TEXAS SYMPOSIUM ON RELATIVISTIC ASTROPHYSICS AND COSMOLOGY

PROCEEDINGS
SUPPLEMENTS
NUCLEAR PHYSICS B

Recently published:

Proceedings of the 7th International Workshop on
Deep Inelastic Scattering and QCD
DESY Zeuthen, Germany
19–23 April, 1999

Proceedings of the XVIIIth International Conference
on Neutrino Physics and Astrophysics
Takayama, Japan
4–9 June 1998

Forthcoming volumes:

Proceedings of the International Workshop on
Particles in Astrophysics and Cosmology
Valencia, Spain
3–8 May, 1999

Proceedings of the International Conference
on the Structure and Interactions of the Photon
Freiburg, Germany
23–27 May 1999

Nuclear Physics B (Proc. Suppl.) 80 (2000) JANUARY 2000

TEXAS SYMPOSIUM ON RELATIVISTIC ASTROPHYSICS AND COSMOLOGY

Proceedings of the XIXth Texas Symposium on Relativistic Astrophysics and Cosmology
"Texas in Paris"

Paris, France 14 – 18 December 1998

Edited by

É. AUBOURG, T. MONTMERLE and J. PAUL
CEA Saclay, Saclay, France

P. PETER
Departement d'Astrophysique Relativiste et de Cosmologie
Observatoire de Paris-Meudon
Meudon, France

NORTH-HOLLAND

FOREWORD

Following a recent tradition, the Texas Symposium crossed the Atlantic. This time, from the banks of the Rio Grande (well, almost) to the banks of the river Seine. From Paris, Texas (well, almost), to Paris, France. This location in the heart of the "old world" made distances shorter and travelling easier for many physicists and astrophysicists eager to exchange ideas, results, and, we believe, culture, sometimes for the first time. Over 900 participants from 51 nations shared emotions and enthusiasm about their passion: the study of the universe, from our Sun to the Big Bang. Supernovae, gamma-ray bursts, black holes, magnetars, neutrinos, gravitational waves, cosmic microwave background, exotic particles, all these topics and more were discussed during a whole week in the "City of Lights", quite an appropriate name for the venue – light is, after all, the privileged messenger from the universe!

"Time flies" ... Relativists know it perhaps more than anyone else! But at the 19th edition of this incontrovertible event of modern astrophysics, we were particularly happy to meet some of the "founding fathers" who gathered in Dallas for the first time 34 years ago, in a simple workshop format, to discuss relativity and cosmology. Unfortunately, others were not with us: we will particularly miss the elegant and friendly silhouette of André Lichnerowicz, who passed away a few weeks before, and Dave Schramm, who died in an accident in December 1997, and whose towering figure will be remembered by the younger generation. May this Symposium be dedicated to their memories.

Non-scientists friends ask: how can you talk to each other if you are nearly a thousand? Well, it's a matter of organization, and atmosphere sometimes. Based on the success of previous meetings, we kept the previous format of invited talks and highlight talks for the general audience, and of mini-symposia for more detailed – and sometimes heated – discussions, reflected in over 600 contributions. As general organizers of the Symposium, we thank our many collaborators for their dedication, patience, and hard work to make this edition another success – at least, this is how we lived it, in a beehive, forum atmosphere.

This book, and the CD-ROM that accompanies it, reflects in a large part what happened (scientifically speaking: we cannot be held responsible for the rest ...). The book contains most of the invited and highlight talks, the CD-ROM (a novelty in the series) contains all the contributions we have received, with nearly 300 papers. A complete index at the end of the book lists them, giving either their page numbers or a code for the CD-ROM. We hope you enjoy both.

We take this opportunity to thank the many institutions which helped us with various forms of support. In France, we must mention first of all the Commissariat à l'Energie Atomique (CEA), via the Service d'Astrophysique, the Service de Physique des Particules, and the Service de Physique Théorique, in Saclay, which bore a large part of the financial burden and provided precious administrative help, the Centre National de la Recherche Scientifique, via its two institutes, Institut National des Sciences de l'Univers (INSU) and Institut National de Physique Nucléaire et de Physique des Particules (IN2P3), and the Centre National d'Etudes Spatiales (CNES). Abroad, we acknowledge the precious support of the International Union for Pure and Applied Physics (IUPAP), of the European Southern Observatory (ESO), of the European Space Agency (ESA), and last but not least, of the National Aeronautics and Space Administration (NASA).

Since such a gathering cannot foster new contacts without the friendly atmosphere of social events, it is also a pleasure to thank Jean Audouze for his welcome at the Palais de la Découverte of which he was the newly appointed director, Patrick Blandin, director of the "Grande galerie de l'Evolution" of the Muséum National d'Histoire Naturelle, for a superb "Noah's Ark" reception, and also the TELESIS Consultants and Moët & Chandon champagne company, which provided the (interstellar?) bubbles.

The Editors

AVANT-PROPOS

Reprenant une tradition instituée récemment, le Texas Symposium a de nouveau traversé l'Atlantique. Cette fois-ci, il a quitté les rives du Rio Grande (si l'on veut) pour les rives de la Seine, et Paris, Texas (enfin presque) pour Paris, France. Ce lieu, au coeur de l'Ancien Monde, a facilité le déplacement de nombreux collègues étrangers, afin de partager nos idées, nos résultats, et, nous le pensons aussi, notre culture. Pour certains, c'en était la toute première occasion. Plus de 900 participants, venus de 51 pays, ont partagé leur passion: l'étude de l'univers, du Soleil au Big Bang. Supernovas, sursauts gamma, trous noirs, magnetars, neutrinos, ondes gravitationnelles, fond diffus cosmologique, particules exotiques, voilà seulement quelques-uns des sujets abordés au cours d'une longue semaine dans la "Ville-lumière", un nom opportunément choisi si l'on songe que la lumière est un messager privilégié de l'univers!

"Comme le temps passe!" Les relativistes le savent mieux que personne. Mais la 19eme édition de ce rendez-vous incontournable de l'astrophysique moderne a été l'occasion de faire la connaissance des "Pères Fondateurs", qui se réunirent pour la première fois à Dallas, Texas, il y a 34 ans, pour une simple réunion de travail autour de la cosmologie et de la relativité. Malheureusement, certains manquaient à l'appel: nous ne verrons plus la silhouette élégante et amicale d'André Lichnerowicz, décédé quelques semaines auparavant, ni celle, combien impressionnante, de Dave Schramm, qui perdit la vie dans un accident en décembre 1997. Nous dédions ce Symposium à leur mémoire.

Nos amis non scientifiques nous ont posé cette question: comment pouvez-vous vous parler si vous êtes près de mille? Eh bien, avec un peu d'organisation, et de la bonne volonté! Sur la base des succès obtenus par les précédentes éditions de la série, nous avons gardé le format des "Exposés invités" et des "Exposés d'actualité" en séances plénières, et des "mini-symposiums" pour les discussions plus détaillées, sinon les plus animées, qui se sont traduites par plus de 600 communications. Ayant supervisé l'organisation du Symposium, nous tenons à remercier nos nombreux collaborateurs pour leur dévouement, leur patience, et la peine qu'ils n'ont pas ménagée, qui ont contribué à ce que nous croyons avoir été un succès, à en juger par l'atmosphère de ruche bourdonnante qui n'a cessé de régner.

Le livre, ainsi que le CD-ROM qui l'accompagne, est un reflet assez fidèle de ce qui s'est passé — enfin, sur le seul plan scientifique! Le livre contient la plus grande partie des exposés invités et d'actualité, et le CD-ROM (une première dans cette série) contient la totalité des contributions reçues, soit en tout près de 300 contributions. Un index en fin de livre permet de s'y retrouver, soit par une page, soit par un numéro de code. Bonne lecture!

Nous saisissons également cette occasion pour remercier les nombreuses institutions qui nous ont aidés sous diverses formes. Nos remerciements vont en premier lieu au Commissariat à l'Energie Atomique (CEA), via ses Services d'Astrophysique, de Physique des Particules, et de Physique Théorique, à Saclay, qui a supporté une partie majeure du poids financier du Symposium, le Centre National de la Recherche Scientifique, par l'intermédiaire de l'Institut National des Sciences de l'Univers (INSU) et de l'Institut National de Physique Nucléaire et de Physique des Particules (IN2P3), ainsi que le Centre National d'Etudes Spatiales (CNES). A l'étranger, nous avons reçu le soutien de l'International Union for Pure and Applied Physics (IUPAP), de l'Observatoire Européen pour l'Hémisphère Austral (ESO), de l'Agence Spatiale Européenne (ESA), et "last but not least", de la National Aeronautics and Space Administration (NASA).

Comme une telle réunion ne peut promouvoir de nouveaux contacts sans la chaleur de son "programme social", c'est avec plaisir que nous remercions pour son accueil dans ses murs le nouveau Directeur du Palais de la Découverte, Jean Audouze, ainsi que Patrick Blandin, Directeur de la Grande Galerie de l'Evolution du Muséum National d'Histoire Naturelle, pour une fabuleuse réception dans son "Arche de Noé", et aussi TELESIS Consultants et la maison Moët & Chandon pour avoir fourni de jolies bulles (interstellaires?)

Les éditeurs

COMMITTEES

International Advisory Committee

M.A. ABRAMOWICZ (Gothenburg)
M. ARNOULD (Bruxelles)
G.V. BICKNELL (Mount Stromlo)
G.F. BIGNAMI (Milano)
R.D. BLANDFORD (CalTech)
S. BONAZZOLA (Paris)
J.R. BOND (Toronto)
B. CARTER (Paris)
C. CÉSARSKY (Saclay)
S. DESER (Waltham)

J. ELLIS (Geneva)
G. FONTAINE (Paris)
J.A. FRIEMAN (Batavia)
R. GIACCONI (Garching)
J. ISERN (Blanes)
S. KATO (Kyoto)
O. LE FÈVRE (Marseille)
M.S. LONGAIR (Cambridge)
L. MAIANI (Roma)
F. MIRABEL (Saclay)
J.V. NARLIKAR (Pune)

P.J.E. PEEBLES (Princeton)
M.J. REES (Cambridge)
B. SADOULET (Berkeley)
J. SILK (Berkeley/Paris)
A.A. STAROBINSKY (Moscow)
R. SUNYAEV (Garching)
Y. TOTSUKA (Tokyo)
J. TRÜMPER (Garching)
J. VAN PARADIJS (Amsterdam)
H.J. VÖLK (Heidelberg)

National Scientific Organizing Committee:

J. PAUL (Saclay), Chair
J. AUDOUZE (Paris)
F. BOUCHET (Paris)
A. BRILLET (Orsay)
M. CASSÉ (Saclay)
B. DEGRANGE (Palaiseau)

B. FORT (Paris)
I. GRENIER (Saclay)
J.-M. HAMEURY (Strasbourg)
J.-P. LASOTA (Paris)
R. MALINA (Marseille)
T. MONTMERLE (Saclay)

L. MOSCOSO (Saclay)
G. PELLETIER (Grenoble)
P. SALATI (Annecy)
R. SCHAEFFER (Saclay)
G. VEDRENNE (Toulouse)

Local Organizing Committee

T. MONTMERLE (Saclay), Chair
É. AUBOURG (Saclay)
J.-Ph. BERNARD (Orsay)
F. BERNARDEAU (Saclay)
M. BORATAV (Paris)
F. COUCHOT (Orsay)

M. DENNEFELD (Paris)
Ph. DUROUCHOUX (Saclay), Co-Chair
D. ELBAZ (Saclay)
R. FERLET (Paris)
P. FLEURY (Palaiseau)

Y. GIRAUD-HERAUD (Paris)
C. LAURENT (Meudon)
A. LEFEBVRE (Orsay)
J. PAUL (Saclay)
P. PETER (Meudon)
D. YVON (Saclay)

David Norman Schramm, October 10, 1945 – December 19, 1997

Something unusual and unfortunate happened at the XIXth Texas Symposium: David Schramm was not attending it. He died one year before its venue, on December 19, 1997, while doing one of the things he loved most – flying his airplane.

Among his other passions was doing research in nuclear and particle astrophysics, and of course in cosmology. He is remembered by his colleagues as a driving force in the latter field, which he used to qualify as "entering its golden age". He was one of the most influential and productive astrophysicists of his generation. His contributions were manifold – meteoritical isotopes, nucleocosmochronology, supernova collapse and neutrinos, Big-Bang nucleosynthesis, neutrino cosmology, the r-process, solar neutrinos, particle dark matter, gamma-ray bursts, structure formation, topological defects, ultra-high energy cosmic rays, astrophysical and cosmological constraints to particle physics, the first paper on quintessence, and on and on. The idea he had pursued all along his career was to relate the results of Big-Bang nucleosynthesis and particle physics – the prediction of neutrino families through the abundance of primordial helium.

Through this prolific work, he was a major figure in the recent merging of particle physics, nuclear physics and astrophysics for the study of the early universe: the cosmology group he built at Chicago is one of the most preeminent, and he co-founded the astrophysics group at Fermilab, one of the first gatherings of particle physicists and astrophysicists. Not only Chicago, but also Pasadena (Caltech) and Austin (University of Texas) owe him a lot of scientific achievements. A list of his academic awards would fill a whole page, and he was rightly very proud to have been elected member of the American Academy of Sciences early in his career. He was recognized worldwide both for his excellence in research and for his leadership skills, whether with students, friends, or colleagues.

As Vice President for Research of the University of Chicago since 1995, he was also an effective advocate for basic research, both to the US government and to the general public. He wrote two popular books, *From Quarks to the Cosmos: Tools of Discovery,* and *The Shadows of Creation: Dark Matter and the Structure of the Universe,* in addition to hundreds of scientific papers and 13 technical books.

He was also a gifted teacher, ready to devote a great deal of time and energy to his students, or to take them with him aboard Big-Bang aviation when going to scientific conferences: in 1994, he received the University's Faculty Award for Excellence in Graduate Teaching.

He was passionate and brilliant in all the things he was doing, either physics, flying, climbing, or wrestling – he was a champion Greco–Roman wrestler who competed in college and graduate school, and was a finalist in the 1968 Olympic trials.

He is missed by his colleagues and former students from France and elsewhere, and we chose to dedicate to his memory this first Texas Symposium he would not illuminate by his presence.

The Editors

CONTENTS

(Abstracted/Indexed in: Current Contents: Physical, Chemical & Earth Sciences/INSPEC)

Opening ceremony

Invited talks

Highlight talks

Summaries of Mini-Symposia

Opening ceremony

ELSEVIER

Nuclear Physics B (Proc. Suppl.) 80 (2000) 3–4

NUCLEAR PHYSICS B
PROCEEDINGS
SUPPLEMENTS

www.elsevier.nl/locate/npe

Physics and astrophysics: the new approach

Richard Bonneville[a]

[a] Direction des Programmes Scientifiques
Centre National d'Etudes Spatiales
2, place Maurice-Quentin
75001 Paris, France

On behalf of the president of CNES, the French space agency, I would like to thank the Organizing Committee for giving me the opportunity to deliver a few words at the beginning of this Texas Symposium.

When Thierry Montmerle asked me to come to this opening session, I wondered what I was to say. And I remembered a science fiction novel from the Polish writer Stanislas Lem. The book is called "Masters Voice". One day, by chance, someone detects a neutrino beam coming ¿from deep space, and this beam seems to exhibit some structure. People get convinced it is a signal sent from an intelligent species far away and an international project named MastersVoice is set up in order to try to decipher it. Billions are spent, but the outputs are not very conclusive. At the end, a brilliant theoretician presents a cosmological model according to which the existence the neutrino beam with its characteristics was a physical necessity of the model. But the narrator ends the story by expressing the opinion that it was actually a signal but that we were not civilized enough to be the actual adressees.

For a long time, cosmology has been a domain only for theoreticians because observations were scarce. Now, with space research, it has come to the status of an experimental science. For instance, in the millimetric and submillimetric range of the electromagnetic spectrum, Planck-Surveyor will accurately measure the high-order anisotropies of the cosmological black-body radiation, which are a key test for the theoretical models. At the other extremity of the spectrum, in the hard X-ray and γ-ray domains, the space observatories provide access to violent events connected with such exotic objects as quasars and black holes, which contain natural laboratories of high energy physics.

In October 1993, on the occasion of a week-long seminar on long-term planning which we organized in order to hear the priorities of the French scientific community, we have introduced fundamental physics in our scientific program as a separate theme besides astrophysics. The following seminar, held in March 1998, confirmed the maturity of this domain and I would like to emphasize three points.

The first one is the increasing convergence between astrophysics and particle physics in the investigation of problems related to the unification of the fundamental interactions, and to the special place of gravity with respect to the other interactions.

Only space provides us with intense and/or rapidly variable gravity fields, which allows to test the gravitation theories in their totality, and not only in the limiting case of weak or slowly varying fields. We are confident that during the first decade of the next century, a new type of space observatories, presently under study on both sides of the Atlantic, will open a new window to research by allowing the direct evidence of gravitational waves in the low-frequency range, as they are emitted in such violent events as the coalescence of binary systems.

Space experiments also allow new tests of General Relativity. Gravity Probe B will be launched soon and at least three space experiments, in the US, in Italy and in France are under preparation, which all aim at testing the Equivalence Principle in low-Earth orbit, with an accuracy improved by 3 to 6 orders of magnitude with respect to the best ground- based experiments. A positive answer would be the signature of new fields and particles which are predicted by the theories aiming at unifying gravity with the other fundamental interactions. A negative answer would strongly constrain the models, but without doubt it would not constrain the imagination of the theoreticians!

Other ideas have been proposed, such as performing redshift experiments and/or angular deflection tests onboard spacecrafts passing close to the Sun. Such experiments would allow us to measure the post- Newtonian parameters with a considerable accuracy (to better than 10^{-6}) and thus to possibly give evidence for traces of exotic fields whose effects have been masked by the cosmological evolution.

A second aspect I would like to underline is the need for an intense effort of technological research to prepare all these projects : efficient drag- free control of the spacecrafts, high stability lasers, highly sensitive accelerometers, ultrastable clocks and accurate time transfer techniques. I will quote for instance the ACES experiment to be flown on the space station that will distribute the best time reference ever.

A third important feature is the increasing need for international cooperation : for instance the Planck-Surveyor satellite which the European Space Agency plans to launch in 2007 has gathered two consortia of European scientists, with US participation, and, in the field of γ-ray astrophysics, we plan to participate to the future GLAST satellite from NASA. On this occasion, I would like to congratulate our Russian colleagues and my friends from CEA and CESR with whom we had developed the SIGMA telescope, since the Symposium occurs at a moment when we have just learnt that the *GRANAT* γ-ray observatory has ended a 9-year lifetime, during which it discovered numerous galactic and extragalactic sources of hard radiation, including black holes.

I thank you for your attention and I express my best wishes for the success of this Symposium.

ELSEVIER

Nuclear Physics B (Proc. Suppl.) 80 (2000) 5–7

NUCLEAR PHYSICS B
**PROCEEDINGS
SUPPLEMENTS**
www.elsevier.nl/locate/npe

Welcome Address on behalf of the *Institut National de Physique Nucléaire et de Physique des Particules*

Gérard Fontaine[a]

[a] IN2P3 – Centre National de la Recherche Scientifique
3, rue Michel-Ange
75016 Paris, France

Dear Colleagues and Friends,

After René Pellat's talk, I also have the pleasure to welcome you for this important symposium on relativistic astrophysics and cosmology gathering about 900 participants in France.

As a particle physicist in charge of particle astrophysics at the French IN2P3-CNRS, I would like, in this welcome address, to emphasize the close relationship between astrophysics and subatomic physics :

This connection was initiated in the 30's, when nuclear physics explained the origin of the energy in stellar systems, and began to address the question of nucleosynthesis.

At about the same period, with the discoveries of the positron, muons, pions, kaons and hyperons in the cosmic radiation from the 30's to the early 50's, particle physics started with cosmic ray observations. Only later on did this physics go its own way using dedicated accelerators of increasingly higher energies.

Cosmology has long been a theoretical science based on general relativity and on particle theories. Inflation, phase transitions, quantum gravity, Planck scale physics and superstrings are concepts of common interest to cosmologists and particle physicists alike. In addition, cosmology has now turned into an observational science which feeds back important constraints to subatomic physics.

The present Universe can be used as an immense laboratory where benefit is taken from the extreme conditions in energy and density that prevail at specific locations. These conditions, which lie far beyond what will probably ever be achieved artificially, allow the observation of the behavior of matter in this extreme environment.

Cosmic particle detectors, derived from accelerator technology are constantly improving in sensitivity. They are opening new windows on the universe, thus founding new astronomies : X and gamma-ray astronomy already exist; neutrino and ultra-high energy charged particle astronomies are probably for tomorrow, and gravitational waves astronomy for the day after tomorrow.

Neutrinos are of special interest to both the astrophysicist and the particle physicist. Their exact nature is yet unknown. In particular, they may have a non zero mass opening the possibility of oscillations. Such an exciting feature, combined with their enormous penetrating power and their important role in the early universe would make them a central figure of the game.

It was therefore natural that researchers originating either from astrophysics or from subatomic physics take interest in each-other scientific field, and join forces to attack the questions at this frontier of knowledge shared by their disciplines. This trend, initiated about 10 to 15 years ago, has led to the birth of a new domain named *particle astrophysics* or *astroparticle physics*. It is a wide field, rich of exciting subjects that cannot be addressed by other means. This domain appears to be currently the focus of an increasing interest worldwide.

From my point of view, the researches in the domain can briefly be classified in four sectors, to which many of the sessions of this symposium can clearly be associated:

* Cosmology : from the primordial universe to the nature of dark matter, many questions are still open and are the subject of vigorous researches. In parallel with the search for anti-matter in space, the quest for dark matter is raging, either in its non-baryonic form, such as Weakly Interacting Massive Particles (detected by their interactions in underground laboratories), or in the form of baryonic matter such as cold fractal molecular clouds or Massive Astrophysical Compact Halo Objects. Observations of the fluctuations of the Cosmic Microwave Background Radiation provide an exceptionally accurate method for the determination of the main cosmological parameters, complementary to the one based on type Ia Supernovae observations. The search for large scale structures should also shed light on the process of the formation of structures in the Universe, and help us to relate the initial density fluctuations to the primordial Universe particle content. Subatomic physicists engaged in this research will have to acquire experience on how to use optical telescopes or develop bolometers, but the rewards will be worth their efforts.

* High energy phenomena, compact objects and cosmic rays or radiation at large : Compact objects (neutron stars, black holes and their environment ranging from accretion disks to active galactic nuclei and jets) may be the sites of the production and the acceleration of particles, both charged and neutrals, up to very large energies. These travel down to us. They are detected either in space (up to a few GeV) or from the ground by using the earth atmosphere as a giant calorimeter (from about 50 GeV to 10^{20} eV). Many of the detection techniques in this sector are familiar to subatomic physicists. On the other hand, their use on board of spacecrafts is certainly new to them.

Compact objects can also be the source of gravitational waves, for instance when binary systems coalesce, and an open question is whether such sources are correlated with γ-ray bursts. Particle physicists on both sides of the Atlantic ocean are engaged in the tracking of these feeble space-time oscillations, although the involved optical techniques are not in their present usual toolbox.

* Nuclear astrophysics is now focused on a deeper understanding of the various processes of nucleosynthesis which explain the elemental abundances in the universe. Astrophysical observations, but also cross-sections measurements with accelerators provide constraints on many processes, ranging from solar physics and steady state stellar nuclear reactions, to explosive phenomena, and eventually to the Big Bang itself, thus coupling to the important questions of the solar neutrino deficit or of the quantity of hadronic dark matter.

* Neutrino sector is by itself a wide field where the main quest concerns the very nature of this intriguing particle. Is it a Dirac or a Majorana system ? Is it massive and does it contribute to the missing mass of the Universe ? Does it oscillate and if so between which states and by which mechanism ? These important questions are attacked by means of experiments using accelerators, nuclear reactors, cosmic rays producing the so called "atmospheric neutrino", as well as solar and supernova neutrinos offering the longest possible baselines for measurement. The recent result obtained by the Superkamiokande collaboration has caught worldwide attention and is triggering a large effort to confirm and improve it by means of new detectors.

In France, research in the field of astroparticle physics is carried out by several hundred of scientists (astrophysicists and subatomic physicists) from various institutions and scientific departments. Thanks to these diverse scientific origins, they bring together different expertise and complementary knowledge. On the other hand, they still have to learn how to take full benefit of the diversity of traditions in their working methods. A similar not fully settled state of the research organization exists in many countries. I am nevertheless convinced that the potential interest of the fields that these two communities are exploring and pioneering together will lead them to overcome any difficulty originating from a difference of culture. The movement has started from the floor and is gaining momentum here as in many places worldwide.

I also want to bear witness to the will of the scientific institutions and funding agencies to support this interdisciplinary domain and to set up adequate structures in order to facilitate its development. Of course, the path ahead of us is not straight. There are many difficulties to overcome : diversity of funding sources, budget restrictions, priorities to establish, etc. You all know too well these contingencies. However within the constraints that they set in the French landscape of research, a scientific policy exists which is determined to build bridges between astrophysics and subatomic physics, and to bring closer all the actors of the play.

When I turn back, and take for instance the time of the Blois particle astrophysics conference in 1992 as a start, and when I consider all that has been achieved since, I feel very optimistic for the long range term, and I foresee a bright future for this field.

On behalf of IN2P3, I wish you an exciting and very productive conference.

Thank you.

ELSEVIER

Nuclear Physics B (Proc. Suppl.) 80 (2000) 9–11

NUCLEAR PHYSICS B
**PROCEEDINGS
SUPPLEMENTS**
www.elsevier.nl/locate/npe

Cosmological Turtles

Hubert Reeves[a]

[a] Service d'Astrophysique
DAPNIA/Sap
CEA Saclay
91191 Gif/Yvette, France

There is a story about an elderly lady who stood up to speak at the end of a public lecture on cosmology. "I have read in a book on Indian mythology", she said, "that the world is an immense turtle swinning in a large basin". "And what supports that basin ?" answers the speaker. -"Another turtle." - "And what about that turtle ?" -"My dear lady, don't try to be smart. I know what you have in mind. It is turtles all the way down !"

This allegory may not really apply to the universe, but, in some sense, it may apply to the theories of the universe. This the subject of my talk.

The arguments in favor of the Big Bang theory are today too numerous and too-well founded to justify their being called into question in a any radical way. On the market place of cosmologocal theories, the Big Bang is, by far, the "best buy" and it has no serious rivals. This, however, should not prevent us from being lucid and aware of its weak points.

1. Cosmologist's nightmares

A number of nightmares haunts the sleep of the cosmologist. They take the form of questions to which he would very much like to have satisfactory answers. There are problems related to cosmology, to physics, and also to structure formation in the early universe. I shall make a separate list of them.

In domain of cosmology, we would like to understand properties of the early universe such as: (a) the very high isothermy (to a few parts in 10^5)

of the fossil radiation (Cosmic Microwave Background, CMB: 3K); (b) its extreme flatness (Euclidean geometry); (c) its very low entropy (sum of thermal plus gravitational entropies); and (d) its very low bulk rotation with respect to the inertial (Galilean) frame. In the realm of physics we ask questions about (e) the nature and numerical value of the cosmological constant. The absence of cosmic antimatter raises the question (f) of the lack of symmetry between matter and antimatter. A related problem is the value of the "baryonic number", i.e., the ratio of nucleons to photons in the universe $(= 3 \times 10^{-10})$. Also problematic is (g) the absence of magnetic monopoles, contrary to standard theoretical predictions. On the subject of the formation of large-scale structures, we would like to understand (h) the nature and (i) the spectrum of the initial perturbations growing into galaxies and clusters of galaxies. Also, we try (j) to resolve the causality paradox attached to their growth in the early times (their volumes were then larger than their sphere of causality). The slow tempo of gravitational contraction processes in the initial (linear) growth is another difficulty (k) of the models. Finally the distribution of full-grown structures, as revealed by the joint analysis of the angular distribution of the CMB and of large-scale surveys of galaxies, is also problematic (l).

2. The inflation turtle

In the "inflation paradigm" resides the most popular answers to a number of these questions. Various phase transitions and the consequent gigantic blowing up of space (episode of inflation) can in principle account for: the initial isothermy, the initial flatness, the low rotation and the low entropy. It accounts, at least partially, for the absence of antimatter and magnetic monopoles, for the nature and spectrum of the initial distribution of fluctuations, and for the causality paradoxes.

According to the standard model of particle physics, phase transitions and episodes of inflation most likely did take place (at least twice) in the early universe. One occurred around 10^{15} K at the electroweak phase transition, when the electromagnetic and weak interaction became dissociated. Another one occurred around 10^{12} K, at the quark-hadron phase transition when quarks combined into nucleons. Other more hypothetical phase transitions are often associated with the end of the so-called Grand Unification period (around 10^{28} K), and also with the Planck era (around 10^{32} K). However, in order to give satisfactory answers to the previously mentioned problems, the phase transitions must necessarily have very specific properties. First and foremost, the growth factors of resulting space inflations must have been enormous: at least 10^{30} ! As a consequence, the present spatial geometry must be flat to within 1 part in 10^4 ($\Omega_{total} = 1 \pm 10^{-4}$). The inflation turtle rests on this new turtle hypothesis. The matter contribution to Ω_{total} does not seem to foot the bill ($\Omega_{matter} = 0.3$). There are indications that the cosmological constant invoked to account for the supernova survey ($\Omega_\Lambda = 0.7$) may save the situation, but this is still uncertain. Recent models, called "open inflationary theories", have been developed, in which the present flatness is not required.

But there are many more requirements on the way to the solutions. Detailed models of the individual episodes show that the forms of the potential energy of the associated Higgs (scalar) fields must be highly "fine-tuned" in order to solve the various problems. The inflation turtle (including its open inflationary version) rests on the hypothesis that the potential energy indeed has these properties. Now we come to problems not (or not entirely) answerable in terms of the inflation paradigm. What is the value of the cosmological constant ? Estimates of Ω_Λ from arguments based on the standard model of particle physics give values of the order of 10^{100}, absurdly larger than the observational upper limit ($\Omega_\Lambda < 1$). There is some hope that supersymmetric theories may solve this problem, but no quantitative estimates are available. As mentioned before, the problem of the absence of antimatter is connected to the value of the baryonic number. Its computed value is strongly dependant on the (largely unknown) properties of particle physics in the early universe.

Finally, we come to the problems of the slow growth of the initial perturbations. The most likely answer is in terms of the presence of dark matter accelerating the rate of growth after the emission of the CMB. The best estimates of the respective densities of luminous matter ($\Omega_{luminous} = 0.005$), of baryonic matter ($\Omega_{baryonic} = 0.02$) and total matter ($\Omega_{matter} = 0.3$) give strong evidence in favor of the existence of (at least) two components of dark matter: one baryonic and one non-baryonic, usually called "exotic". The nature and properties of this exotic matter are still unknown. The best accounts of the distribution of large-scale structures invoke the presence of mostly "cold" dark matter (non relativistic at recombination time), but also some "hot" dark matter (relativistic at recombination time) is required. These are, for the moment, free parameters of the models.

And now we come to our (presently !) ultimate turtle: the "fundamental physics" required to justify all these assumptions, i.e., to support the precarious scaffolding of turtles.

The one main problem remaining to be solved in physics is the quantification of gravity, which, we hope ! will lead to a unified theory of the four interactions. The main contender to this theory is the superstring theory. Despite its beauty and seductive power for theorists, it has made, up to now, very little quantitative predictions testable by observations. Needless to say, this theory is hardly of any practical help in the solutions of our various problems. Furthermore, there is no guarantee that it will form a stable basis for the scaffolding. That it will not itself rely on new assumptions... which would themselves...

ELSEVIER

Nuclear Physics B (Proc. Suppl.) 80 (2000) 13–14

NUCLEAR PHYSICS B
PROCEEDINGS
SUPPLEMENTS

www.elsevier.nl/locate/npe

In Remembrance of David Schramm

Michel Cassé[a]

[a] Service d'Astrophysique
DAPNIA/Sap
CEA Saclay
91191 Gif/Yvette, France

On behalf of the French astrophysical, cosmological and physical community I would like to express a few sentences in remembrance of David Schramm, a friend and colleague who died by accident a year ago. We have followed his brillant carrere and recognize his mastership of cosmology, and all aspects of nucleosynthesis. He completed his undergraduate training at MIT in 1967 and received his PhD from the California Institute of Technology in 1971.

During 23 years the dynamic presence of David Schramm enriched the life of the University of Chicago. His vision of cosmology and particle physics led to the formation of an astrophysical group in the Fermi National Accelerator Laboratory. He lectured frequently worldwide at scientific conferences, schools, workshops, colloquia and public symposia.

Dave was an imaginative and brilliant scientist whose research affected much of our understanding of contemporary astrophysics, from the "Dynamical r-process Nucleosynthesis in Supernova" to dark matter, through the age of the universe. He is best known for his work unifying the fields of Big Bang cosmology and elementary particle physics.

Elected to the National Academy of Science of the United States in 1986, he is the recipient of numerous awards. His most oustanding accomplishment was in astroparticles, his leadership in this fast growing domain is aknowledged all around the world. He became an effective spokesman for basic research in America and in the whole world.

Through various seminars at Moriond, at CERN, and cristal clear review talks, he convinced French particle physicists of the interest of astrophysics and cosmology to provide constraints on the properties of elementary particles. More specifically, he showed how the observation of neutrinos from SN 1987a in the Large Magellanic Cloud gives limits on their mass and lifetime.

He convinced skeptical physicists that non baryonic dark matter was unescapable. He was influential in the search of baryonic dark matter in the halo of our galaxy by microlensing techniques and served as co-advisor of a thesis in Saclay. The merging of high energy physics and cosmology owes him very much. One can say, according to Michel Spiro, that he has put in perspective a standard model for particle physics and cosmology that serves today as a guiding line for both community.

On the front of nucleosynthesis, his pioneering efforts are of great interest in showing how elemental abundances can be used in cosmology. He has collaborated with many of us on light element nucleosynthesis, starting from Hubert Reeves and Jean Audouze, on deuterium, then Elisabeth Vangioni-Flam and myself on Helium-3 and finally Martin Lemoine on lithium. All of us were impressed by his exceptional talent and vitality. He lived and worked twice the normal rate.

Beyond his exceptional professional qualities, he has left the memory of a gentleman. He was particularly kind to young students and to the general public, tirelessly explaining and explaining again. He wrote several books of exceptional clarity for the general public. A translation of his book by Elisabeth Vangioni-Flam from Intitut d'Astrophysique de Paris "Shadows of the Universe" has just appeared under the title "Les mirages de la création".

He was also an expert mountain climber, having scaled the highest peaks on five continents. Himself, he was big in every sense of the word, bigger than life. He achieved so much, and was active in so many arenas, that he has left many unfillable graps. It is still impossible to fully grasp the scope of his disparition. He is not the sort of person one would ever forget. The 19th Texas Symposium is dedicated to David Norman Schramm.

Invited talks

ELSEVIER

Nuclear Physics B (Proc. Suppl.) 80 (2000) 17–32

NUCLEAR PHYSICS B
**PROCEEDINGS
SUPPLEMENTS**

www.elsevier.nl/locate/npe

Oscillation Solutions to Solar Neutrino Problem

V. Berezinsky[a]

[a]INFN, Laboratori Nazionali del Gran Sasso, I–67010 Assergi (AQ), Italy
and Institute for Nuclear Research, Moscow, Russia

The current status of oscillation solutions to the Solar Neutrino Problem is reviewed. Four oscillation solutions are discussed in the light of 708d Superkamiokande data: MSW, Just-So VO, VO with Energy-Independent Suppression (EIS) and Resonant-Spin-Flavor-Precession (RSFP). Only EIS VO is strongly disfavoured by the global rates, mostly due to the Homestake data. Vacuum oscillations give an interesting solution which explains high-energy excess of events observed by Superkamiokande and predicts *semi-annual* seasonal variation of *Be*-neutrino flux. There are indications to these variations in the GALLEX and Homestake data. No direct evidence for oscillation is found yet.

1. Introduction

With Superkamiokande in operation and SNO in preparation the study of solar neutrinos has entered a decisive stage, when neutrino oscillations can be directly discovered. At present we know almost certainly that something happens to neutrinos either inside the Sun or on the way between the Sun and Earth. This knowledge has been provided mainly by solar-neutrino experiments, by helioseismic observations and by theoretical analysis.

1. *Helioseismic data* confirm the Standard Solar Models (SSM's) with precision sufficient for reliable prediction of solar-neutrino fluxes. The seismic data (in agreement within a fraction of percent with the SSM predictions) are valid down to radial distance $0.05R_\odot$, where production of B- and Be - neutrinos has maximum, while the other neutrinos are mostly produced at larger distances. For the smaller radial distances, where production of neutrinos falls down due to decreasing of volume, the seismic data still exist, though with worse precision.

Acoustic frequencies comprise the set of seismic data. Nothing in physics is measured with greater precision than frequencies. This is why seismic measurements give the super-precise data on density and sound speed inside the Sun. Within fraction of percent seismically measured density and sound speed are different from the SSM predic-

tions (especially at distance $0.7R_\odot$) and this difference is statisticlally significant. It might imply that some physical processes are not included in SSM's, and they are of great interest for physics of the Sun. But not for solar neutrinos! This statistically significant difference, *e.g.* in measured and predicted sound speed, produces negligible difference in neutrino fluxes, which is out of interest for any present (and most probably for any future) solar-neutrino experiment.

Almost for half century we thought that solar neutrinos with their tremendous penetrating power give us the best way to look inside the Sun. We see now that seismic observations do it with higher precision, while solar neutrinos provide us with the unique information about neutrino properties.

2. *Nuclear cross-sections* is now the dominat source of uncertainties in the calculated solar-neutrino fluxes. The impressive progress exists here too. In the LUNA experiment at Gran Sasso the cross-section of one of the most intriguing reaction,$^3He +^3He \rightarrow ^4He + 2p$, was measured at energy corresponding to maximum of the Gamow peak in the Sun. The famous speculations about solving or ameliorating the SNP due to increase of this cross-section at very low energy, have been now honorably buried. In the nearest future most of cross-sections relevant to SNP will be measured in the LUNA experiment at very low energy. There is also considerable progress

in calculations of cross-sections and screening of nuclear reactions in the solar plasma. A rather exceptional case is cross-section of Hep reaction $p + {}^3He \rightarrow {}^4He + e^+ + \nu$, in which neutrinos with the highest energies are produced. Uncertainties in calculation of this cross-section are very large.

3. *Solar Neutrino Problem (SNP)* is a deficit of neutrino fluxes (as compared to the SSM prediction) detected in all solar-neutrino experiments (Homestake, SAGE, GALLEX, Kamiokande and Superkamiokande). This deficit is described by factor ~ 3 for Homestake and by factor ~ 2 for all other experiments.

4. *Astrophysical solution* to SNP is strongly disfavoured by combination of any two solar-neutrino experiments, *e.g.* the boron and chlorine experiments (Superkamiokande and Homestake) or the gallium and boron experiments (GALLEX/SAGE and Superkamiokande). The ratio of *Be* to *B* neutrino fluxes, extracted from each pair of experiments mentioned above, is negative (or too small). This is the essence of failure of astrophysical solution. The arbitrary variation of temperature and unknown cross-sections do not solve a problem of *Be/B* ratio.

Solar neutrino experiments have a status of disappearance oscillation experiment.

But solar-neutrino oscillations are not proved yet. In this paper I will discuss the status of different oscillation solutions to SNP.

2. Status of Astrophysical Solution to SNP

The global rates of four solar-neutrino experiments [1–4], as reported up to 1999, are listed in Table 1 and compared with calculations of Bahcall and Pinsonneault 1998 [5].

The deficit of detected neutrino fluxes seen in the last column of Table 1 is impossible to explain by astrophysics or/and nuclear physics. This conclusion is based on the following.

(i) Compatibility of the boron (Superkamiokande) and chlorine (Homestake) signals or boron and gallium (GALLEX/SAGE) signals results in unphysically small ratio $\Phi_\nu(Be)/\Phi_\nu(B)$ [6]-[15]. The best fit value of this ratio is nega-

Table 1
The solar-neutrino data of 1998 compared with the SSM prediction , Bahcall and Pinnsoneault 1998 [5]. The data of Superkamiokande are given in units $10^6 \ cm^{-2}s^{-1}$.

	DATA	SSM [5]	DATA/SSM
SUPERK	2.42 ± 0.08	5.15	0.47 ± 0.02
GALLEX (SNU)	77.5 ± 7.7	129	0.60 ± 0.06
SAGE (SNU)	66.6 ± 8.0	129	0.52 ± 0.06
HOMEST (SNU)	2.56 ± 0.23	7.7	0.33 ± 0.03

tive. The statement above is model-independent.

(ii) The arbitrary variation of unknown nuclear cross-sections and the central temperature cannot bring $\Phi_\nu(Be)/\Phi_\nu(B)$ ratio in agreement with observations [16].

(iii) Seismic observations of the density and sound speed radial profiles confirm SSM at distances down to $0.05R_\odot$ with accuracy better than fraction of percent [17]-[20] and at the center with accuracy better than 4% for sound speed [21].

(iv) As minimum, SSM is a good approximation to realistic model of the Sun. In this case there must be a track, when changing the parameters of SSM and/or introducing the new physical phenomena, one arrives from SSM neutrino fluxes to the observed ones. Such a track does not exist [15].

We shall analyze the last item at some details. But two comments are in order now.

There are two other, more model-dependent arguments against astrophysical solution to SNP.

If one takes three major components of neutrino fluxes (*pp*, *Be* and *B*) as independent and positive, and the CNO neutrino flux (which gives much smaller contribution) - according to SSM, then arbitrary variation of those three fluxes do not give acceptable fit to the observational data at 99.99%CL [22].

The deficit of *B* neutrinos seen in the Superkamiokande data (Table 1) is another problem for astrophysical solution. Some time ago many people thought that with extreme and correlated uncertainties in pBe-cross-section and in the central temperature T_c this discrepancy can be eliminated. Now the situation looks like follows. In the helioseismically constrained solar models (HCSM) [23] the central temperature $T_c = 1.58 \cdot 10^7$ K within maximum uncertainty $\Delta T_c/T_c = 1.4\%$. Taking T_c 1.4% lower and S_{17} 40% lower we obtain the minimum B-neutrino flux $3.0 \cdot 10^6$ $cm^{-2}s^{-1}$, i.e. 7.4σ higher than the measured one. The most recent attempt [24–26]

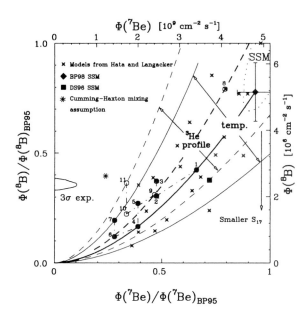

Figure 1. *Neutrino fluxes allowed by arbitrary ^3He mixing accompanied by independent variations of temperature, S_{34} and S_{17}. The solid lines limit the allowed region in case ^3He radial profile is the same as in SSM, and the dashed lines limit the region when both, temperature and ^3He profiles are varying. Some trajectories from the SSM allowed regions are shown for illustration. The best fit is given by point 11, separated by more than 6σ from experimentally allowed region (shown as "3σ exp." in the figure).*

to reconcile the astrophysical solution with mea-

sured neutrino fluxes involves an old idea of ^3He-mixing in the solar core. In SSM's ^3He-abundance is very low in the Be, B-neutrino production zone. ^3He is accumulated at much larger distance $r \sim 0.3R_\odot$. It is assumed that due to some process (it could be gravity wave induced diffusion [25] or non-linear instability [24]) the "fresh" ^3He is brought into solar core. It could happen as the short repeating episodes. Then neutrinoless channel in nuclear reactions, $^3He + ^3He \rightarrow ^4He + 2p$, is enhanced and the central temperature T_c decreases too.

A general analysis of astrophysical solution, which includes arbitrary ^3He-mixing has been recently performed in ref.([27]). The ^3He-mixing was assumed not to be accompanied by mixing of other elements. Additionally all other relevant parameters in the neutrino production zone were varying within wide range: T_c - within $\pm 5\%$, S_{17} - within $\pm 40\%$ and S_{34} - in the range $(-20\% + 40\%)$. The temperature and ^3He radial profiles were also varying. The results are presented in Fig.1 as allowed regions between two limiting curves, thin solid ones ("temp.") and two broken ones ("^3He profile"). The best fit is at least 6σ away from observationally allowed region. It can be interpreted as well that there is no allowed track from the SSM's region ("SSM") to the observationally allowed region (see Fig.1). A trajectory with variation of temperature T_c is shown by thick solid line ("temp.")

3. Oscillation Solutions

Due to oscillations, electron neutrino emitted from the Sun can be found at the Earth as neutrino with another flavor: muon, tau or sterile neutrino. These neutrinos either do not give a signal in the detector (e.g. muon neutrinos in gallium or chlorine detectors) or interact weaker due to NC (e.g. muon- or tau-neutrinos in Superkamiokande). I will not discuss in this review sterile neutrinos. Atmospheric neutrino oscillations imply that ν_μ and ν_τ neutrinos are maximally mixed. In this case solar ν_e neutrino oscillates with equal probability to each of those neutrinos.

The probability to find emitted electron neutrino in the same flavor state in the detector $P_{\nu_e \to \nu_e}$ is called survival probability or (less precisely) suppression factor. In general case survival probability depends on energy, $P_{\nu_e \to \nu_e}(E)$, *i.e.* solar neutrinos are suppressed in energy-dependent way and actually this property allows to solve SNP with help of oscillations.

Four oscillation solutions are currently discussed in the literature (see Table 2.)

Table 2
Oscillation Solutions to SNP.

	$\sin^2 2\theta$ best fit	$\Delta m^2 \ (eV^2)$ best fit
MSW		
SMA	$6.0 \cdot 10^{-3}$	$5.4 \cdot 10^{-6}$
LMA	0.76	$1.8 \cdot 10^{-5}$
LOW	0.96	$7.9 \cdot 10^{-8}$
VO		
Just-so	0.75	$8.0 \cdot 10^{-11}$
EIS	~ 1	$10^{-9} - 10^{-3}$
RSFP		
	small	$10^{-8} - 10^{-7}$

1. *MSW solution* [28].
MSW effect in the Sun is a resonance conversion of ν_e into ν_μ or ν_τ. For neutrino energies at interest it occurs in the narrow layer, $\Delta R \sim 0.01 R_\odot$, at the distance $R \sim 0.1 R_\odot$ from the center of the Sun. There are three MSW solutions to SNP, which explain the global rates in all four solar-neutrino experiments: Small Mixing Angle (SMA) MSW, Large Mixing Angle (LMA) MSW and LOW solution with low probability (it appears only at 99%CL). The best fits of these solutions to the rates are reported in Table 2.

2. *Vacuum Oscillations (VO)*
The concept of vacuum oscillations was first put forward by B.Pontecorvo [29] (for a review see [30]). The survival probability for ν_e neutrino with energy E at distance r is given by

$$P_{\nu_e \to \nu_e} = 1 - \sin^2 2\theta \sin^2 \left(\frac{\Delta m^2}{4E} r \right), \qquad (1)$$

where $l_\nu = 4\pi E / \Delta m^2$ is the vacuum oscillation length. At $\Delta m^2 = 8 \cdot 10^{-11} \ eV^2$ (the best fit) the oscillation length of neutrino with energy $E_\nu \sim$

$3 \ MeV$ is $l_\nu \sim 1 \cdot 10^{13} \ cm$, *i.e.* of order of distance between the Sun and Earth. That is why this VO solution is called *just-so*. Since observational data need large suppression of neutrino flux, by factor ~ 2, $\sin^2 2\theta \sim 1$ is needed: see Eq.(1). Thus just-so VO solution must be large mixing angle solution. The best fit values are given in Table 2.

3. *EIS VO solution*
VO with Energy Independent Suppression (EIS) occurs when $\Delta m^2 \gg 10^{-10} \ eV^2$. In this case the oscillation length is much smaller than distance between the Sun and the Earth. Oscillatory function in Eq.(1) is averaged to factor $1/2$ and hence suppression factor is energy independent and equals to $1 - \frac{1}{2} \sin^2 2\theta$. Since this suppression should be of order 0.5, $\sin^2 2\theta \sim 1$ is needed. On the other hand one must assume $\Delta m^2 \ll 10^{-3} \ eV^2$ because of non-observation of ν_e oscillation in the atmospheric neutrinos.

The energy independent suppression is excluded by observed rates at 99.8%CL [22]. However, the Homestake data give the dominant contribution to this conclusion. If these data are arbitrarily excluded from analysis, EIS VO survives. I will not give more attention to discussion of EIS VO solution. Further details a reader can find in references [31–34].

4. *RSFP solution*
The Resonant Spin-Flavor Precession (RSFP) describes two physical effects working simultaneously: the spin-flavor precession, when neutrino spin (coupled to magnetic moment) precesses around magnetic field, changing simultaneously neutrino flavor, and the resonant, density-dependent effect, which produces difference in potential energy of neutrinos with different flavors (similar to the MSW effect). This complex transition occurs in the external magnetic field due to presence of non-diagonal (transition) neutrino magnetic moments. The RSFP was first recognized in ref.'s [35,36]. For excellent review see [37].

This theory had a predecessor. The precession of neutrino magnetic moment around magnetic field converts left-chiral electron neutrino ν_{eL} into sterile right component ν_{eR}, suppressing thus ν_e-flux[38]. However, the suppression effect in this case is energy-independent and thus contradicts to the observed solar-neutrino rates. In ref's [38,39] the matter effect was included and in [40] spin-flavor precession was discovered. The observed rates in solar-neutrino experiments can be explained only by the RSFP, because only this type of precession give the energy-dependent suppression factor. Majorana neutrino can have only transition magnetic moment. RSFP induces the transition ν_{eL} to $\bar{\nu}_{\mu R}$, i.e. electron neutrino to muon antineutrino, which can scatter off the electron due to NC. The survival probability is similar to that of SMA MSW (see Fig.3 from [41]). Neutrino mixing is not needed directly for RSFP effect, but it is needed indirectly to provide the transition magnetic moment of the Majorana neutrino. To be a solution to SNP, RSFP needs a transition magnetic moment $\mu \sim 10^{-11}\mu_B$, magnetic field in the resonance layer of the Sun, $B \sim 20 - 100\ kG$ and Δm^2 in the range $10^{-8} - 10^{-7}\ eV^2$ (see Section 6).

4. Signatures of Oscillation Solutions

A common signature of most neutrino oscillation solutions is distortion of B-neutrino spectrum. The survival probabilities for SMA MSW, LMA MSW and just-so VO are shown in Fig.2. The survival probabilities for RSFP are similar to SMA MSW and shown in Fig.3. One can see there that LMA MSW (and LOW too) predicts small distortion of B-neutrino spectrum spectrum in the region of observation $5 - 15\ MeV$. For EIS VO the distortion is absent. The strongest spectrum distortion one can expect for SMA MSW and just-so VO. However, spectrum of recoil electrons are distorted weaker than that of neutrinos, because of cross-section and averaging over energy bins in observations (*e.g.* see Fig.8). *The absence of distortion of neutrino or recoil-electron spectra is not a general argument against neutrino oscillations.*

Anomalous NC/CC ratio is another common signature of neutrino oscillations which can be observed in SNO. The NC events will be seen there by detection of neutrons produced in $\nu + D \rightarrow p + n + \nu$ reaction. Oscillation $\nu_e \rightarrow \nu_\mu(\nu_\tau)$ does not change NC interaction but changes CC

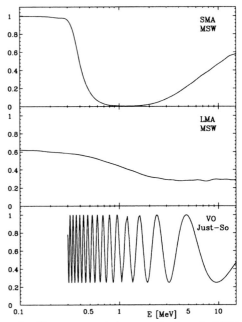

Figure 2. *Electron neutrino survival probabilities.*

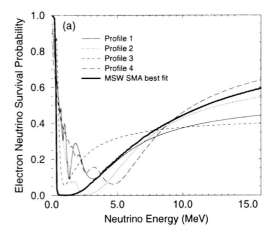

Figure 3. *RSFP survival probabilities from [41] compared with SMA MSW (best fit) survival probability, shown by thick solid line. The four other curves correspond to different radial profiles of magnetic field.*

interaction and thus the ratio of NC/CC rate. In case of oscillation to sterile neutrino the NC/CC ratio is not changed. Therefore, *the normal ratio NC/CC is not a general argument against neutrino oscillations.*

MSW solutions have very specific signatures which are unique. They are day/night effect, zenith angle dependence of solar-neutrino flux and difference in day/night neutrino spectra. All these effects are caused by the MSW matter effect in the Earth.

The signature of *just-so VO* is anomalous seasonal variation of neutrino flux [42]. Due to ellipticity of the Earth's orbit, the distance between the Sun and Earth changes with time, causing the universal 7% variation of the flux due to r^{-2} effect ("geometrical" seasonal variation). As a result of just-so VO the flux of ν_e neutrinos changes additionally due to dependence of survival probability (1) on distance. This effect is absent for MSW solution, because vacuum oscillation length is too small. As follows from Eq.(1) neutrinos with different E have different phases and it weakens the observed effect, since finite energy interval ΔE is used in measurements. In case of monochromatic Be neutrinos the time variations are strongest. For the detailed calculations of anomalous seasonal variations see [43] and for the recent calculations [44–48].

For *EIS VO solution* the anomalous seasonal variations are absent, because the oscillation length is too small. It results in a signature, which can be observed by BOREXINO: Be-neutrino flux is suppressed by a factor ~ 2, but does not show anomalous seasonal variations.

RSFP has two signatures. As a result of RSFP electron neutrinos oscillate inside the Sun into muon/tau antineutrinos. Due to vacuum oscillations on the way to the Earth these neutrinos oscillate to electron antineutrinos. The latter oscillations are suppressed by mixing angle, which is small in case of RSFP. However, even small fluxes of electron antineutrino can be reliably detected (*e.g.* by KamLand [49]). The second signature is prediction of 11-year periodicity for Be-neutrino flux [37]. RSFP occurs in the resonant layers, which are located at different distances for neutrinos of different energies: for pp and B neutri-

nos the resonant layers are located near the solar center and at the periphery, respectively, where magnetic field is weak and RSFP too. The resonant layer for Be neutrinos is located at intermediate distance, where magnetic field is large and RSFP is strongest. 11-year variations of neutrino flux is caused by periodic variation of toroidal magnetic field at the bottom of convective zone. The magnetic activity of the sun exhibits quasiperiodic time variations with the mean period 11 yr. Taking into account changing of magnetic polarity, one can argue for 22 yr as a basic period for large-scale magnetic field. This periodicity is thought to be originated due to toroidal field, generated in so-called *overshoot layer* by dynamo mechanism and located near the bottom of convective zone. Theoretically, magnetic field there can reach 100 kG. This field rises through convective zone to the surface of the sun. The 11 yr periodicity should be observed most effectively by neutrino detectors sensitive to Be-neutrinos: Homestake, GALLEX and BOREXINO. In particular (see Fig.3) when toroidal magnetic field disappears (due to change of magnetic polarity) survival probability increases from ~ 0.1 to ~ 1. Since B-neutrino flux is also suppressed by factor $\sim 0.4 - 0.6$, 11 yr variations should be seen in the combined Kamiokande and Superkamiokande data.

5. 708-day Superkamiokande Data

After 708 days of solar-neutrino observations Superkamiokande has not found direct evidences for neutrino oscillations. There are only some indications to the distortion of the spectrum of recoil electrons, which will be discussed in this Section.

The spectrum of the recoil electrons (708 d) is shown in Fig.4 [1,50] as the ratio to (undistorted) spectrum calculated in BP98 SSM [5]. The spectrum is suppressed by overall factor 0.47, but there is no distortion of the spectrum, except the high energy excess at $E_e \geq 13 \, MeV$. In principle, this excess can be a result of low statistics or small systematic errors at the end of the boron neutrino spectrum. For example, due to very steep end of the electron spectrum, even small systematic error in electron energy (e.g. due to calibration) could enhance the number of events in the highest energy bins.

Another possible explanation of this excess [51] is that the Hep neutrino flux might be significantly larger (about a factor 10–20) than the SSM prediction. The Hep flux depends on solar properties, such as ^{3}He abundance and the temperature, and on S_{13}, the zero-energy astrophysical S-factor of the $p + {}^3He \rightarrow {}^4He + e^+ + \nu$ reaction. Both SSM based [51] and model-independent [52] approaches give a robust prediction for the ratio $\Phi_\nu(Hep)/S_{13}$. Therefore, this scenario implies a cross-section larger by a factor 10–20 than the present calculations. Such a large correction to the calculation does not seem likely, though is not excluded. The signature of Hep neutrinos, the presence of electrons above the maximum boron neutrino energy, can be tested by the SNO experiment.

The observed excess is difficult to explain by neutrino oscillations. The oscillation parameters $\sin^2 2\theta$ and Δm^2, which correspond to the allowed global rates in four neutrino experiments, result in the recoil electron spectra in bad agreement with the excess. The spectra for the best fit MSW solutions (LMA shown by short-dash lines and SMA – by long-dash lines) are displayed in Fig.4 (calculations by K.Inoue).

Night/Day excess is found to be small, consistent with zero within 1.7σ. The excess after 201.6, 504 and 708 days of observation, respectively, looks as follows:

$$\frac{N - D}{N + D} \times 100 = -0.4 \pm 3.1 \pm 1.7 \quad (201.6d)$$

$$\frac{N - D}{N + D} \times 100 = +2.3 \pm 2.0 \pm 1.4 \quad (504d)$$

$$\frac{N - D}{N + D} \times 100 = +2.9 \pm 1.7 \pm 0.39 \quad (708d)$$

The observed (708 d) excess has the right sign, but is statistically insignificant (1.7σ). Note that systematic error is much smaller than observed effect. Statistics, if increased by factor 5, can make the effect statistically significant. *The absence of day/night effect does not rule out MSW solution.*

The zenith-angle dependence is not seen in Superkamiokande data.

The Superkamiokande data for day/night effect, zenith-angle dependence and recoil-electron energy spectrum (especially absence of distortion in most part of the spectrum) have the great restriction power for the oscillation solutions.

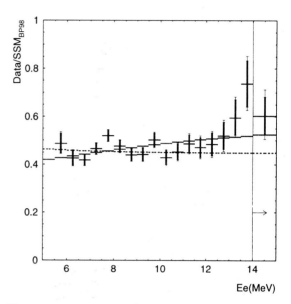

Figure 4. *Energy spectrum of recoil electrons from 708d of Superkamiokande data [1,50]. Plotted is ratio of the observed spectrum to one predicted by BP SSM [5]. The long- and short- dash lines show SMA MSW and LMA MSW spectra, respectively, for best-fit rates (calculations by K.Inoue).*

6. Status of Oscillation Solutions

I will discuss here the status of MSW, VO and RSFP solutions in the light of 708d Superkamiokande data.

6.1 MSW solutions

The regions in oscillation parameter space allowed by global rates do not explain the high energy excess in the recoil-electron spectrum [53] (see Fig.4). MSW solutions need an alternative explanation of this excess, *e.g.* by Hep neutrinos (see Section 5). The status of MSW solutions is determined by *combined* restrictions due to global rates, day/night effect, zenith-angle flux dependence and the energy spectrum (under assumption of arbitrary S_{13}). The result of such analysis is expressed in goodness of the total fit χ^2.

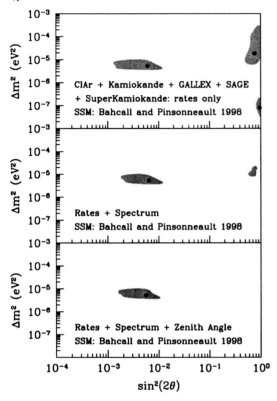

Figure 5. *Status of MSW oscillation solutions after 504 days of Superkamiokande data [22].*

The data of Superkamiokande for 504 days disfavoured LMA MSW solution [22]. In Fig.5 the upper panel shows the regions allowed (at 99% CL) by the global rates. The middle panel includes additionally restrictions due to spectrum and the low one includes three restrictions (rates, spectrum and zenith-angle dependence). All regions are shown at 99% CL. One can see how LMA and LOW solutions disappear.

In Fig.6 the 708d data of Superkamiokande are shown as allowed by rates (upper panel), the regions excluded by day/night effect (middle panel) and excluded by spectrum (low panel). Note, that in the middle panel the regions above and below the central one are allowed at 68% CL.

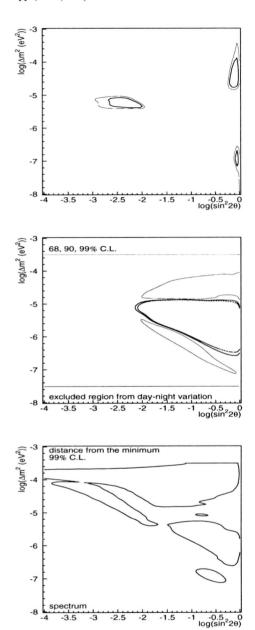

Figure 6. *Status of MSW solutions after 708 days of Superkamiokande data (courtesy of Y. Totsuka)*

The visual inspection shows that all three MSW solutions are allowed.

However, the quantitative analysis of 708d data presented by K.Inoue [1] shows that the SMA MSW solution is not acceptable at 90% CL if day/night effect and spectrum (with free Hep flux) are included in the analysis simultaneously. LMA MSW solution is more favourable.

In conclusion, exclusion of any MSW solution looks unstable and statistics-dependent. I mean that conclusions are changing too drastically with accumulation of data and with method of analysis (inclusion or not day and night spectra, inclusion Hep flux as a free parameter *etc*). Inclusion of too many data together may be misleading if the data are partially inconsistent. Finally, I would like to remind a reader that the data of Superkamiokande are still preliminary, and conclude that it is premature to speak of exclusion of any MSW solutions.

6.2 VO Solutions

If high energy excess in the spectrum is due to Hep neutrinos, just-so VO solution fits the rates and the spectrum. In case the excess is due to oscillations, the regions in oscillation parameter space allowed by the rates (Fig.7, upper panel) are excluded by energy spectrum (low panel). The spectrum is well fitted by vacuum oscillations with $\Delta m^2 = 4.2 \cdot 10^{-10}$ eV^2 and $sin^2 2\theta = 0.93$ [53], but this point is located outside the regions allowed by the rates (Fig.7), *i.e.* it does not represent the SNP solution. The status of this point has been further analysed in [54].

To explain both the excess and the rates it was assumed that boron neutrino flux is 15–20% smaller than the SSM prediction, and that the chlorine signal is about 30% larger than the Homestake observation. This assumed 3.4σ increase of the chlorine signal could have a combined statistical and systematic origin. In practice, the SSM boron neutrino flux and the Homestake signal were rescaled with help of parameters f_B and f_{Cl}, as $\Phi_B = f_B \Phi_B^{SSM}$ and $R_{Cl} = 2.56 f_{Cl}$ SNU, where 2.56 SNU is the Homestake signal.

For each pair f_B and f_{Cl} there were found the parameters $(\Delta m^2, sin^2 2\theta)$, that explain the observed rates, and B-neutrino spectrum was calculated for these parameter values.

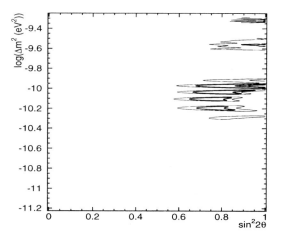

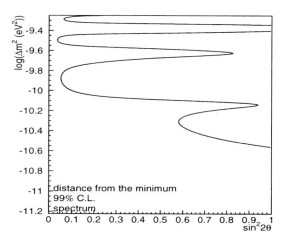

Figure 7. *Just-so VO solution: regions allowed by rates (upper panel) and excluded by spectrum (low panel)– courtesy of Y.Totsuka.*

In particular, for $f_B = 0.8$ and $f_{Cl} = 1.3$ the oscillation parameters ($\Delta m^2 = 4.2 \cdot 10^{-10}$ eV^2, $\sin^2 2\theta = 0.93$) give a good fit to all rates (χ^2/d.o.f. $= 3.0/3$) and to the spectrum with the excess (see Fig.8). More generally, the oscillation parameters give rates in agreement with the experiments at the 2σ level when $0.77 \leq f_B \leq 0.83$ and $1.3 \leq f_{Cl} \leq 1.55$.

These VO solutions will be referred to as HEE VO (with HEE for High Energy Excess) to distinguish them from ordinary just-so VO solutions.

The anomalous seasonal variations are rather

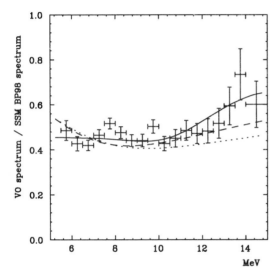

Figure 8. *Ratio of the vacuum oscillation spectra to the SSM spectrum. The solid curve corresponds to the HEE VO solution with $\Delta m^2 = 4.2 \cdot 10^{-10}$ eV^2 and $\sin^2 2\theta = 0.93$. The dashed and dotted curves correspond to the VO solutions of Refs. [15] and [22], respectively.*

unusual in the HEE VO solution. They are described by time dependence of survival probability for the electron neutrino: $P_{\nu_e \to \nu_e}(t)$. In particular, for Be-neutrinos with energy $E = 0.862$ MeV it equals to

$$P(t) = 1 - \sin^2 2\theta \sin^2 \left(\frac{\Delta m^2 a}{4E} \left(1 + e \cos \frac{2\pi t}{T}\right) \right) \quad (2)$$

where $a = 1.496 \cdot 10^{13}$ cm is the semi-major axis, $e = 0.01675$ is the eccentricity of the Earth's orbit, and $T = 1$ yr is the orbital period.

As seen in Fig.9, the case of the HEE VO (solid curve) is dramatically different from the just-so VO case: there are two maxima and minima during one year and the survival probability oscillates between $1 - \sin^2 2\theta \approx 0.14$ and 1. The explanation is obvious: the HEE VO solution has a large Δm^2, which results in a phase $\Delta m^2 a/(4E) \approx 93$, large enough to produce two full harmonics during one year, when the phase changes by about 3% due to the factor $(1 + e \cos 2\pi t/T)$.

The HEE VO solution predicts (see Fig.9) that Be electron neutrinos should arrive almost unsuppressed during about four months a year!

According to the SSM, beryllium neutrinos contribute 34.4 SNU out of the total gallium signal of 129 SNU. Therefore, the strong ^{7}Be neutrino oscillation predicted by the HEE VO solution also implies an appreciable variation of total gallium signal. In Fig.9 the dotted curve shows this variation for the HEE VO solution, which can be compared with the weaker variation in the just-so VO solution (dashed-dotted curve).

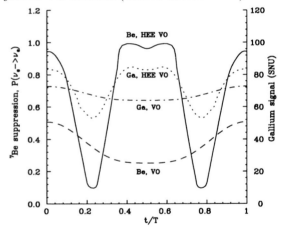

Figure 9. *Anomalous seasonal variations of the beryllium neutrino flux and gallium signal for the VO and HEE VO solutions. The survival probability $P_{\nu_e \to \nu_e}$ for Be neutrinos is given for the HEE VO (solid curve) and for just-so VO (dashed curve) as function of time (T is an orbital period). The dotted (dash-dotted) curve shows the time variation of gallium signal in SNU for the HEE VO and for just-so VO [22] solutions.*

In Figs. 10 - 15 the predictions of HEE VO solution for seasonal time variations are compared with observations of GALLEX, SAGE, Homestake and Superkamiokande. While the agreement of each of the observational data with the HEE VO solution might appear accidental and not sta-

tistically significant, the combined agreement between this model and experiment looks like indication in favour of the HEE VO solution.

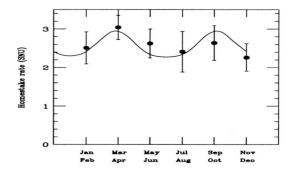

Figure 12. *Seasonal variations predicted by the HEE VO in comparison with the Homestake data [4]. The fit with the HEE VO gives $\chi^2/d.o.f.=1.4/5$, while the fit with constant (no-oscillation) gives $\chi^2/d.o.f=3.1/5$.*

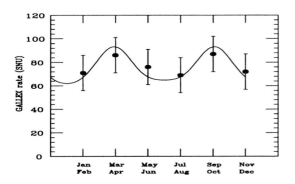

Figure 10. *Seasonal variations predicted by the HEE VO in comparison with the GALLEX data [2]. The fit with the HEE VO has $\chi^2/d.o.f.=0.87/4$, while a time independent fit gives $\chi^2/d.o.f.=1.36/5$.*

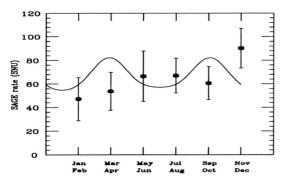

Figure 11. *Seasonal variations predicted by the HEE VO in comparison with the SAGE preliminary data [3]. The fit with the HEE VO has $\chi^2/d.o.f.=8.9/5$, while a time independent fit gives $\chi^2/d.o.f.=3.8/5$.*

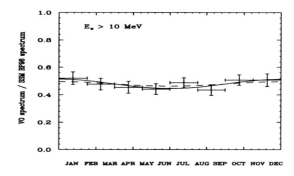

Figure 13. *Seasonal variations predicted by the HEE VO for $E_e > 10$ MeV in comparison with the Superkamiokande data at the same energies. The fit with the HEE VO gives $\chi^2/d.o.f.=2.7/7$, while the one with the geometrical effect only gives $\chi^2/d.o.f.=2.3/7$.*

The anomalous seasonal variation of ν_{Be}- flux predicted by the HEE VO will be reliably tested by BOREXINO and LENS.

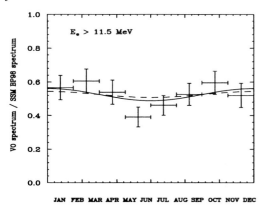

Figure 14. *The same as in Fig.13 for $E_e >$ 11.5 MeV. The fit with the HEE VO gives $\chi^2/d.o.f.=2.7/7$, while the one with the geometrical effect only gives $\chi^2/d.o.f.=2.3/7$.*

6.3 RSFP solution

As was recently demonstrated [41], the RSFP solution can successfully explain the rates (see also [55,56] for early calculations) and high-energy excess in the Superkamiokande spectrum.

This solution has more free parameters to fit the data. For the Majorana neutrino they are: Δm^2, transition magnetic moment μ_ν, scale of toroidal magnetic field in the convective zone, B, and radial profile for magnetic field, $B(r)$, in the wide range of distances. The mixing angle is an arbitrary parameter in the RSFP solution which determines the magnetic moment, but it must be small enough, $sin2\theta < 0.25$ [37].

In Fig.15 the calculated recoil-electron spectra (for four magnetic radial profiles) are compared with the 504d Superkamiokande data. The agreement is reasonably good, though from deflections of the individual points one can guess that χ^2 is not very small. In Fig.16 the regions explaining the rates in four solar-neutrino experiments are shown in parameter space Δm^2 and the mean magnetic field $< B >$ for two magnetic radial

profiles [41]. One can see there the allowed range of parameters. One of the signatures of the RSFP solution, the time-variation of the neutrino signal, is probably testable now. There are two widely discussed effects: 11 year periodicity and (June+December)/(March+Sept) ratio of fluxes. There are some indications to 11yr periodicity in the Homestake signal, especially when correlations with various solar phenomena (sun spot number, green coronal line, surface magnetic field *etc*) are included. My personal opinion is that correlation is allowed as an argument, if the the time variability of the signal is established. Such a lesson was taught us by a bitter experience with Cyg X-3, when correlation with X-ray variability was used as a proof of high-energy gamma-ray signal. Meanwhile, the Homestake signal is perfectly compatible [57] with constant flux ($\chi^2/d.o.f. = 0.6$). The data of all other detectors are also consistent with time-independent flux.

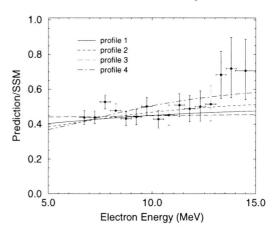

Figure 15. *RSFP spectra for recoil electrons for four different magnetic radial profiles, compared with 504d Superkamiokande data.*

The suppression of neutrino flux in the RSFP model disappears when magnetic field vanishes. It happens in two cases: when polarity of magnetic field in the Sun changes and when neutrino flux arrives, propagating in the plane of solar equator (June 5 and December 5). This effect is strongest for Be-neutrinos (see Fig.3).

The GALLEX data do not show an excess of the rate in June and December or the deficit in March and September. Using three month intervals centered at June 5 and December 5 ("high" rates) and at March 5 and September 5 ("low" rates) the GALLEX collaboration has obtained as a mean rate $78.5 \pm 12\,SNU$ for 20 "high" runs and $90 \pm 12\,SNU$ for 19 "low" runs, *i.e.* within limited statistics the wrong-sign effect (T.Kirsten, private communication). In Figs. 13–14 one can see a similar wrong-sign effect in Superkamiokande data. On the other hand there are no accurate model calculations of this effect in the RSFP models, to compare with the data above.

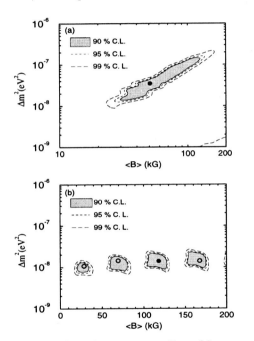

Figure 16. *The RSFP regions allowed by rates for two magnetic field radial profiles [41]*

7. Future Experiments

SNO, e.g. [58], is 1 kt heavy water detector, which will start to operate this year. In contrast to Superkamiokande, electron neutrinos will be directly seen here in the CC current reaction $\nu_e + D \rightarrow p + p + e^-$. Thanks to the large cross-section, event rate is close to that of Superkamiokande. Neutrino energy is given by electron energy and mass difference of D and H. The SNO data can be helpful in detection of Hep neutrinos above the end of B-neutrino spectrum. The NC reactions $\nu_x + D \rightarrow \nu_x + p + n$, seen by presence of neutrons result in anomalous NC/CC ratio in case of oscillation of ν_e to active neutrino component. This is a signature of neutrino oscillation. SNO is more sensitive than Superkamiokande to the day/night effect, which is a "smoking gun" of MSW effect. This is because ν_e neutrinos are directly measured in CC reaction. Detection of day/night effect in Superkamiokande, SNO and probably ICARUS is the last hope for this effect, because all other planned now detectors are not sensitive to it.

ICARUS [59] is a liquid argon detector. Detection of solar neutrinos is based on CC-reaction $\nu_e + {}^{40}Ar \rightarrow {}^{40}K + e^-$ and νe scattering. With excellent energy resolution (about 5%) and low threshold of electron detection (about 5 MeV), ICARUS has great potential for super-precise measurement of electron spectrum and flux of Hep neutrinos.

KamLand [49] is 1kt liquid scintillator detector for $\bar{\nu}_e$ neutrinos, based on the Reines reaction $\bar{\nu}_e + p \rightarrow e^+ + n$. Solar $\bar{\nu}_e$ can be detected, though without clear signature of solar origin (*e.g.* no directionality). Detection of $\bar{\nu}_e$ neutrinos with $E_\nu \sim 3\,MeV$ from a nuclear reactor at distance $L \sim 100\,km$ can test the oscillations with $\Delta m^2 \sim E_\nu/L \sim 3 \cdot 10^{-6}\,eV^2$, *i.e.* close to that of LMA MSW solution.

HELLAZ [60] is a low temperature and high pressure hellium detector, registering neutrinos in νe scattering. The recoil electron energy and scattering angle are measured with high precision and thus the energy of neutrino is known with comparable accuracy. This detector is designed for pp neutrinos, but probably Be neutrinos can be registered too.

BOREXINO and LENS are two low-energy neutrino detectors, complementary in physical interpretation of the results. BOREXINO will start to operate in the beginning of the next millennium. LENS is a new proposal to the Gran Sasso Laboratory based on the recent idea put forward by R.S.Raghavan [62].

BOREXINO at Gran Sasso is 300t liquid scintillator detector for registering *Be*-neutrinos due to νe scattering. It measures CC+NC signal from ν_e neutrinos together with NC signal from ν_μ and ν_τ, in case of oscillation to active components. LENS is a liquid scintillator detector loaded by Yb or Gd nuclei. Neutrinos are detected due to reactions $\nu_e + {}^{176}Yb \rightarrow {}^{176}Lu^* + e^-$ (or $\nu_e + {}^{160}Gd \rightarrow {}^{160}Tb^* + e^-$) with a threshold 244 keV (300 keV). The prompt signal from electron is accompanied by a delayed signal from photon or conversion electron from an excited nucleus. This strongly reduces the background. LENS will detect pp and Be ν_e-neutrinos. The combination of the BOREXINO and LENS data will provide us with the following physical information. (i) With pp- and Be- neutrino fluxes measured separately, the whole neutrino spectroscopy of the Sun will be completed. The suppression of neutrino fluxes at different energies will be explicitly known. (ii) With Be ν_e-neutrino flux known from LENS, BOREXINO will give a signal from ν_μ and ν_τ, in case of ν_e oscillation to active neutrinos. Thus, the combination of both experiments have a status of *appearance oscillation experiment*. In case of ν_e oscillation to sterile neutrino, BOREXINO should not show the signal in excess of that predicted by LENS. (iii) Just-so VO and HEE VO solutions predict strong seasonal variation of Be neutrino flux. The BOREXINO/LENS observations will confirm or reject these models. EIS VO model predicts absence of anomalous seasonal variation accompanied by suppression (by factor ~ 2) of both pp and Be ν_e neutrinos. This is also can be tested by the combined BOREXINO and LENS data. (iv) Both detectors can observe 11yr periodicity in Be-neutrino flux and measure (June+Dec)/(March+Sept) ratio, which are signatures of the RSFP solution.

8. Conclusions

Solar Neutrino Problem (SNP) is deficit of neutrino fluxes as compared to the SSM predictions detected in all solar-neutrino experiments. The astrophysical (including nuclear physics) solution to SNP is excluded or strongly disfavoured. SNP has a status of disappearance oscillation experiment. No direct signature of oscillations has been found yet.

Currently there are six oscillation solutions to SNP: SMA MSW, LMA MSW, LOW MSW, Just-so VO, EIS VO, and RSFP. Two of them are disfavoured: EIS VO (vacuum oscillations with energy-independent suppression) is excluded by observed rates at 99.8% CL and can survive if the Homestake result is excluded from analysis; LOW MSW is seen only at $\geq 99\%$ CL.

Distortion of B-neutrino spectrum (as compared with the SSM specrum) is a common signature of oscillation solutions (which is absent only in EIS VO and weak in the LMA MSW). The ratio of the observed electron spectrum (708d of Superkamiokande data) to that of predicted by the SSM model is flat at $5.5~MeV \leq E_e \leq 13~MeV$ and has an excess at $E_e \geq 13~MeV$. This excess cannot be explained by the MSW solutions and by those just-so VO solutions, which explain the rates. It is not excluded that this excess is due to Hep neutrinos or small systematic experimental error.

Day/Night effect and zenith-angle dependence of neutrino flux is a signature of MSW solutions. After 708d of Superkamiokande observations this effect (in percent) is $2.9 \pm 1.7 \pm 0.30$, i.e. consistent with zero at 1.7σ. Statistics, if increased by factor 5, might make this effect statistically significant. SNO, in the operation soon, is more sensitive than Superkamiokande to day/night effect (due to CC events). There are still chances that day/night effect will be discovered in this round of observations. If not, the future detectors planned at present (BOREXINO, LENS and HELLAZ), will not also be able to see it.

The HEE VO solution with $\Delta m^2 = 4.2 \cdot 10^{-10}$ eV^2 and $\sin^2 2\theta = 0.93$ explains the spectrum with high energy excess and the rates, if B-neutrino flux is assumed to be $15 - 20\%$ smaller than in SSM and if the chlorine signal is about 30% larger than in the Homestake observations. This solution predicts high amplitude semi-annual time variation of Be-neutrino flux, that can be reliably observed by BOREXINO.

Another oscillation solution which explains all rates and Superkamiokande spectrum (including high energy excess) is the RSFP model. An open problem for this model is prediction of 11yr (or 22yr) variations and (June+Dec)/(March+Sept) excess, that are not observed.

Future low-energy neutrino detectors, BOREXINO and LENS, are very sensitive to VO solutions and they will either confirm or reject them.

Acknowledgments

I am grateful to Gianni Fiorentini and Marcello Lissia for enjoyable permanent collaboration and discussions. I am very much indebted to Yoji Totsuka and Kunio Inoue who provided me with the Superkamiokande data in the form of ps-files. I have learned much about Superkamiokande data from discussions with Kunio Inoue. I would like to thank Plamen Krastev for preparing a compilation of figures from Ref.[22] and for discussions. Sandro Bettini and Till Kirsten are thanked for discussions and useful remarks. I am honoured and grateful to the organizers of 19th Texas Symposium for inviting me for a plenary talk. I appreciate very much their efforts to the excellent organization of the conference.

REFERENCES

1. K.Inoue, Talk at 8th Int. Workshop "Neutrino Telescopes", Venice, 23 - 28 February, 1999.
2. T.Kirsten, Proc. of the 18th Int. Conf. "Neutrino 98", 4 - 9 June 1998, to be published in Nucl. Phys. B (Proc. Suppl) 1999.
3. V.Gavrin, Proc. of the 18th Int. Conf. "Neutrino 98", 4 - 9 June 1998, to be published in Nucl. Phys. B (Proc. Suppl) 1999.
4. Homestake collaboration, B .T. Cleveland et al, Ap.J., **496** (1998) 505.
5. J.N.Bahcall, S.Basu and M.H.Pinsonneault, Phys. Lett. **B 433** (1998) 1.
6. J.N.Bahcall and H.A.Bethe, Phys. Rev. Lett. **65**, (1990) 2233.
7. S.Bludman, N.Hata, D.Kennedy and P.Langacker, Phys. Rev. **D 47** (1993) 2220.
8. V.Castellani, S.Degl'Innocenti, G.Fiorentini, Astron. Astrophys. **271** (1993) 601.
9. N.Hata, S.Bludman and P.Langacker, Phys.Rev.**D 49** (1994) 3622.
10. V.Berezinsky, Comm. Nucl. Part. Phys., **21** (1994) 249.
11. J.Bahcall, Phys. Lett., **B 338** (1994) 276.
12. W.Kwong and S.P.Rosen, Phys. Rev. Lett. **73** (1994) 369.
13. S.Degl'Innocenti, G.Fiorentini and M.Lissia, **43** (1995) 66.
14. V.Castellani, S.Degl'Innocenti, G.Fiorentini, M.Lissia and B.Ricci, Phys. Rep. **281** (1997) 309.
15. N.Hata and P.Langacker, Phys. Rev. **D56** (1997) 6107.
16. V.Berezinsky, G.Fiorentini and M.Lissia, Phys. lett. **B365** (1996) 185.
17. W.A.Dziembowski, Bull. Astron. Soc. India, **24** (1996) 133.
18. S.Degl'Innocenti, W.A.Dziembowski, G.Fiorentini and B.Ricci B., Astrop. Phys. **7** (1997) 77.
19. J.N.Bahcall, M.H.Pinsonneault, S.Basu, and Christensen-Dalsgaard J., Phys. Rev. Lett. **78** (1997) 171.
20. A.S.Brun, S.Turck-Chieze and P.Morel, astro-ph/9806272.
21. V.Castellani, S.Degl'Innocenti, W.A.Dziembowski, G.Fiorentini and B.Ricci, Nucl. Phys. B (Proc. Suppl) **70** (1999) 301
22. J.N.Bahcall, P.I.Krastev and A.Yu.Smirnov, Phys. Rev. **D58** (1998) 096016.
23. B.Ricci, V.Berezinsky, S.Degl'Innocenti, W.A.Dziembowski, and G.Fiorentini, Phys. Lett. **B407** (1997) 155.
24. D.Gough, Annals N.Y. Academy Sci., **647** (1992) 199.

25. E.Schatzman, in: Proc. 4th Int. Solar Neutrino Conf., Heidelberg, 4 - 11 April 1997 (ed. W.Hampel) p.21 (1997).
26. A.Cumming and W.C.Haxton, Phys.Rev.Lett. **77** (1996) 4286.
27. V.Berezinsky, G.Fiorentini and M.Lissia, astro-ph/9902222.
28. S.P.Mikheyev and A.Yu.Smirnov, Nuovo Cim. **9 C** (1986) 17,
    S.P.Mikheyev and A.Yu.Smirnov A.Yu., Sov. Phys. Uspekhi **30** (1986) 759,
    L. Wolfenstein, Phys.Rev. **D17** (1978) 2369.
29. B.Pontecorvo, ZhETP **33** (1957) 549.
30. S.M.Bilenky and S.T.Petcov, Rev. Mod. Physics, **59**, (1987) 671.
31. P.F.Harrison, D.H.Perkins, and W.G.Scott, Phys. Lett. **B 349** (1995) 137; Phys, Lett. **B 396** (1997) 186.
32. R.Foot and R.R.Volkas, Phys.Rev. **D 52** (1995) 6595.
33. G.Comforto, A.Marcionni, F.N.Martelli, F.Vetrano, M.Lanfranchi, and G.Torricelli-Ciamponi, Astrop. Phys. **5** (1996) 147.
34. A.J.Baltz, A.S.Goldhaber, and M.Goldhaber, Phys. Rev. Lett. **81** (1998) 5730.
35. E.Akhmedov, Phys. Lett. **B 213** (1998) 64.
36. C.S.Lim and W.J.Marciano, Phys. Rev. **D 37** (1988) 1368.
37. E.Kh.Akhmedov, in: Proc. 4th Int. Solar Neutrino Conf., Heidelberg, (ed. W.Hampel) p.388 (1997).
38. M.B.Voloshin, M.I.Vysotsky, and L.B.Okun, Sov. J. Nucl. Phys. **44** (1986) 845.
39. R.Barbieri and G.Fiorentini, Nucl. Phys. **B 304** (1998) 909.
40. J.Schechter and J.W.F.Valle, Phys. Rev. **D 24** (1981) 1883.
41. M.M.Guzzo and H.Nunokawa, hep-ph/9810408.
42. I.Ya.Pomeranchuk (cited in [30]).
43. P.I.Krastev and S.T.Petcov, Nucl. Phys. **B449** (1995) 605.
44. S.L.Glashow, P.J.Kerman, and L.M.Kraus, Phys. Lett. **B190** (1998) 412.
45. S.P.Mikheyev and A.Yu.Smirnov, Phys. Lett. **B 429** (1998) 343.
46. J.M.Gelb and S.P.Rosen, hep-ph/9809508 (1998).
47. A.Yu.Smirnov, Nucl. Phys. B (Proc. Suppl.) **70** (1999) 324.
48. M.Maris and S.T.Petcov, hep-ph/9903303.
49. A.Suzuki, Talk at "Neutrino 98", Takayama,4 - 9 June 1998, to be published in Nucl. Phys. B (Proc. Suppl.) 1999.
50. Y.Totsuka, Talk at 19th Texas Symposium on Relativistic Astrophysics, Paris, 14 - 18 December, 1998.
51. J.N.Bahcall and P.I.Krastev, Phys.Lett. **B 436** (1998) 243.
52. G.Fiorentini, V.Berezinsky, S. Degl'Innocenti, and B.Ricci, Phys. Lett. **B 444** (1998) 387.
53. Y.Suzuki, Talk at "Neutrino 98", Takayama,4 - 9 June 1998, to be published in Nucl. Phys. B (Proc. Suppl.) 1999.
54. V.Berezinsky, G.Fiorentini, and M.Lissia, hep-ph/9811352
55. K.S.Babu, R.N.Mohapatra and I.Z.Rothstein, Phys. Rev. **D 44** (1991) 2265.
56. E.Kh. Akhmedov, A.Lanza, and S.T.Petcov, Phys. Lett. **B 303** (1993) 85.
57. M.Lissia, Proc. of 18th Texas Symposium on Relativistic Astrophysics, Chicago 15 - 20 December 1996, World Scientific, eds A.V.Olinto, J.A.Frieman and D.N.Schramm, (1998) 706.
58. A.B.MacDonald, Nucl. Phys. B (Proc. Suppl.) **35** (1994) 340.
59. ICARUS collaboration, Proposal v.1 and v.2, Laboratori Nazionali del Gran Sasso , 1993
60. T.Ypsilantis, Proc. 4th Int. Workshop "Neutrino Telescopes", ed M. Baldo-Ceolin (1992) 289.
61. BOREXINO collaboration, Proposal, Univ. of Milan, 1991.
62. R.S.Raghavan, Phys. Rev. Lett. **78** (1997) 3618.

ELSEVIER

Nuclear Physics B (Proc. Suppl.) 80 (2000) 33–40

NUCLEAR PHYSICS B
PROCEEDINGS SUPPLEMENTS

www.elsevier.nl/locate/npe

Extremely high energy cosmic rays

James W. Cronin[a]

[a]Enrico Fermi Institute, University of Chicago,
5640 S. Ellis Ave., Chicago, IL, 60637, USA

The evidence for the existence of cosmic rays with energies in excess of 10^{20} eV is now overwhelming. There is so far no indication of the GZK cutoff in the energy spectrum at 5×10^{19} eV. This conclusion is not firm for lack of statistics. A cutoff would be expected if the sources of the cosmic rays were distributed uniformly throughout the cosmos. The sources of cosmic rays with energy above the GZK cutoff must be at a distance ≤ 100 Mpc, and if they are protons they are very likely to point to these sources. There are no easy explanations how known astrophysical objects can accelerate protons (or atomic nuclei) to these energies. This difficulty has led to speculation that there may be exotic sources such as topological defects which produce these energetic cosmic rays directly along with a copious supply of neutrinos of similar energy. The fluxes of these cosmic rays is very low and large instruments are required to observe them even with modest statistics. One such instrument, the Pierre Auger Observatory, is described. It is designed for all-sky coverage and the construction of its southern site will begin in Argentina in 1999.

1. The cosmic ray energy spectrum above 10^{18} eV

In recent years the interest in extremely high energy cosmic rays (EHECR), those with energy $\geq 10^{18}$ eV (EeV), has revived because of a number of discoveries. Therefore there are many excellent reviews, books, and conference proceedings to which the reader is referred [1]. The energy spectrum of cosmic rays is quite well measured up to 10^{19} eV. Above the knee (3×10^{15} eV) it falls as a power law in energy, $dN/dE \sim E^{-\alpha}$, with an index $\alpha = 3$.

Above 10^{17} eV the cosmic ray spectrum has significant structure, which is displayed in Fig 1, where the differential spectrum has been multiplied by E^3 to better display the observed structures. These data are the contribution of four experiments which have operated over the past 20 years. These experiments observe the cosmic rays indirectly by means of the air showers they produce. They are from the Haverah Park surface array in England [2], the Yakutsk surface array in Siberia [3], the Fly's Eye fluorescence detector in Utah [4], and the AGASA surface array in Japan [5]. Before plotting, the energy scale of each experiment has been adjusted by amounts $\leq 20\%$ to show most clearly the common features. The

method of energy determination in each of these experiments is quite different, and the fact that they agree within 20% is remarkable.

The spectrum continues with an index of 3.0 until about 5×10^{17} eV where it steepens with an index of about 3.3. Above an energy of 10^{18} eV it is difficult for the galaxy to contain even iron nuclei and galactic accelerators that can produce such energies cannot be imagined. If cosmic rays at these energies continue to be produced in the galaxy, they should show a strong anisotropy which correlates with the distribution of matter in our galaxy. Above this energy such a correlation is not observed probably due to lack of statistics. Above 5×10^{18} eV the spectrum hardens to a spectral index of 2.7. This hardening of the spectrum may be due to a new component that is extragalactic.

The composition of the cosmic rays is notoriously difficult to measure with the indirect air shower methods. Such evidence as does exist suggests that the composition is moving towards a lower mean atomic number as the energy increases from 10^{17} eV to 10^{19} eV [6].

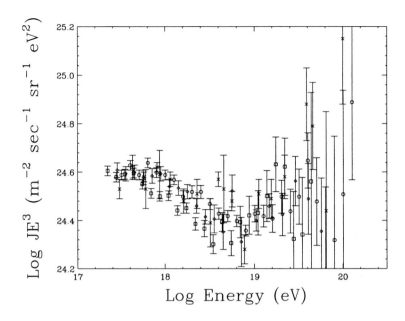

Figure 1. Upper end of the cosmic ray spectrum. Haverah Park [2] points (crosses) serve as a reference. Yakutsk [3] points [diamonds] have been reduced by 20%. Fly's Eye [4] points (squares) have been increased by 10%. Agasa [5] points (circles) have been reduced by 10%.

2. The Difficulty of Acceleration

Above 10^{19} eV the precision of the spectrum measurement suffers from lack of statistics. There have been about 60 events recorded with energy greater than 5×10^{19} eV. Yet it is above this energy that the scientific mystery is the greatest. There is little understanding how known astrophysical objects can produce particles of such energy. At the most primitive level, a necessary condition for the acceleration of a proton to an energy of 10^{20} eV requires that the product of the magnetic field B and the size of the region R be much larger than 3×10^{17} gauss-cm. This value is appropriate for a perfect accelerator such as might be scaled up from the Tevatron at Fermilab. The Tevatron has a BR=3×10^9 gauss-cm and accelerates protons to 10^{12} eV. The possibility of acceleration of cosmic rays to energies above 10^{19} eV seems difficult and the literature is filled with speculations. Two reviews which discuss the

basic requirements are given by Greisen [7] and Hillas [8]. While these were written some time ago, they are excellent in outlining the basic problem of cosmic ray acceleration. Biermann [9] has recently reviewed all the ideas offered to achieve these high energies. Hillas in his outstanding review of 1984 presented a plot which graphically shows the difficulty of cosmic ray acceleration to 10^{20} eV. Figure 2 is a reproduction of his figure. Plotted are the size and strength of possible acceleration sites. The upper limit on the energy is given by:

$$E_{18} \leq 0.5\beta ZB_{\mu g}L_{kpc}.$$

Here the E_{18} is the maximum energy measured in units of 10^{18} eV. L_{kpc} is the size of the accelerating region in units of kilo-parsec, and $B_{\mu g}$ is the magnetic field in μgauss. The factor β was introduced by Greisen to account for the fact that the effective magnetic field in the accelerator analogy is much less than the ambient field. The factor β in the Hillas discussion is the velocity of the shock wave (relative to c) which provides the acceleration. Lines corresponding to a 10^{20} eV proton with $\beta=1$ and 1/300 are plotted. A line is also plotted for iron nuclei ($\beta=1$). With Z=26, iron is in principle easier to accelerate, but in a realistic situation it is difficult to avoid the disintegration of the nucleus during the acceleration process. Real proton accelerators should lie *well* above the solid line. The figure is also relevant for "one shot" acceleration as it represents the emf induced in a conductor of length L moving with a velocity β through a uniform magnetic field B. Synchrotron energy loss is also important. For protons the synchrotron loss rate at 10^{20} eV requires that the magnetic field be less than 0.1 gauss for slow acceleration (the accelerator analogy)[7]. The conclusion from this figure is that the acceleration of cosmic rays to 10^{20} eV is not a simple matter. Because of this difficulty some authors have seriously postulated that the cosmic rays are not accelerated but are directly produced by "top down" processes. Defects in the fabric of space-time can have huge energy content, and can release this energy in the form of high energy cosmic rays [10].

3. Diagnostic Tools

There are some natural diagnostic tools which make the analysis of the cosmic rays above 5×10^{19} eV easier than at lower energies. The first of these is the 2.7K Cosmic Background Radiation (CBR). Greisen [11] and Zatsepin and Kuz'min [12] pointed out that protons, photons, and nuclei all interact strongly with this radiation (GZK effect).

The collision of a proton of 10^{20} eV colliding with a CBR photon of 10^{-3} eV produces several hundred MeV in the center of mass system. The cross section for pion production is quite large so that collisions are quite likely, resulting in a loss of energy for the primary proton. In Fig. 3 we plot the probability for selected proton energies that the source is further away than the indicated distance. This probability assumes that the sources are distributed uniformly and the spectrum has a powerlaw dependence with $\alpha=2.5$.

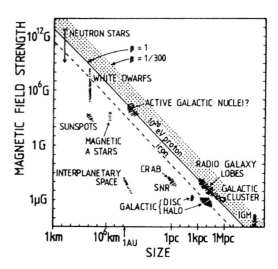

Figure 2. Plot adapted from Hillas [8]. Size and magnetic field of possible sites of acceleration. Objects below the solid line cannot accelerate protons to 10^{20} eV.

Almost independent of the initial energy of a proton, it will be found with less than 10^{20} eV after propagating through a distance of 100 Mpc (3×10^8 light years). Thus the observation of a cosmic ray proton with energy greater than 10^{20} eV implies that its distance of travel is less than 100 Mpc and that its initial energy at its source had to have been much greater. This distance corresponds to a red shift of ~ 0.025 and is small compared to the size of the universe. Similar arguments can be made for nuclei or photons in the energy range considered. There are a limited number of possible sources which fit the Hillas criteria (Fig. 2) within a volume of radius 100 Mpc about the earth. The fact that the cosmic rays, if protons, will be little deflected by galactic and extragalactic magnetic fields serves as a second diagnostic tool [13]. The deflection of protons of energy 5×10^{19} eV by the galactic magnetic field (~ 2 μgauss) and the intergalactic magnetic fields ($\leq 10^{-9}$ gauss) is only a few degrees, so that above 5×10^{19} eV it is possible that the cosmic

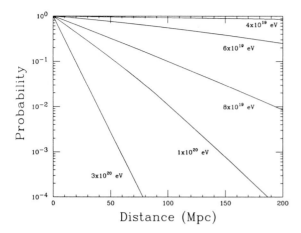

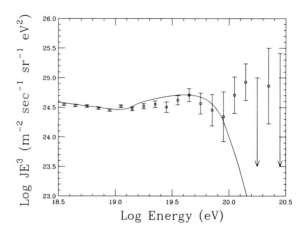

Figure 3. Probability that cosmic ray proton has traveled further than the indicated distance. Sources are assumed to be distributed uniformly throughout space and emit cosmic rays with a differential spectral index $\alpha=2.5$. This figure is based on calculations by Paul Sommers, University of Utah.

Figure 4. Cosmic ray spectrum observed by the AGASA experiment during seven years of operation [14] [18]. The solid line is the spectrum expected for a universal distribution of sources showing the GZK cutoff, including the effects of instrument resolution.

rays will point to their sources. We thus approach an astronomy where the "light" consists of cosmic rays and the sources visible are dominated by those less than 100 Mpc away.

4. Cosmic Ray Astronomy

The energy 5×10^{19} eV represents a lower limit for which the notion of an astronomy of charged particles from "local" sources can be applied. For a universal distribution of sources with energy $\geq 5 \times 10^{19}$ eV, about half of the events will come from a distance of less than 100 Mpc due to attenuation of the more distant sources by the GZK effect. Above this energy the deflection of protons by galactic and extragalactic magnetic fields is expected to be no more than a few degrees. Recently the AGASA group [14] has published the spectrum of EHECR based on seven years of operation of their 100 km^2 surface array. Their spectrum is plotted in Fig 4. There is no evidence of a GZK cutoff in this spectrum. To date about 13 events worldwide have been observed with en-

ergies in excess of 10^{20} eV. Ninty percent of these must have come from a distance less than 50 Mpc. Of these events, two particularly stand out with energies reported to be 2×10^{20} eV by the AGASA experiment [15] and 3×10^{20} eV by the Fly's Eye experiment [16].

The events above 5×10^{19} eV are too few to derive a spectral index. It is not clear that a single spectrum is even the proper way to characterize these events. Since they must come from "nearby" the actual number of sources may not form an effective continuum in space, so the spectrum observed may vary with direction. The distribution of matter within 100 Mpc is not uniformly distributed over the sky. It is probably more fruitful to take an astronomical approach and plot the arrival directions of these events on the sky in galactic coordinates.

Arrival direction data are available for the Haverah Park experiment [17] and the AGASA experiment [18]. In Fig. 5 we plot the arrival directions of 27 AGASA events and 16 Haverah Park events. The size of the symbols corresponds to the angular resolution. What is remarkable in

AGASA (smaller ellipses) Haverah Park (larger ellipses)

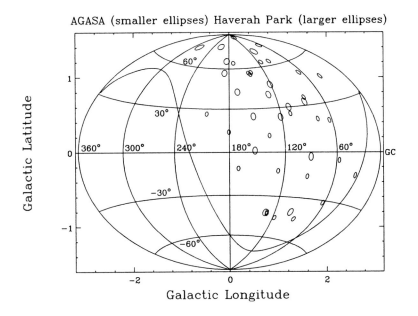

Figure 5. Plot in galactic coordinates of arrival directions of cosmic rays with energy $\geq 5 \times 10^{19}$ eV. Large symbols, Haverah Park [17]; small symbols, AGASA [18]. The size of the symbols indicate the resolution of the experiments (63% of the events within the symbol). The empty region bounded by the solid line is the part of the sky not seen by the experiments which are located in the Northern Hemisphere.

this figure is the number of coincidences of cosmic rays coming from the same direction in the sky. Of the 43 events reported by AGASA and Haverah Park, there are two triplets. The probability of a chance coincidence for this occurance is about 0.1%. From astrophysical experience this small probabiliy is not sufficient to rule out chance, but the possibility that these clusters may be real cannot not be ignored. The cluster in the southern galactic hemisphere contains the largest AGASA event of 2×10^{20} eV, a Haverah Park event of about 1×10^{20} eV, and an AGASA event of 5×10^{19} eV. This cluster is particularly interesting as it contains cosmic rays separated by a factor of four in energy. The intervening magnetic fields have not separated the cosmic rays in space by more than a few degrees. This is an encouraging prospect for future experiments where, with many more events, one may observe point sources, clusters, and larger scale anisotropies in

the sky. The crucial questions will be: Does the distribution of cosmic rays in the sky follow the distribution of matter within our galaxy or the distribution of "nearby" extragalactic matter, or is there no relation to the distribution of matter? Are there point sources or very tight clusters? What is the energy distribution of events from these clusters? Are these clusters associated with specific astrophysical objects? If there is no spatial modulation or no correlation with observed matter, what is the spectrum? This situation would imply an entirely different class of sources which are visible only in the "light" of cosmic rays with energy $\geq 5 \times 10^{19}$ eV. Of course there may be a combination of these possibilities. If even crude data on primary composition is available, it can be divided into catagories of light and heavy components which may have different distributions. Crucial to these considerations is uniform exposure over the whole sky. And a final and perhaps most fundamental question is: Is there an end to the cosmic ray spectrum?

5. Speculations on the nature and origin of the cosmic rays with energy $\geq 10^{20}$ eV.

When the first observation of a cosmic ray with energy $\geq 10^{20}$ eV was made some 36 years ago [19], it was considered a curiosity and attracted very little attention beyond the cosmic ray community. With the more recent results from Haverah Park, Yakutsk, Fly's Eye and particularly AGASA, the existance of these extraordinary cosmic rays cannot be ignored. There is a very rich literature that has emerged in the last few years giving a variety of explanations for these high energy particles. The fact that there is such a variety is the best evidence that we are dealing with a real mystery. We will briefly review the various categories of speculations. We cannot be exhaustive here and will refer the reader to some recent reviews that are more comprehensive.

As in all scientific mysteries it is reasonable to seek the explanation in terms of known astrophysical objects and known laws of physics. As we pointed out in section 2, acceleration by "conventional" means is difficult. The possible mechanisms for acceleration have recently been discussed by Blandford [20]. While he cannot conclude that it is impossible to accelerate cosmic rays to 10^{20} eV, he points out the extreme difficulty to achieve those energies. There is also the question of whether there are a sufficient number of qualifying sources within the distances limited by the GZK effect.

These difficulties have led to speculations that the cosmic rays are produced by direct creation at the highest energy; no acceleration is required. These sources are generally explained in terms of topological defects. These "top down" processes have been extensively discussed in several recent reviews. [21] [22] [23]. The characteristics of topological defect models are the emission of large fluxes of neutrinos and γ-rays along with the protons. The energy spectrum of the γ-rays can be degraded depending on the strength of random magnetic fields between the source and the earth. While most predictions normalize the

density of defects to the observed cosmic rays, the relic densities that are required are often difficult to explain as having evolved from the early universe. One of the speculations concerns massive heavy relics which collect about the galactic center [24]. Decays of these particles will produce a spectrum of energies unaffected by the GZK effect since they are galactic in origin and there will be a prenounced enhancement towards the galactic center.

There has been extensive work on the observational consequences of bursting sources, which could be either topological defects or γ-ray bursts [25]. A time delay is expected between the most energetic cosmic ray and lower energies, with the highest energy arriving first. The scale of the time delay depends on the strength of the random magnetic fields and the distance to the source. For B=10^{-9} μ gauss and 30 Mpc distance typical time delays between the most energetic and least energetic cosmic ray in the triplet observed in southern galactic hemisphere is 2×10^5 years. The observed triplets are most likely from a continuous source or a tight cluster of independent sources.

There have been a number of speculations on how to evade the GZK cutoff. These do not address the question of how the enormous energies are produced, but seek to have the cosmic rays transported to the earth from the entire universe. Farrar and Bierman [26] suggest that distant quasars ($z \sim 1$) produce neutral (SUSY) particles that thread their way undeflected through the CBR. A failure of Lorentz invariance can alter the kinematics of the collisions between the cosmic ray proton and the cosmic background photons so that the pion production is below threshold [27] [28]. It has been suggested that cosmologically produced neutrinos collide with nearby relic neutrinos producing Z bosons whose decays are the observed cosmic rays [29]. This idea requires that at least one of the relic neutrinos have a mass of the order of a few eV. At present measurements of the electron neutrino mass give the unphysical result that the square of the mass is negative. Taking this to be the truth, an explanation of many of the observed facts concerning cosmic rays including evasion of the GZK cutoff has been crafted [30].

Many of these speculations discussed above seem farfetched and many will be discarded when a much larger sample of data can be collected. To achieve this goal a number of new projects are underway.

6. New detectors for EHECR

The discussion so far makes clear that the EHECR's are a mystery that will require even larger detectors than the present AGASA detector which has an aperture of ~ 125 km^2-sr. The rate of cosmic rays with energy $\geq 5 \times 10^{19}$ eV is about 4 km^{-2}-sr^{-1}-century^{-1} and 1 km^{-2}-sr^{-1}-century^{-1} for energies $\geq 10^{20}$ eV. Thus detectors with very large acceptances are required to gather even modest statistics.

The next improvement in the sensitivity of cosmic ray detectors will be the High Resolution Fly's Eye (HiRes) [31]. It anticipates a time-averaged aperture of 300 km^2-sr at 10^{19} eV and 1000 km^2-sr at 10^{20} eV. The HiRes experiment is expected to begin operation at the end of 1999.

We will briefly describe a more ambitious approach to the problem, the Pièrre Auger Observatories named after the French physicist who discovered extensive air showers [32]. In 1938 Auger demonstrated that particles were arriving from outside the earth with energies $\geq 10^{15}$ eV[33].

The Auger project is a comprehensive experiment designed to study cosmic rays with energy $\geq 10^{19}$ eV with the least possible bias concerning theories of their origin. Since the cosmic rays are likely to point to the sources, a comprehensive study requires that the entire celestial sphere be observed.

Two instruments will be built at mid-latitude sites in the southern and northern hemispheres. Each instrument will observe cosmic rays at zenith angles up to 60°, so that as the earth turns the whole sky is nearly uniformly observed. Each instrument is a hybrid consisting of a surface array to measure the lateral distribution of the shower particles on the ground, and a fluorescence detector to measure the longitudinal development of the shower. The surface array consists of 1600 water tanks (10 m^2 x 1.2 m deep) spread over 3000 km^2. The configuration of the fluorescence detectors is such that when conditions permit their operation (dark, moonless nights) they will register $\geq 90\%$ of the showers which trigger the surface array. Approximately 10% of the showers will be observed by both detectors. This subset of events will permit a cross check of the energy and provide the maximum possible information on the composition of the primary.

An international collaboration consisting of 19 countries is pooling resources to build two cosmic ray observatories each with an aperture of 7000 km^2-sr for energies $\geq 10^{19}$ eV. The observatories are to be built in Mendoza Province, Argentina (35.2° S, 69.2° W, altitude 1400m) and the state of Utah in the United States (39.1°, 112.6° W, altitude 1400m). Construction has begun at the southern site and is expected to be complete by 2003. Construction is expected to begin in the northern site in 2002.

REFERENCES

1. Yoshida, S. and Dai H., 1998, J. Phys. **G 24**, 905; Sokolsky, P., P. Sommers, and B. R. Dawson, 1992 Physics Rep. **217**, 225; *Proceedings of the Paris Workshop on the Highest Energy Cosmic Rays*, 1992, Nucl. Phys. lo(Proc. Suppl.) **B 28**, 213; *Proceedings of the International Symposium on Extremely High Energy Cosmic Rays: Astrophysics and Future Observations*, 1996, ed M. Nagano, (Institute for Cosmic Ray Research, University of Tokyo); Swordy, S., rapporteur talk, 1994, *Proceedings of the 23rd International Cosmic Ray Conference, (Calgary)* 243; Watson, A. A., 1991, Nucl. Phys. (Proc. Suppl.) **B 22**, 116; V. S. Berezinskiĭ, S. V. Bulanov, V. A. Dogiel, V. L. Ginzburg (editor) and V. S. Ptuskin, 1990, *Astrophysics of Cosmic Rays*, North Holland, Elsevier Science Publishers, The Netherlands;Bhattachargee, P. and G. Sigl, astro-ph/9811011, submitted to Physics Reports.

2. Lawrence, M. A., et al., 1991, J. Phys. **G 17**, 773.

3. Afanasiev, B. N., et al., 1995, *Proceedings of the 24th International Cosmic Ray Confer-*

ence (Rome) **2**, 756.

4. Bird, D. J., et al., 1994, Ap. J. **424**, 491.

5. Yoshida S., et al., 1995, Astroparticle Physics **3**, 105.

6. Yoshida, S. and H. Dai , 1998, J. Phys. **G 24**, 905.

7. Greisen, K., 1965, *Proceedings of the 9th International Cosmic Ray Conference (London)* **2**, 609.

8. Hillas, A. M., 1984, Ann. Rev. Astron. Astrophys. **22**, 425.

9. Biermann, P., 1997, J. Phys. **G 23**, 1.

10. Bhattacharjee, P., C. T. Hill, and D. N. Schramm, 1992, Phys. Rev. Letters **69**, 567.

11. Greisen, K., 1966, Phys. Rev. Letters **16**, 748.

12. Zatsepin, G. T. and V. A. Kuz'min, 1966, JETP Letters **4**, 78.

13. Kronberg, P. P., 1994, Rep. Prog. Phys. **57**, 325; Kronberg, P. P., 1994, Nature **370**, 179; Cole, P., Comments Astrophys. **16**, 1992, 45.

14. Takeda, M., et al., 1998, Phys. Rev. Letters, **81**, 1163.

15. Hayashida, N., et al., 1994, Phys. Rev. Letters, **73**, 3491.

16. Bird, D. J., et al., 1995, Ap. J., **441**, 144; Elbert, J. W. and P. Sommers, 1995, Ap. J., **441**, 151.

17. Watson, A. A., 1997, University of Leeds, private communication.

18. Takeda, M. et al., 1999, astro-ph/9902239, to be published in Astrophys. J.

19. Linsley, J., 1962, Phys. Rev. Letters, **10**, 146

20. Blandford, R., *Acceleration of Ultra High Energy Cosmic Rays* 1998, paper presented at Nobel Symposium, August 1998, Stockholm, Sweden.

21. Berezinsky, V., P. Blasi, and A. Vilenkin, 1998, astro-ph/9803271.

22. Bhattachargee, P., and G. Sigl, 1998, astro-ph/9811011.

23. Hillas M., 1998, Nature, **395**, 15.

24. Berezinsky, V., 1997, Phys. Rev. Letters, **79**, 4302.

25. Waxman, E., and J. Miralda-Escudé, 1996, Astrophys. J. **473** L89; Lemoine, M., et al., 1997, Astrophys. J., **486**, L115.

26. Farrar, G. R. and P. L. Bierman, 1998, Phys. Rev. Letters, **81**, 3579.

27. Glashow, S. L. and S. Coleman, 1997, Phys. Letters, **B405**, 249; hep-ph/9703240.

28. Gonzalez-Mestres, L., 1997, in *Proceedings of Workshop on Observing Giant Cosmic Airshowers from $\geq 10^{20}$ eV Particles from Space: AIP conf. proc. 433* ed J. Krizmanic, et al., p 148.

29. Weiler T., 1997, astro-ph/9807324.

30. Ehrlich, R., 1998, astro-ph/9807324.

31. Abu-Zayyad, T., et al., 1997, in *Proceedings of the 25th International Cosmic Ray Conference*, Durban, edited by M. S. Potgeiter, B. C. Raubenheimer, and D. J. van der Walt (World Scientific, Singapore), Vol 5, p321; this paper and the eleven that immediately follow describe various aspects of the HiRes detector.

32. Pièrre Auger Design Report, 2nd ed. March 1997, Fermilab; Design Report and more than 200 technical notes available at Auger web site, www.auger.org.

33. Auger, P., et al., 1938, Comptes Rendus **206**, 1721; Auger, P., 1939, Rev. Mod. Phys. **11**, 288.

ELSEVIER

Nuclear Physics B (Proc. Suppl.) 80 (2000) 41–50

NUCLEAR PHYSICS B
PROCEEDINGS
SUPPLEMENTS

www.elsevier.nl/locate/npe

Experimental Tests of Relativistic Gravity

Thibault DAMOUR[a]

[a]Institut des Hautes Etudes Scientifiques, 91440 Bures-sur-Yvette, France
and DARC, CNRS - Observatoire de Paris, 92195 Meudon Cedex, France

The confrontation between Einstein's gravitation theory and experimental results, notably binary pulsar data, is summarized and its significance discussed. Experiment and theory agree at the 10^{-3} level or better. All the basic structures of Einstein's theory (coupling of gravity to matter; propagation and self-interaction of the gravitational field, including in strong-field conditions) have been verified. However, the theoretical possibility that scalar couplings be naturally driven toward zero by the cosmological expansion suggests that the present agreement between Einstein's theory and experiment might be compatible with the existence of a long-range scalar contribution to gravity (such as the dilaton field, or a moduli field, of string theory). This provides a new theoretical paradigm, and new motivations for improving the experimental tests of gravity.

1. Introduction

Einstein's gravitation theory can be thought of as defined by two postulates. One postulate states that the action functional describing the propagation and self-interaction of the gravitational field is

$$S_{\text{gravitation}} = \frac{c^4}{16\pi\,G} \int \frac{d^4x}{c}\ \sqrt{g}\ R(g). \qquad (1)$$

A second postulate states that the action functional describing the coupling of all the fields describing matter and its electro-weak and strong interactions (leptons and quarks, gauge and Higgs bosons) is a (minimal) deformation of the special relativistic action functional used by particle physicists (the so called "Standard Model"), obtained by replacing everywhere the flat Minkowski metric $\eta_{\mu\nu} = \text{diag}(-1, +1, +1, +1)$ by $g_{\mu\nu}(x^\lambda)$ and the partial derivatives $\partial_\mu \equiv \partial/\partial x^\mu$ by g-covariant derivatives ∇_μ. Schematically, one has

$$S_{\text{matter}} = \int \frac{d^4x}{c}\ \sqrt{g}\ \mathcal{L}_{\text{matter}}\ [\psi, A_\mu, H; g_{\mu\nu}]. \qquad (2)$$

Einstein's theory of gravitation is then defined by extremizing the total action functional, $S_{\text{tot}}\ [g, \psi, A, H] = S_{\text{gravitation}}\ [g] + S_{\text{matter}}\ [\psi, A, H, g]$.

Although, seen from a wider perspective, the two postulates (1) and (2) follow from the unique requirement that the gravitational interaction be mediated only by massless spin-2 excitations [1], the decomposition in two postulates is convenient for discussing the theoretical significance of various tests of General Relativity. Let us discuss in turn the experimental tests of the coupling of matter to gravity (postulate (2)), and the experimental tests of the dynamics of the gravitational field (postulate (1)). For more details and references we refer the reader to [2] or [3].

2. Experimental tests of the coupling between matter and gravity

The fact that the matter Lagrangian depends only on a symmetric tensor $g_{\mu\nu}(x)$ and its first derivatives (i.e. the postulate of a universal "metric coupling" between matter and gravity) is a strong assumption (often referred to as the "equivalence principle") which has many observable consequences for the behaviour of localized test systems embedded in given, external gravitational fields. In particular, it predicts the constancy of the "constants" (the outcome of local non-gravitational experiments, referred to local standards, depends only on the values of the coupling constants and mass scales entering the Standard Model) and the universality of free fall (two test bodies dropped at the same location and with the same velocity in an external gravitational field fall in the same way, independently of their masses and compositions).

Many sorts of data (from spectral lines in distant galaxies to a natural fission reactor phenomenon which took place at Oklo, Gabon, two billion years ago) have been used to set limits on a possible time variation of the basic coupling constants of the Standard Model. The best results concern the electromagnetic coupling, i.e. the fine-structure constant α_{em}. A recent reanalysis of the Oklo phenomenon gives a conservative upper bound [4]

$$-6.7 \times 10^{-17}\,\mathrm{yr}^{-1} < \frac{\dot{\alpha}_{em}}{\alpha_{em}} < 5.0 \times 10^{-17}\,\mathrm{yr}^{-1}, (3)$$

which is much smaller than the cosmological time scale $\sim 10^{-10}$ yr^{-1}. It would be interesting to confirm and/or improve the limit (3) by direct laboratory measurements comparing clocks based on atomic transitions having different dependences on α_{em}. [Current atomic clock tests of the constancy of α_{em} give the limit $|\dot{\alpha}_{em}/\alpha_{em}| < 3.7 \times 10^{-14}$ yr^{-1} [5].]

The universality of free fall has been verified at the 10^{-12} level both for laboratory bodies [6], e.g. (from the last reference in [6])

$$\left(\frac{\Delta a}{a}\right)_{\mathrm{Be\,Cu}} = (-1.9 \pm 2.5) \times 10^{-12}, \qquad (4)$$

and for the gravitational accelerations of the Moon and the Earth toward the Sun [7],

$$\left(\frac{\Delta a}{a}\right)_{\mathrm{Moon\,Earth}} = (-3.2 \pm 4.6) \times 10^{-13}. \qquad (5)$$

In conclusion, the main observable consequences of the Einsteinian postulate (2) concerning the coupling between matter and gravity ("equivalence principle") have been verified with high precision by all experiments to date (see Refs. [2], [3] for discussions of other tests of the equivalence principle). The traditional paradigm (first put forward by Fierz [8]) is that the extremely high precision of free fall experiments (10^{-12} level) strongly suggests that the coupling between matter and gravity is exactly of the "metric" form (2), but leaves open possibilities more general than eq. (1) for the spin-content and dynamics of the fields mediating the gravitational interaction. We shall provisionally adopt this paradigm to discuss the tests of the other Einsteinian postulate, eq. (1). However, we shall emphasize at the end that recent theoretical findings suggest a new paradigm.

3. Tests of the dynamics of the gravitational field in the weak field regime

Let us now consider the experimental tests of the dynamics of the gravitational field, defined in General Relativity by the action functional (1). Following first the traditional paradigm, it is convenient to enlarge our framework by embedding General Relativity within the class of the most natural relativistic theories of gravitation which satisfy exactly the matter-coupling tests discussed above while differing in the description of the degrees of freedom of the gravitational field. This class of theories are the metrically-coupled tensor-scalar theories, first introduced by Fierz [8] in a work where he noticed that the class of non-metrically-coupled tensor-scalar theories previously introduced by Jordan [9] would generically entail unacceptably large violations of the equivalence principle. The metrically-coupled (or equivalence-principle respecting) tensor-scalar theories are defined by keeping the postulate (2), but replacing the postulate (1) by demanding that the "physical" metric $g_{\mu\nu}$ (coupled to ordinary matter) be a composite object of the form

$$g_{\mu\nu} = A^2(\varphi)\, g^*_{\mu\nu}, \qquad (6)$$

where the dynamics of the "Einstein" metric $g^*_{\mu\nu}$ is defined by the action functional (1) (written with the replacement $g_{\mu\nu} \to g^*_{\mu\nu}$) and where φ is a massless scalar field. [More generally, one can consider several massless scalar fields, with an action functional of the form of a general nonlinear σ model [10]]. In other words, the action functional describing the dynamics of the spin 2 and spin 0 degrees of freedom contained in this generalized theory of gravitation reads

$$S_{\mathrm{gravitational}}[g^*_{\mu\nu}, \varphi] = \frac{c^4}{16\pi G_*} \int \frac{d^4x}{c}\sqrt{g_*}$$
$$\times [R(g_*) - 2g_*^{\mu\nu}\, \partial_\mu\, \varphi\, \partial_\nu\, \varphi]. (7)$$

Here, G_* denotes some bare gravitational coupling constant. This class of theories contains an arbitrary function, the "coupling function" $A(\varphi)$. When $A(\varphi) = $ const., the scalar field is not coupled to matter and one falls back (with suitable boundary conditions) on Einstein's theory. The simple, one-parameter subclass $A(\varphi) = \exp(\alpha_0 \, \varphi)$ with $\alpha_0 \in \mathbb{R}$ is the Jordan-Fierz-Brans-Dicke theory [8], [11], [12]. In the general case, one can define the (field-dependent) coupling strength of φ to matter by

$$\alpha(\varphi) \equiv \frac{\partial \ln A(\varphi)}{\partial \varphi}. \tag{8}$$

It is possible to work out in detail the observable consequences of tensor-scalar theories and to contrast them with the general relativistic case (see, e.g., ref. [10]).

Let us now consider the experimental tests of the dynamics of the gravitational field that can be performed in the solar system. Because the planets move with slow velocities ($v/c \sim 10^{-4}$) in a very weak gravitational potential ($U/c^2 \sim (v/c)^2 \sim 10^{-8}$), solar system tests allow us only to probe the quasi-static, weak-field regime of relativistic gravity (technically described by the so-called "post-Newtonian" expansion). In the limit where one keeps only the first relativistic corrections to Newton's gravity (first post-Newtonian approximation), all solar-system gravitational experiments, interpreted within tensor-scalar theories, differ from Einstein's predictions only through the appearance of two "post-Einstein" parameters $\overline{\gamma}$ and $\overline{\beta}$ (related to the usually considered Eddington parameters γ and β through $\overline{\gamma} \equiv \gamma - 1$, $\overline{\beta} \equiv \beta - 1$). The parameters $\overline{\gamma}$ and $\overline{\beta}$ vanish in General Relativity, and are given in tensor-scalar theories by

$$\overline{\gamma} = -2 \, \frac{\alpha_0^2}{1+\alpha_0^2}, \tag{9}$$

$$\overline{\beta} = +\frac{1}{2} \, \frac{\beta_0 \, \alpha_0^2}{(1+\alpha_0^2)^2}, \tag{10}$$

where $\alpha_0 \equiv \alpha(\varphi_0)$, $\beta_0 \equiv \partial\alpha(\varphi_0)/\partial\varphi_0$; φ_0 denoting the cosmologically-determined value of the scalar field far away from the solar system. Essentially, the parameter $\overline{\gamma}$ depends only on the

linearized structure of the gravitational theory (and is a direct measure of its field content, i.e. whether it is pure spin 2 or contains an admixture of spin 0), while the parameter $\overline{\beta}$ parametrizes some of the quadratic nonlinearities in the field equations (cubic vertex of the gravitational field).

All currently performed gravitational experiments in the solar system, including perihelion advances of planetary orbits, the bending and delay of electromagnetic signals passing near the Sun, and very accurate range data to the Moon obtained by laser echoes, are compatible with the general relativistic predictions $\overline{\gamma} = 0 = \overline{\beta}$ and give upper bounds on both $|\overline{\gamma}|$ and $|\overline{\beta}|$ (i.e. on possible fractional deviations from General Relativity). The best current limits come from: (i) VLBI measurements of the deflection of radio waves by the Sun, giving [13]: $-3.8 \times 10^{-4} < \overline{\gamma} < 2.6 \times 10^{-4}$, and (ii) Lunar Laser Ranging measurements of a possible polarization of the orbit of the Moon toward the Sun ("Nordtvedt effect" [14]) giving [7]: $4\overline{\beta} - \overline{\gamma} = -0.0007 \pm 0.0010$.

The corresponding bounds on the scalar coupling parameters α_0 and β_0 are: $\alpha_0^2 < 1.9 \times 10^{-4}$, $-8.5 \times 10^{-4} < (1 + \beta_0)\alpha_0^2 < 1.5 \times 10^{-4}$. Note that if one were working in the more general (and more plausible; see below) framework of theories where the scalar couplings violate the equivalence principle one would get much stronger constraints on the basic coupling parameter α_0 of order $\alpha_0^2 \lesssim 10^{-7}$ [15].

The parametrization of the weak-field deviations between generic tensor-scalar theories and Einstein's theory has been extended to the second post-Newtonian order [16]. Only two post-post-Einstein parameters, ε and ζ, representing a deeper layer of structure of the gravitational interaction, show up. These parameters have been shown to be already significantly constrained by binary-pulsar data: $|\varepsilon| < 7 \times 10^{-2}$, $|\zeta| < 6 \times 10^{-3}$.

4. Tests of the dynamics of the gravitational field in the strong field regime

In spite of the diversity, number and often high precision of solar system tests, they have an important qualitative weakness : they probe neither the radiation properties nor the strong-field aspects of relativistic gravity. Fortunately, the discovery [17] and continuous observational study of pulsars in gravitationally bound binary orbits has opened up an entirely new testing ground for relativistic gravity, giving us an experimental handle on the regime of strong and/or radiative gravitational fields.

The fact that binary pulsar data allow one to probe the propagation properties of the gravitational field is well known. This comes directly from the fact that the finite velocity of propagation of the gravitational interaction between the pulsar and its companion generates damping-like terms in the equations of motion, i.e. terms which are directed against the velocities. [This can be understood heuristically by considering that the finite velocity of propagation must cause the gravitational force on the pulsar to make an angle with the instantaneous position of the companion [18], and was verified by a careful derivation of the general relativistic equations of motion of binary systems of compact objects [19]]. These damping forces cause the binary orbit to shrink and its orbital period P_b to decrease. The measurement, in some binary pulsar systems, of the secular orbital period decay $\dot{P}_b \equiv dP_b/dt$ [20] thereby gives us a direct experimental probe of the damping terms present in the equations of motion.

The fact that binary pulsar data allow one to probe strong-field aspects of relativistic gravity is less well known. The a priori reason for saying that they should is that the surface gravitational potential of a neutron star $Gm/c^2R \simeq 0.2$ is a mere factor 2.5 below the black hole limit (and a factor $\sim 10^8$ above the surface potential of the Earth). Due to the peculiar "effacement" properties of strong-field effects taking place in General Relativity [19], the fact that pulsar data probe the strong-gravitational-field regime can only be seen when contrasting Einstein's theory with more general theories. In particular, it has been found in tensor-scalar theories [21] that a self-gravity as strong as that of a neutron star can naturally (i.e. without fine tuning of parameters) induce order-unity deviations from general relativistic predictions in the orbital dynamics of a binary pulsar thanks to the existence of nonperturbative strong-field effects. [The adjective

"nonperturbative" refers here to the fact that this phenomenon is nonanalytic in the coupling strength of the scalar field, eq. (8), which can be as small as wished in the weak-field limit]. As far as we know, this is the first example where large deviations from General Relativity, induced by strong self-gravity effects, occur in a theory which contains only positive energy excitations and whose post-Newtonian limit can be arbitrarily close to that of General Relativity.

A comprehensive account of the use of binary pulsars as laboratories for testing strong-field gravity will be found in ref. [22]. Two complementary approaches can be pursued : a phenomenological one ("Parametrized Post-Keplerian" formalism), or a theory-dependent one [10], [22], [23].

The phenomenological analysis of binary pulsar timing data consists in fitting the observed sequence of pulse arrival times to the generic DD timing formula [24] whose functional form has been shown to be common to the whole class of tensor-multi-scalar theories. The least-squares fit between the timing data and the parameter-dependent DD timing formula allows one to measure, besides some "Keplerian" parameters ("orbital period" P_b, "eccentricity" e,...), a maximum of eight "post-Keplerian" parameters: $k, \gamma, \dot{P}_b, r, s, \delta_\theta, \dot{e}$ and $\dot{x}$. Here, $k \equiv \dot{\omega}P_b/2\pi$ is the fractional periastron advance per orbit, γ a time dilation parameter (not to be confused with its post-Newtonian namesake), $\dot{P}_b$ the orbital period derivative mentioned above, and r and s the "range" and "shape" parameters of the gravitational ("Shapiro") time delay caused by the companion. The important point is that the post-Keplerian parameters can be measured without assuming any specific theory of gravity. Now, each specific relativistic theory of gravity predicts that, for instance, $k, \gamma, \dot{P}_b, r$ and s (to quote parameters that have been successfully measured from some binary pulsar data) are some theory-dependent functions of the (unknown) masses m_1, m_2 of the pulsar and its companion. Therefore, in our example, the five simultaneous phenomenological measurements of $k, \gamma, \dot{P}_b, r$ and s determine, for each given theory, five corresponding theory-dependent curves in the $m_1 - m_2$ plane (through the 5 equations $k^{\text{measured}} = k^{\text{theory}}(m_1, m_2)$, etc...). This yields three $(3 = 5 - 2)$ tests of the specified theory, according to whether the five curves meet at one point in the mass plane, as they should. [In the most general (and optimistic) case, discussed in [22], one can phenomenologically analyze both timing data and pulse-structure data (pulse shape and polarization) to extract up to nineteen post-Keplerian parameters.] The theoretical significance of these tests depends upon the physics lying behind the post-Keplerian parameters involved in the tests. For instance, as we said above, a test involving $\dot{P}_b$ probes the propagation (and helicity) properties of the gravitational interaction. But a test involving, say, k, γ, r or s probes (as shown by combining the results of [10] and [21]) strong self-gravity effects independently of radiative effects.

Besides the phenomenological analysis of binary pulsar data, one can also adopt a theory-dependent methodology [10], [22], [23]. The idea here is to work from the start within a certain finite-dimensional "space of theories", i.e. within a specific class of gravitational theories labelled by some theory parameters. Then by fitting the raw pulsar data to the predictions of the considered class of theories, one can determine which regions of theory-space are compatible (at say the 90% confidence level) with the available experimental data. This method can be viewed as a strong-field generalization of the parametrized post-Newtonian formalism [2] used to analyze solar-system experiments. When non-perturbative strong-field effects are absent one can parametrize strong-gravity effects in neutron stars by using an expansion in powers of the "compactness" $c_A \equiv -2\, \partial \ln m_A / \partial \ln G \sim G\, m_A / c^2\, R_A$. Ref. [10] has then shown that the observable predictions of generic tensor-multi-scalar theories could be parametrized by a sequence of "theory parameters", $\overline{\gamma}$, $\overline{\beta}$, β_2, β', β'', β_3, $(\beta\beta')$... representing deeper and deeper layers of structure of the relativistic gravitational interaction beyond the first-order post-Newtonian level parametrized by $\overline{\gamma}$ and $\overline{\beta}$. When non-perturbative strong-field effects develop, one cannot use the

multi-parameter approach just mentioned. A useful alternative approach is then to work within specific, low-dimensional "mini-spaces of theories". Of particular interest is the two-dimensional mini-space of tensor-scalar theories defined by the coupling function $A(\varphi) = \exp\left(\alpha_0 \varphi + \frac{1}{2}\beta_0 \varphi^2\right)$. The predictions of this family of theories (parametrized by α_0 and β_0) are analytically described, in weak-field contexts, by the post-Einstein parameter (9), and can be studied in strong-field contexts by combining analytical and numerical methods [23].

Let us now briefly summarize the current experimental situation. Concerning the first discovered binary pulsar PSR1913 + 16 [17], it has been possible to measure with accuracy the three post-Keplerian parameters k, γ and $\dot{P}_b$. From what was said above, these three simultaneous measurements yield *one* test of gravitation theories. After subtracting a small ($\sim 10^{-14}$ level in $\dot{P}_b$!), but significant, perturbing effect caused by the Galaxy [25], one finds that General Relativity passes this $(k - \gamma - \dot{P}_b)_{1913+16}$ test with complete success at the 10^{-3} level. More precisely, one finds [26], [20]

$$\left[\frac{\dot{P}_b^{\text{obs}} - \dot{P}_b^{\text{galactic}}}{\dot{P}_b^{\text{GR}}[k^{\text{obs}}, \gamma^{\text{obs}}]}\right]_{1913+16} = 1.0032 \pm 0.0023(\text{obs})$$
$$\pm 0.0026(\text{galactic})$$
$$= 1.0032 \pm 0.0035\,, \quad (11)$$

where $\dot{P}_b^{\text{GR}}[k^{\text{obs}}, \gamma^{\text{obs}}]$ is the GR prediction for the orbital period decay computed from the observed values of the other two post-Keplerian parameters k and γ.

This beautiful confirmation of General Relativity is an embarrassment of riches in that it probes, at the same time, the propagation *and* strong-field properties of relativistic gravity ! If the timing accuracy of PSR1913 + 16 could improve by a significant factor two more post-Keplerian parameters (r and s) would become measurable and would allow one to probe separately the propagation and strong-field aspects [26]. Fortunately, the discovery of the binary pulsar PSR1534 + 12 [27] (which is significantly stronger than PSR1913+16 and has a more favourably oriented orbit) has opened a new testing ground, in which it has been possible to probe strong-field gravity independently of radiative effects. A phenomenological analysis of the timing data of PSR1534 + 12 has allowed one to measure the four post-Keplerian parameters k, γ, r and s [26]. From what was said above, these four simultaneous measurements yield *two* tests of strong-field gravity, without mixing of radiative effects. General Relativity is found to pass these tests with complete success within the measurement accuracy [26], [20]. The most precise of these new, pure strong-field tests is the one obtained by combining the measurements of k, γ and s. Using the most recent data [28] one finds agreement at the 1% level:

$$\left[\frac{s^{\mathrm{obs}}}{s^{\mathrm{GR}}[k^{\mathrm{obs}}, \gamma^{\mathrm{obs}}]} \right]_{1534+12} = 1.007 \pm 0.008 . \quad (12)$$

Recently, it has been possible to extract also the "radiative" parameter $\dot{P}_b$ from the timing data of PSR1534+12. Again, General Relativity is found to be fully consistent (at the $\sim 15\%$ level) with the additional test provided by the $\dot{P}_b$ measurement [28]. Note that this gives our second direct experimental confirmation that the gravitational interaction propagates as predicted by Einstein's theory.

More recently, measurements of the pulse shape of PSR 1913 + 16 [29], [30] have detected a time variation of the pulse shape compatible with the prediction [31], [32] that the general relativistic spin-orbit coupling should cause a secular change in the orientation of the pulsar beam with respect to the line of sight ("geodetic precession"). As envisaged long ago [31] this precession will cause the pulsar to disappear (around 2035) and to remain invisible for hundreds of years [29], [30].

A theory-dependent analysis of the published pulsar data on PSRs 1913 + 16, 1534 + 12 and 0655 + 64 (a dissymetric system constraining the existence of dipolar radiation [33]) has been recently performed within the (α_0, β_0)-space of tensor-scalar theories introduced above [23]. This analysis proves that binary-pulsar data exclude large regions of theory-space which are compatible with solar-system experiments. This is illustrated in Fig. 9 of Ref. [23] which shows that β_0 must be larger than about -5, while any value of β_0 is compatible with weak-field tests as long as α_0 is small enough.

5. Was Einstein 100% right ?

Summarizing the experimental evidence discussed above, we can say that Einstein's postulate of a pure metric coupling between matter and gravity ("equivalence principle") appears to be, at least, 99.999 999 999 9% right (because of universality-of-free-fall experiments), while Einstein's postulate (1) for the field content and dynamics of the gravitational field appears to be, at least, 99.9% correct both in the quasi-static-weak-field limit appropriate to solar-system experiments, and in the radiative-strong-field regime explored by binary pulsar experiments. Should one apply Occam's razor and decide that Einstein must have been 100% right, and then stop testing General Relativity ? My answer is definitely, no !

First, one should continue testing a basic physical theory such as General Relativity to the utmost precision available simply because it is one of the essential pillars of the framework of physics. This is the fundamental justification of an experiment such as Gravity Probe B (the Stanford gyroscope experiment), which will advance by one order of magnitude our experimental knowledge of post-Newtonian gravity.

Second, some very crucial qualitative features of General Relativity have not yet been verified : in particular the existence of black holes, and the direct detection on Earth of gravitational waves. Hopefully, the LIGO/VIRGO network of interferometric detectors will observe gravitational waves early in the next century.

Last, some theoretical findings suggest that the current level of precision of the experimental tests of gravity might be naturally (i.e. without fine tuning of parameters) compatible with Einstein being actually only 50% right ! By this we mean that the correct theory of gravity could involve, on the same fundamental level as the Einsteinian tensor field $g^*_{\mu\nu}$, a massless scalar field φ.

Let us first question the traditional paradigm [8], [2] according to which special attention should be given to tensor-scalar theories respecting the equivalence principle. This class of theories was, in fact, introduced in a purely *ad hoc* way so as to prevent too violent a contradiction with experiment. However, it is important to notice that the scalar couplings which arise naturally in theories unifying gravity with the other interactions systematically violate the equivalence principle. This is true both in Kaluza-Klein theories (which were the starting point of Jordan's theory) and in string theories. In particular, it is striking that (as first noted by Scherk and Schwarz [34]) the dilaton field Φ, which plays an essential role in string theory, appears as a necessary partner of the graviton field $g_{\mu\nu}$ in all string models. Let us recall that $g_s = e^\Phi$ is the basic string coupling constant (measuring the weight of successive string loop contributions) which determines, together with other scalar fields (the moduli), the values of all the coupling constants of the low-energy world. This means, for instance, that the fine-structure constant $\alpha_{\rm em}$ is a function of Φ (and possibly of other moduli fields). In intuitive terms, while Einstein proposed a framework where geometry and gravitation were united as a dynamical field $g_{\mu\nu}(x)$, i.e. a soft structure influenced by the presence of matter, string theory extends this idea by proposing a framework where geometry, gravitation, gauge couplings and gravitational couplings all become soft structures described by interrelated dynamical fields. Symbolically, one has $g_{\mu\nu}(x) \sim g^2(x) \sim G(x)$. This spatiotemporal variability of coupling constants entails a clear violation of the equivalence principle. In particular, $\alpha_{\rm em}$ would be expected to vary on the Hubble time scale (in contradiction with the limit (3) above), and materials of different compositions would be expected to fall with different accelerations (in contradiction with the limits (4), (5) above).

The most popular idea for reconciling gravitational experiments with the existence, at a fundamental level, of scalar partners of $g_{\mu\nu}$ is to assume that all these scalar fields (which are massless before supersymmetry breaking) will acquire a mass after supersymmetry breaking. Typically one expects this mass m to be in the TeV range [35]. This would ensure that scalar exchange brings only negligible, exponentially small corrections $\propto \exp(-mr/\hbar c)$ to the general relativistic predictions concerning low-energy gravitational effects. However, the interesting possibility exists that the mass m be in the milli eV range, corresponding to observable deviations from usual gravity below one millimeter [36], [37], [38].

But, the idea of endowing the scalar partners of $g_{\mu\nu}$ with a non zero mass is fraught with many cosmological difficulties [39], [40], [41]. Though these cosmological difficulties might be solved by a combination of ad hoc solutions (e.g. introducing a secondary stage of inflation to dilute previously produced dilatons [42], [43]), a more radical solution to the problem of reconciling the existence of the dilaton (or any moduli field) with experimental tests and cosmological data has been proposed [44] (see also [45] which considered an equivalence-principle-respecting scalar field). The main idea of Ref. [44] is that string-loop effects (i.e. corrections depending upon $g_s = e^\Phi$ induced by worldsheets of arbitrary genus in intermediate string states) may modify the low-energy, Kaluza-Klein type matter couplings ($\propto e^{-2\Phi} F_{\mu\nu} F^{\mu\nu}$) of the dilaton (or moduli) in such a manner that the VEV of Φ be cosmologically driven toward a finite value Φ_m where it decouples from matter. For such a "least coupling principle" to hold, the loop-modified coupling functions of the dilaton, $B_i(\Phi) = e^{-2\Phi} + c_0 + c_1 e^{2\Phi} + \cdots +$ (nonperturbative terms), must exhibit extrema for finite values of Φ, and these extrema must have certain universality properties. A natural way in which the required conditions could be satisfied is through the existence of a discrete symmetry in scalar space. [For instance, a symmetry under $\Phi \to -\Phi$ would guarantee that all the scalar coupling functions reach an extremum at the self-dual point $\Phi_m = 0$.]

A study of the efficiency of this mechanism of cosmological attraction of φ towards φ_m (φ denoting the canonically normalized scalar field in the Einstein frame, see Eq. (7)) estimates that the present vacuum expectation value φ_0 of the scalar field would differ (in a rms sense) from φ_m by

$$\varphi_0 - \varphi_m \sim 2.75 \times 10^{-9} \times \kappa^{-3} \Omega_m^{-3/4} \Delta\varphi . \quad (13)$$

Here κ denotes the curvature of the gauge coupling function $\ln B_F(\varphi)$ around the maximum

φ_m, Ω_m denotes the present cosmological matter density in units of $10^{-29} g$ cm^{-3}, and $\Delta\varphi$ the deviation $\varphi - \varphi_m$ at the beginning of the (classical) radiation era. Equation (13) predicts (when $\Delta\varphi$ is of order unity[1]) the existence, at the present cosmological epoch, of many small, but not unmeasurably small, deviations from General Relativity proportional to the *square* of $\varphi_0 - \varphi_m$. This provides a new incentive for trying to improve by several orders of magnitude the various experimental tests of Einstein's equivalence principle. The most sensitive way to look for a small residual violation of the equivalence principle is to perform improved tests of the universality of free fall. The mechanism of Ref. [44] suggests a specific composition-dependence of the residual differential acceleration of free fall and estimates that a non-zero signal could exist at the very small level

$$\left(\frac{\Delta a}{a}\right)^{\max}_{\mathrm{rms}} \sim 1.36 \times 10^{-18}\,\kappa^{-4}\,\Omega_m^{-3/2}\,(\Delta\varphi)^2, \quad (14)$$

where κ is expected to be of order unity (or smaller, leading to a larger signal, in the case where φ is a modulus rather than the dilaton).

Let us emphasize that the strength of the cosmological scenario considered here as counterargument to applying Occam's razor lies in the fact that the very small number on the right-hand side of eq. (14) has been derived without any fine tuning or use of small parameters, and turns out to be naturally smaller than the 10^{-12} level presently tested by equivalence-principle experiments (see equations (4), (5)). The estimate (14) gives added significance to the project of a Satellite Test of the Equivalence Principle (nicknamed STEP, and currently studied by NASA, ESA and CNES) which aims at probing the universality of free fall of pairs of test masses orbiting the Earth at the 10^{-18} level [47].

REFERENCES

1. R.P. Feynman, F.B. Morinigo and W.G. Wagner, *Feynman Lectures on Gravitation*, edited by Brian Hatfield (Addison-Wesley, Reading, 1995);
    S. Weinberg, Phys. Rev. **138** (1965) B988,
    V.I. Ogievetsky and I.V. Polubarinov, Ann. Phys. N.Y. **35** (1965) 167;
    W. Wyss, Helv. Phys. Acta **38** (1965) 469;
    S. Deser, Gen. Rel. Grav. **1** (1970) 9;
    D.G. Boulware and S. Deser, Ann. Phys. N.Y. **89** (1975) 193;
    J. Fang and C. Fronsdal, J. Math. Phys. **20** (1979) 2264;
    R.M. Wald, Phys. Rev. D **33** (1986) 3613;
    C. Cutler and R.M. Wald, Class. Quantum Grav. **4** (1987) 1267;
    R.M. Wald, Class. Quantum Grav. **4** (1987) 1279.
2. C.M. Will, *Theory and Experiment in Gravitational Physics*, 2nd edition (Cambridge University Press, Cambridge, 1993); see also gr-qc/9811036.
3. T. Damour, *Gravitation and Experiment* in *Gravitation and Quantizations*, eds B. Julia and J. Zinn-Justin, Les Houches, Session LVII (Elsevier, Amsterdam, 1995), pp 1-61; see also gr-qc/9711061.
4. T. Damour and F. Dyson, Nucl. Phys. B **480** (1996) 37; hep-ph/9606486.
5. J.D. Prestage, R.L. Tjoelker and L. Maleki, Phys. Rev. Lett. **74** (1995) 3511.
6. P.G. Roll, R. Krotkov and R.H. Dicke, Ann. Phys. (N.Y.) **26** (1964) 442;
    V.B. Braginsky and V.I. Panov, Sov. Phys. JETP **34** (1972) 463;
    Y. Su et al., Phys. Rev. D **50** (1994) 3614.
7. J.O. Dickey et al., Science **265** (1994) 482;
    J.G. Williams, X.X. Newhall and J.O. Dickey, Phys. Rev. D **53** (1996) 6730.
8. M. Fierz, Helv. Phys. Acta **29** (1956) 128.
9. P. Jordan, Nature **164** (1949) 637; *Schwerkraft und Weltall* (Vieweg, Braunschweig, 1955).
10. T. Damour and G. Esposito-Farèse, Class. Quant. Grav. **9** (1992) 2093.
11. P. Jordan, Z. Phys. **157** (1959) 112.
12. C. Brans and R.H. Dicke, Phys. Rev. **124** (1961) 925.
13. T.M. Eubanks, D.N. Matsakis, J.O. Martin, B.A. Archinal, D.D. McCarthy, S.A. Klioner,

---

[1] However, $\Delta\varphi$ could be $\ll 1$ if the attractor mechanism already applies during an early stage of potential-driven inflation [46].

S. Shapiro and I.I. Shapiro, Bull. Am. Phys. Soc., Abstract # K 11.05 (1997).

14. K. Nordtvedt, Phys. Rev. **170** (1968) 1186.

15. T. Damour and D. Vokrouhlicky, Phys. Rev. D **53** (1996) 4177.

16. T. Damour and G. Esposito-Farèse, Phys. Rev. D **53** (1996) 5541.

17. R.A. Hulse and J.H. Taylor, Astrophys. J. Lett. **195** (1975) L51; see also the 1993 Nobel lectures in physics of Hulse (pp. 699-710) and Taylor (pp. 711-719) in Rev. Mod. Phys. **66**, n^{0}3 (1994).

18. P.S. Laplace, *Traité de Mécanique Céleste*, (Courcier, Paris, 1798-1825), Second part : book 10, chapter 7.

19. T. Damour and N. Deruelle, Phys. Lett. A **87** (1981) 81;
T. Damour, C.R. Acad. Sci. Paris **294** (1982) 1335;
T. Damour, in *Gravitational Radiation*, eds N. Deruelle and T. Piran (North-Holland, Amsterdam, 1983) pp 59-144.

20. J.H. Taylor, Class. Quant. Grav. **10** (1993) S167 (Supplement 1993) and references therein; see also J.H. Taylor's Nobel lecture quoted in [17].

21. T. Damour and G. Esposito-Farèse, Phys. Rev. Lett. **70** (1993) 2220.

22. T. Damour and J.H. Taylor, Phys. Rev. D. **45** (1992) 1840.

23. T. Damour and G. Esposito-Farèse, Phys. Rev. D **54** (1996) 1474.

24. T. Damour and N. Deruelle, Ann. Inst. H. Poincaré **43** (1985) 107 and **44** (1986) 263.

25. T. Damour and J.H. Taylor, Astrophys. J. **366** (1991) 501.

26. J.H. Taylor, A. Wolszczan, T. Damour and J.M. Weisberg, Nature **355** (1992) 132.

27. A. Wolszczan, Nature **350** (1991) 688.

28. I.H. Stairs et al., Astrophys. J. **505** (1998) 352;
I.H. Stairs et al., contribution to Moriond 1999, astro-ph/9903289.

29. M. Kramer, Astrophys. J., in press, astro-ph/9808127.

30. J.H. Taylor and J.M. Weisberg, in preparation;
J.H. Taylor, contribution to Moriond 1999.

31. T. Damour and R. Ruffini, C.R. Acad. Sc. Paris, Série A, **279** (1974) 971.

32. B.M. Barker and R.F. O'Connel, Phys. Rev. D **12** (1975) 329.

33. C.M. Will and H.W. Zaglauer, Astrophys. J. **346** (1989) 366.

34. J. Scherk and J.H. Schwarz, Nucl. Phys. B **81** (1974) 118; Phys. Lett. B **52** (1974) 347.

35. B. de Carlos, J.A. Casas, F. Quevedo and E. Roulet, Phys. Lett. B **318** (1993) 447.

36. I. Antoniadis, Phys. Lett. B **246** (1990) 377;
I. Antoniadis, S. Dimopoulos and G. Dvali, Nucl. Phys. **B516** (1998) 70; hep-ph/9710204.

37. S. Ferrara, C. Kounnas and F. Zwirner, Nucl. Phys. B **429** (1994) 589.

38. N. Arkani-Hamed, S. Dimopoulos and G. Dvali, Phys. Lett. **B429** (1998) 263 and Phys. Rev. **D59** (1999) 086004; hep-ph/9803315 and 9807344.

39. R. Brustein and P.J. Steinhardt, Phys. Lett. **B302** (1993) 196.

40. G.D. Coughlan et al., Phys. Lett. **B131** (1983) 59;
J. Ellis, D.V. Nanopoulos and M. Quiros, Phys. Lett. **B174** (1986) 176;
T. Banks, D.B. Kaplan and A.E. Nelson, Phys. Rev. **D49** (1994) 779.

41. T. Damour and A. Vilenkin, Phys. Rev. Lett. **78** (1997) 2288.

42. L. Randall and S. Thomas, Nucl. Phys. B **449** (1995) 229.

43. D.H. Lyth and E.D. Stewart, Phys. Rev. Lett. **75** (1995) 201; Phys. Rev. D **53** (1996) 1784.

44. T. Damour and A.M. Polyakov, Nucl. Phys. B **423** (1994) 532; Gen. Rel. Grav. **26** (1994), 1171.

45. T. Damour and K. Nordtvedt, Phys. Rev. Lett. **70** (1993) 2217; Phys. Rev. D **48** (1993) 3436.

46. T. Damour and A. Vilenkin, Phys. Rev. **D53** (1996) 2981.

47. P.W. Worden, in *Near Zero : New Frontiers of Physics*, eds J.D. Fairbank et al. (Freeman, San Francisco, 1988) p. 766;
J.P. Blaser et al., *STEP, Report on the Phase A Study*, ESA document SCI (96)5, March 1996;

GEOSTEP Project, CNES report
DPI/SC/FJC-N^0 96/058, April 1996;
Fundamental Physics in Space, special issue
of Class. Quant. Grav. **13** (1996).
MiniSTEP, NASA-ESA report, December
1996 (second issue).

ELSEVIER

Nuclear Physics B (Proc. Suppl.) 80 (2000) 51–61

NUCLEAR PHYSICS B
PROCEEDINGS
SUPPLEMENTS

www.elsevier.nl/locate/npe

Astrophysical MHD Jets

J. Heyvaerts[a]

[a]Observatoire Astronomique, 11 rue de l'Université, F-67000 Strasbourg, France

1. THE JET SOURCE

It is observed that many objects, like active galactic nuclei, micro-quasars, young stellar objects, possibly γ-ray burst sources [1], suffer mass loss in the form of collimated flows, called jets, some of which reach relativistic bulk velocities. This raises a number of questions: what is the propulsion mechanism of these flows, how can it boost them to relativistic speeds, what causes the, sometimes extreme, focusing observed, are such flows stable or internally turbulent and how is their radiation emitted and transfered? In the past fifteen years, a general concept, though not the only conceivable one, has emerged, which figures such flows as rotating MHD winds emanating from accreting objects or/and from the accretion disk that surrounds them. This review will address the above questions in that general framework.

A typical source of a jet would consist of an accreting, possibly collapsed, object, which may be protected by a closed magnetosphere, and may be surrounded by an accretion disk. A magnetic field threads at least part of the different elements of this system. It may be of external origin or have its source in one of those, possibly by dynamo action in the accreting object or in the accretion disk. Electric currents in the wind zone certainly contribute to it and partly determine the geometry of the wind. The material of the accretion disk orbits the accreting object, which may be spinning fast too. The accretion disk and possibly also the accreting object radiate. The accretion flow in the disk must be dissipative if stationary and therefore the disk suffers heating. Rotation, pressure forces and radiation may be the accelerating agents for the jet outflow. By contrast, the flow in the wind zone need not, but may be dissipative. Dissipation seems to actually occur since the particles which radiate in the jet by synchrotron emission need permanent reacceleration. Dissipation occurs also of course in the shock system which terminates the jet.

2. STATIONARY WINDS

The jet can, as a first approximation, be described as stationary, axisymmetric and dissipationless. Though idealized, this picture suffices for a first approach of such questions as global wind dynamics and collimation. The assumption that the plasma pressure is given by a polytropic law of index γ is sometimes added. We adopt MKSA units and cylindrical coordinates (r, ϕ, z) with unit vectors $\vec{e}_r, \vec{e}_\phi, \vec{e}_z$. Under axisymmetry, any vector quantity is the sum of a toroidal, or azimutal, part and of a poloidal part (in the meridian plane). The poloidal magnetic field, $\vec{B}_P$, is conveniently represented in terms of a flux function $a(r, z)$, by $\vec{B}_P = (\vec{\nabla}a \times \vec{e}_\phi)/r$. Surfaces of constant $a(r, z)$ are the magnetic surfaces, generated by the rotation about the axis of the different field lines. The assumptions of stationarity and axisymmetry result, in perfect MHD, in the existence of a number of quantities conserved following the flow. It can be shown, in particular, that the fluid velocity may be written as

$$\rho\vec{v} = \alpha(a)\vec{B} + \rho r\Omega(a)\vec{e}_\phi \qquad (1)$$

where ρ is the mass density and α and Ω are two quantities which are constant on a magnetic surface, though their value may be different on two different such surfaces. The flow can be viewed as if infinitesimal flux tubes were pipes entrained in rotation at angular velocity Ω, in which plasma flows with a mass flux to magnetic flux ratio α.

This implies that flow surfaces are magnetic surfaces. The azimutal component of the equation of motion integrates as

$$rv_\phi - \frac{rB_\phi}{\mu_0\alpha} = L(a) \qquad (2)$$

where L is a conserved specific angular momentum of the escaping fluid, which consists of a matter and a magnetic part: due to torques exerted by MHD forces, the angular momentum of the fluid is not conserved following the motion, but the combination in eq.(2) is. The polytropic relation can be written as $P = Q(a)\rho^\gamma$, where Q is a "specific entropy" conserved following the motion. The component of the equation of motion along the poloidal field integrates as a law of conservation of the specific energy of the escaping fluid as:

$$\frac{v_P^2 + v_\phi^2}{2} + \frac{\gamma Q\rho^{\gamma-1}}{\gamma - 1} - \frac{GM_\star}{R} - \frac{r\Omega B_\phi}{\mu_0\alpha} = E(a) \quad (3)$$

where it has been asumed that the gravity is that of a point-mass. Note that the specific energy E contains a kinetic, a thermal, a gravitational and a magnetic part. The latter in fact represents the ratio of the Poynting energy flux to the mass flux. These three equations express in integrated form all MHD equations but for the component of the equation of motion normal to magnetic surfaces. Toroidal components of the velocity and magnetic field can thus be expressed in terms of poloidal variables. It then appears that for regularity of the resulting expressions r must assume a value r_A defined by $r_A^2 = L(a)/\Omega(a)$ wherever the density ρ assumes the value $\rho_A = \mu_0\alpha^2$. The quantity r_A is the Alfvén radius and ρ_A is the alfvénic density. They both depend on magnetic surface. The point at $r = r_A$ on a given field line is its Alfvén point, named that way because at this position the poloidal fluid velocity equals the Alfvén speed associated with the poloidal field, defined by $v_{PA}^2 = B_P^2/\mu_0\rho_A$. The locus of all Alfvén points is the Alfvén surface.

For polytropic thermodynamics, the flow is then characterized by five "first integrals", α, L, E, Ω and Q, all depending on magnetic surface, i.e. on a. All unknowns can be expressed in terms of the function a, of the first integrals and of

ρ. The MHD problem reduces to the determination of $\rho(r, z)$ and of the flux function $a(r, z)$, supplemented by the determination of the functions $\alpha(a)$, $L(a)$ and $E(a)$, which, unlike $Q(a)$ and $\Omega(a)$, are not given by boundary conditions at the origin of the wind. The functions $a(r, z)$ and $\rho(r, z)$ are eventually shown to satisfy the system which consists of the Bernoulli equation (3) rewritten as

$$\frac{1}{2}\frac{\alpha^2|\nabla a|^2}{\rho^2 r^2} = E + \frac{GM_\star}{\sqrt{r^2 + z^2}} - \frac{\gamma}{\gamma - 1}Q\rho^{\gamma-1}$$

$$+\rho\Omega^2\frac{r_A^2 - r^2}{\rho_A - \rho} - \frac{1}{2}\frac{\Omega^2 r_A^4}{r^2}\left(1 + \frac{\rho(r_A^2 - r^2)}{r_A^2(\rho_A - \rho)}\right)^2 \quad (4)$$

and of the transfield equation, which is the projection perpendicular to the magnetic field of the poloidal part of the equation of motion. This latter equation is a second order partial differential equation for the flux function $a(r, z)$:

$$\frac{\alpha}{\rho r}\left(\frac{\partial}{\partial z}\frac{\alpha}{\rho r}\frac{\partial a}{\partial z} + \frac{\partial}{\partial r}\frac{\alpha}{\rho r}\frac{\partial a}{\partial r}\right)$$

$$-\frac{1}{\mu_0\rho r}\left(\frac{\partial}{\partial z}\frac{1}{r}\frac{\partial a}{\partial z}\frac{\partial}{\partial r}\frac{1}{r}\frac{\partial a}{\partial r}\right) =$$

$$E' - \frac{Q'\rho^{\gamma-1}}{\gamma - 1} + \frac{\alpha'}{\alpha}\frac{\mu_0\alpha^2\rho}{r^2}\left(\frac{L - r^2\Omega}{\mu_0\alpha^2 - \rho}\right)^2$$

$$-\frac{\rho}{r^2}\frac{(L - r^2\Omega)(L' - r^2\Omega')}{\mu_0\alpha^2 - \rho} - \frac{LL'}{r^2} \quad (5)$$

3. PROPERTIES of the BERNOULLI and TRANSFIELD SYSTEM

The coupled system of equations (4) and (5) describes both the field-aligned dynamics, by the Bernoulli equation (4), and the geometry of the rotating MHD wind, by the transfield equation (5). The latter is a quasi-linear partial differential equation because the highest derivative terms are linear, while the unknown function a enters nonlinearly as the argument of the first integrals and of their derivatives with respect to a, like α', Q' and E', which appear on its r.h.s.

Similar Bernoulli-Transfield systems have been obtained for special-relativistic winds [2,3] and for general relativistic motion in the field of a Kerr black hole [4-7].

If the shape of magnetic surfaces, that is the function $a(r, z)$, is assumed to be known, the

Bernoulli equation (4) reduces to an algebraic equation for the density ρ, or for any equivalent variable, such as the poloidal velocity or the alfvénic Mach number M, related to ρ by $M^2 = \mu_0 \alpha^2 / \rho$. Expressing eq. (4) as $\mathcal{B}(M, r) = E$, it can be turned by differentiation into a differential equation for $M(r)$ at given a:

$$\frac{dM}{dr} = -\frac{\partial \mathcal{B}/\partial r}{\partial \mathcal{B}/\partial M} \qquad (6)$$

It turns out that the denominator vanishes whenever the poloidal flow velocity equals one of the poloidal MHD mode speeds. If the flow velocity passes any of these critical speeds, it must do so at a point where the numerator simultaneously vanishes, for otherwise the solution for $M(r)$ would not be regular. Such points where the differential of $\mathcal{B}$ vanish are called critical points. On each field line there may exist a slow critical point (associated to the slow mode speed) and a fast critical point (associated to the fast mode speed). The locus of these points constitute the slow and fast critical surfaces. The positions $r_s(a)$ and $r_f(a)$ of the critical points on magnetic surface a, and the associated alfvénic Mach numbers $M_s(a)$ and $M_f(a)$ depend on a by the values of the constants of the motion, α, Q, L and Ω, which appear in equation (6), but not of E, which does not. Demanding that the Bernoulli equation (4) be satisfied at both critical points, we obtain two relations between the first integrals, which impose two constraints on the three first integrals which are not determined from boundary conditions:

$$E = \mathcal{B}(M_s, r_s) = \mathcal{B}_s(\alpha, \Omega, Q, L) \qquad (7)$$

$$E = \mathcal{B}(M_f, r_f) = \mathcal{B}_f(\alpha, \Omega, Q, L) \qquad (8)$$

A regularity condition must also be imposed to the transfield equation. A glance at equation (5) indicates that it loses its highest order derivative terms at the Alfvén point, a property that remains when the Bernoulli equation is used to eliminate ρ. Another way to appreciate this is to convert the transfield equation into one that gives the curvature $d\psi/ds$ of poloidal field lines. The result can be written as [8]:

$$(M - 1)(d\psi/ds) = \mathcal{R} \qquad (9)$$

where $\mathcal{R}$ is a function of α, E, L, Ω, Q, α', E', L', Ω', Q', ρ, r, z, $|\nabla a|$. At $M = 1$ the right hand side of this equation must vanish, since the left hand side does, a condition which can be expressed as an ordinary differential equation relating the first integrals [8]. Solutions which would not satisfy this regularity condition would have infinite curvature of poloidal field lines at the Alfvén surface, that is, magnetic surfaces would have a sharp kink there, which would be supported by a current sheet standing at rest in the flow as a so-called rotational discontinuity, i.e. a large amplitude Alfvén wave. Imposition of the Alfvén regularity condition $\mathcal{R} = 0$ at the Alfvén surface provides a relation which, together with the Bernoulli regularity conditions, determines the unknown first integrals to within an integration constant. Since however the shape of magnetic surfaces is not a-priori known, but must be the solution of problem (4)-(5), the determination of the first integrals is not direct but results at best from an iterative solution scheme. It is not certain that the flux function which results as a limit of such an iteration sequence would have all the regularity properties of the functions of the sequence themselves.

Though this is not possible in practice, the density ρ could conceivably be obtained in terms of the first integrals and of $|\nabla a|$ by solving the Bernoulli equation. One would insert the result in the transfield equation, obtaining an equation for $a(r, z)$ alone, the "wind equation". The wind equation, which can rarely be explicitly written down, is nevertheless known to be a quasi-linear partial differential equation for $a(r, z)$ of a mixed elliptic/hyperbolic type [9]. It is elliptic very near the wind source, changes to hyperbolic at the cusp surface (defined below) changes again to elliptic at the slow critical surface and to hyperbolic at the fast mode critical surface. It is singular at the Alfvén surface. The cusp surface is the locus of points where the poloidal speed equals the cusp speed, c_c defined by $c_c^{-2} = c_s^{-2} + c_f^{-2}$, which is the minimum group velocity of slow mode waves.

The slow and fast critical surfaces are not the surfaces where causal connection by MHD modes with the upstream or downstream flow is lost. To see this it is possible to follow the characteristics of the considered MHD modes in the hyperbolic regions. In each hyperbolic region, one of them reaches an asymptote, the so-called limiting separatrix [10,11]. The surfaces generated by rotation of the fast and slow limiting separatrices are the actual borders of causal connection for the considered MHD modes. The plasma upstream from the slow limiting separatrix does not communicate by slow mode waves with the plasma downstream of it, while the plasma downstream from the fast limiting separatrix does not communicate by fast mode waves with the plasma upstream of it. There is ongoing discussion as to whether the limiting separatrices are singular surfaces for the wind equation, and whether or not there is enough degrees of freedom to guarantee that the solution to the wind equation is continuously derivable [10,12].

4. COLLIMATION

4.1. Hoop Stress

Flows collimating on a scale length larger than the characteristic size of the near-source environment may be forced into that regime by MHD stresses, i.e. by the so-called hoop-stress, or pinching force. Ampere's theorem

$$2\pi r B_\phi = \mu_0 I(r, z) \tag{10}$$

gives the electric current $I(r, z)$ flowing through a circle of radius r centred on the axis at altitude z. One component of the radial Lorentz force is

$$(\vec{j}_z \times \vec{B}_\phi) \cdot \vec{e}_r = -\frac{\mu_0}{2\pi r} I \, j_z \tag{11}$$

Whenever I and j_z are of the same sign it exerts pinching towards the axis. Since j_z is the surface density of I, this must necessarily happen somewhere. This force therefore has the potential to focus the wind axially into a jet if it dominates opposing defocusing forces. Is it so? The issue of magnetic collimation, in 2D axisymmetric and stationary flows, reduces to answering this question.

4.2. General Asymptotic Properties

It has been shown [8] that, because the Poynting flux escaping in the wind cannot exceed the total energy flux, the amount of vertical current

I enclosed in a given magnetic surface a at large z, $I_\infty(a)$, is bounded by

$$I_\infty \le 2\pi \, \frac{\alpha(a) E(a)}{\Omega(a)} \tag{12}$$

The boundedness of I_∞ implyes that asymptotic flattening of magnetic surfaces to the equator can be excluded for polytropic flows [8]. It has also been shown that if the vertical current through magnetic surface a approaches 0 as z approaches infinity (a limit noted as $\lim_{\infty a}$) we have

$$\lim_{\infty a} \frac{z(a, r)}{r} = \infty \tag{13}$$

$$\lim_{\infty a} r = \infty \tag{14}$$

These results have been extended to the special-relativistic situation [13]. Since they are of asymptotic validity no general relativistic treatment is needed.

On those magnetic surfaces along which either $\lim_{\infty a}(r) = \infty$ or $\lim_{\infty a}(r) \gg r_A$, the asymptotic form of the transfield equation reduces simply to

$$\frac{\alpha(a)}{v_\infty(a)} \frac{d}{da} \left(I_\infty^2(a) \right) = 0 \tag{15}$$

except in those exceptional regions near which I_∞ vanishes. For current closure, such regions must exist. In their vicinity in particular near the polar axis, other forces become competitive with the pinching force. Asymptotically, such regions in the flow form boundary layers. The above asymptotic form of the transfield equation, eq.(15), applies between them. If $I_\infty(a)$ does not vanish, it must be constant in any region in which eq.(15) applies, that is, it must be piecewise constant in the asymptotic region. As a result magnetic surfaces which enclose the polar axis and are conically radially diverging must enclose a common non-zero current, which means that no poloidal current flows between them, so that $| j_P | \to 0$. If this situation is to occur at all, this implies that the non-zero polar asymptotic current which these surfaces enclose must flow in a set of magnetic surfaces of a finite asymptotic radius. Thus there is in this case a core of magnetic surfaces surrounding the pole which asymptotically become cylindrical, with a finite radius

r_∞. If on the other hand $I_\infty(a) = 0$ the asymptotic structure consists of nested paraboloids as mentioned. The simple asymptotic analysis cannot tell whether I_∞ vanishes or not. Note that current closure is satisfied, since the current returning from the circumpolar regions flows in boundary layers, the position of which depend on the magnetic polarity on the source object. The boundary layer which carries the return current is an equatorial sheet when the wind source has dipolar polarity symmetry.

4.3. Self Similar Models

A number of authors have derived exact solutions for simplified models, either self-similar or in which variables are somehow separated. In spherical self-similarity [14], the distance R to the center of a point at colatitude θ on magnetic surface a is written as

$$R(a, \theta) = r_0(a)g(\theta) \qquad (16)$$

In [14] only cold classical winds have been considered but similar analysis has been extended to relativistic winds [15–17] and to non-zero pressure [18,19]. Cylindrically self similar solutions [20] have also been explored, where

$$r(a, z) = r_0(a)f(z) \qquad (17)$$

as well as latitudinally self similar solutions [21] where separabilty in spherical coordinates is assumed:

$$r(\theta, R) = sin^2\theta \; h(R) \qquad (18)$$

All these solutions exhibit collimation, at least in some part of the parameter space, sometimes with oscillations about the asymptotic radius due to the interaction of the magnetic pinching force with the centrifugal force near the axis.

Self similar models are great because they yield exact solutions but they are poor in matching boundary conditions. For example, in radial or spherical self-similarity, the sign of the electric current leaving the source normal to it has no possibility of changing and the return current concentrates on an axial singularity or is rejected to infinity. Also, the source is necessarily plane, of infinite extent and may be of infinite total flux. In models like [21] no a-priori thermodynamics

can be imposed. Some of the differences in these different results are in fact due to the difficulty of controlling these aspects [17].

4.4. Sketchy Models

Some authors attempted to trade exactness against generality by developing sketchy models in which the geometry of the flow is assumed to be known in some region close to the source, but on which rather general boundary conditions can be imposed [22]. Some transfield equilibrium is still imposed, in that the Alfvén regularity condition is enforced, and Bernoulli regularity conditions at the slow and fast surface are also solved for, yielding all the necessary first integrals. The solution is not an exact one, though, because the shape of the flow is not self consistently solved for, transfield equilibrium being imposed at a few points only. The asymptotic structure of the jet can be determined assuming an external uniform confining pressure, P_{ext}, which can be made to approach zero. The most significant result of such calculations is that the total vertical current in the jet continuously decreases as P_{ext} is reduced, the limit of a constant current being never achieved, though a sizeable one is maintained, even at very low pressure, for fast rotators. Since the model is not self-consistent, this cannot be taken as a proof that unconfined systems with non-zero asymptotic poloidal current do not exist or exist only in limit cases, but it certainly puts a question mark on this.

4.5. Explicit Asymptotics

Explicit calculation of the structure of the flow in the asymptotic domain is possible if the five first integrals are known as a function of the flux function a [23]. For special relativistic winds, the transfield equation outside of the return-current boundary layers mentioned above, reduces to

$$\frac{\vec{\nabla}a}{|\nabla a|} \cdot \left(r^2 B_\phi^2 - r^2 B_P^2 \frac{\Omega^2 r^2}{c^2} \right) = 0 \qquad (19)$$

and integrates as

$$\lim_{\infty a} \left(r^2 B_\phi^2 - r^2 B_P^2 \frac{\Omega^2 r^2}{c^2} \right) = K_\infty(b) \qquad (20)$$

where the constant K_∞ depends on the particular orthogonal trajectory to magnetic surfaces, labeled by b, along which the integration is conducted. This parameter may, but need not, be the value of z at its intersection with the polar axis. Again, this solution holds piecewise true in regions limited by boundary layers where the hoop stress ceases to be the unique potentially dominating force, the polar axis being one such region. If $\hat{K} = \lim_{\infty b}(K_\infty(b))$ is non-zero, the 2D asymptotic problem reduces to an Hamilton Jacobi equation that can be solved [23], showing that the asymptotic structure then consists of nested cones and cylinders. The solution outside of the polar and other boundary layers gives the complement of the semi-opening angle of the asymptotically conical magnetic surfaces $\psi_\infty(a)$ by

$$tan\ \psi_\infty(a) = sinh\left(X(a)/(\mu_0 c \hat{K})\right) \qquad (21)$$

$$X(a) = \int_a^A da' \frac{\Omega(c^2 + \hat{K}\Omega/\alpha)}{\left(E^2 - (c^2 + \hat{K}\Omega/\alpha)^2\right)^{1/2}}$$

while in the cylindrical region nested inside it, the asymptotic radius, r_∞, if much larger than r_A, is given by:

$$r_\infty(a) = r_0\ exp\left(Y(a)/(\mu_0 c \hat{K})\right) \qquad (22)$$

$$Y(a) = \int_0^a da' \frac{\Omega(c^2 + \hat{K}\Omega/\alpha)}{\left(E^2 - (c^2 + \hat{K}\Omega/\alpha)^2\right)^{1/2}}$$

A smooth match between both types of asymptotics requires that at the "regime-change" value of a, a_*, the angle ψ_∞ approaches $\pi/2$ while r_∞ approaches infinity, which is possible only when the integrals X and Y both diverge at a_*. This requires that the function

$$f(a) = \frac{\alpha(a)(E(a) - c^2)}{\Omega(a)} - \hat{K} \qquad (23)$$

has a second order zero, reached by positive values, at $a = a_*$. So,

$$\hat{K} = inf\left(\frac{\alpha(a)(E(a) - c^2)}{\Omega(a)}\right) \qquad (24)$$

This minimum, however, cannot be on the polar axis, because any zero of $f(a)$ there would usually be a first order one. We conclude that if

$\alpha(E - c^2)/\Omega$ increases from pole to equator, the constant $\hat{K}$ must be zero, and the asymptotics must be parabolic. Note that the option $\hat{K} = 0$ remains open when $\alpha(E - c^2)/\Omega$ has a lower minimum at some a_*. If $\hat{K} \neq 0$, the asymptotic wind velocity then has to vanish at the regime-change magnetic surface. The cylindrical jet and the conical wind appear in this case as forming two separate entities. The vanishing of the wind velocity at the regime-change magnetic surface then raises the question: is it possible that the wind velocity approaches zero, while remaining super-fast-mode? Considering for simplicity the case of newtonian motion, it can be shown that as $\rho \to 0$, the fast mode speed is bounded from below by

$$c_F^2 \geq I_\infty \Omega/\alpha \qquad (25)$$

which must also vanish at the regime-change surface for the flow to be asymptotically super-fast-mode implying either that I_∞ vanishes (in contradiction with the assumption) or that α is infinite. An infinite α is possible for a finite mass outflow if B_P vanishes, which happens on magnetic surfaces rooted at a polarity reversal. So, either I_∞ vanishes everywhere and we have parabolic asymptotics, all the energy flux being converted into kinetic form, or the wind chokes off at those magnetic surfaces which are rooted in the wind source at polarity reversals. At such surfaces B_ϕ also vanishes because it is created from B_P by rotation. To sum up these special magnetic surfaces are in fact non-magnetic. As a result I_∞ vanishes and so does the hoop stress, which cannot dominate over other forces locally in the vicinity of these special places. Hence, the regime-change surface is a boundary layer of the asymptotic flow at which there is return-current flow. The livy parts of these MHD flows must reside there and in the vicinity of the polar axis, another region where poloidal currents still flow at large distance from the source. These boundary-layer regions separate vast asymptotic regions devoid of poloidal current. In contrast to other boundary layers which have the structure of sheet pinches, the polar one is needle-shaped and has the structure of a cylindrical pinch. In [23] a matched asymptotics expansion is performed which provides a solution both in this boundary layer and intermediate zones. The main result of such a solution is a relation between the asymptotic density on the axis of the jet, ρ_0, and the total asymptotic current that is conveyed in it. In the newtonian case this relation is

$$\frac{\gamma}{\gamma - 1} Q_0 \rho_0^{\gamma - 1} = \frac{I_\infty \Omega_0}{\alpha_0} \qquad (26)$$

This analysis has been extended to the case where the poloidal current asymptotically vanishes, the variation in z being dealt with by a WKB approximation technique. The variation of the axial density $\rho_0(R)$ with distance R in the dimming jet has been determined. It declines very slowly indeed, as:

$$\rho_0 \sim \left(log(\frac{R}{l}) + \frac{2 - \gamma}{2(\gamma - 1)} log \; log(\frac{R}{l}) \right)^{-\frac{1}{\gamma - 1}} \qquad (27)$$

Therefore the current-supporting core of the flow stands out on the environment with a high contrast, though its central density decreases very slowly. The shape of paraxial magnetic surfaces has been found to be paraboloids $r \propto z^{q(a)}$ with exponent $q(a)$ given by:

$$q(a) = \frac{\int_0^a \Omega(a')E(a')^{-1/2}da'}{\int_0^A \Omega(a')E(a')^{-1/2}da'} \qquad (28)$$

Cylindrical pinch structures have been discussed in the force free approximation assuming a given, special, profile for the electric current $j_z(a)$ [24,25]. In [26] exact one-dimensional solutions to the Grad Shafranoff equation have been obtained for black hole environments in the presence of an externally applied uniform field.

A particle-dominated core surrounded by a force free envelope appeared in these solutions, just as in the ones described above. The visible jet may indeed just be this particle-dominated inner core of a flow which otherwise extends on much larger scales and broader angles.

5. WIND PROPULSION

5.1. Two Forces

Two forces may accelerate a cold wind: the centrifugal force [14], a part of the $(\vec{v} \cdot \vec{\nabla})\vec{v}$ inertial force, and the Lorentz force $\vec{j} \times \vec{B}$. Propulsion

of the lumps of frozen-in plasma by centrifugal effect can be compared to the motion of beads on a wire, the role of the latter being played by the magnetic line [14]. For this image to make sense the field structure should not be destroyed by the flow. So this process operates mainly inside the Alfvén surface. For centrifugal propulsion to occur, the angle between the polar axis and the poloidal field at the source surface must be larger than 30 degrees [14].

The Lorentz force can also assume the role of a propeller. Indeed it can be split as follows, separating poloidal and toroidal parts of the current and field:

$$\vec{j} \times \vec{B} = (\vec{j}_P + \vec{j}_\phi) \times (\vec{B}_P + \vec{B}_\phi) =$$
$$\vec{j}_P \times \vec{B}_P + \vec{j}_\phi \times \vec{B}_P + \vec{j}_P \times \vec{B}_\phi \qquad (29)$$

The first of the three terms in the last sum is toroidal. It exerts a torque which accelerates or decelerates the rotation of the plasma according to the sign of this vector product. The second force, which is poloidal magnetic pressure and tension, is normal to the magnetic surfaces and does not exert propulsion. The third force has a component perpendicular to $\vec{B}_P$ which is just the collimating pinching force and a component aligned to $\vec{B}_P$ that can boost the plasma if its sign is adequate. A moment's thought shows that this force exerts a propulsion effect if B_ϕ decreases with height z along the magnetic surface. This is the plasma gun effect [27] or sweeping twist mechanism [28]: when there is tighter coiling of the magnetic field at lower altitudes, the magnetic field lines act as a spring that is willing to uncoil by pushing plasma out [28]. Strong coiling is only possible after the Alfvén surface. So, unless the field is rather weak, this propulsion mechanism would preferentially act far away from the wind source.

Note that these forces exist even in the presence of dissipation and that they also drive the dynamics of the plasma at the beginning of its expansion, inside the accretion disk.

5.2. Lifting off from Accretion Disk

Cold plasma propulsion towards positive z in the upper layers of an accretion disk must be due to one of these two forces having a positive vertical component. At the disk equator the torque exerted by the force $\vec{j}_P \times \vec{B}_P$ is normally decelerating, extracting angular momentum from the disk. This promotes accretion. Then, the equatorial rotation is slightly sub-keplerian. At the disk surface the plasma must be spun up if it is indeed to be centrifugally ejected. Rotation must then have become slightly super-keplerian there, which can be achieved by magnetic thetering to innermore disk regions which rotate faster. Resistivity degrades the quality of this coupling. The change of sign of the electromagnetic torque between the equatorial plane and the surface implies a reversal of the sign of the radial electric current density over a disk thickness. Plasma expulsion off the disk surface then requires a delicate unbalance between actual and keplerian rotation.

In a stationary state there should be a balance between advection and diffusion of the magnetic field in the resistive disk. Therefore the plasma should not be too diffusive, for otherwise the minimum angle between the poloidal field and the axial direction that is necessary for centrifugal plasma expulsion could not be maintained. Precisely, the effective magnetic Reynolds number based on radial velocity and disk thickness should be of order unity. This condition is not realized in the Shakura-Sunyaev regime [29], the effective viscous Reynolds number based on radial velocity and disk radius being of order unity while the effective magnetic Prandtl number for 3D turbulence is also of order unity. Therefore the effective magnetic Reynolds number mentioned above is in this regime of order (h/r), much less than the needed value of unity. It is concluded that jetting disks in which the plasma propulsion is achieved by centrifugal forces can only operate in a regime vastly different from the Shakura Sunyaev one. As shown by various authors [30,31], this is however totally unlikely in weakly magnetized disks subject to the magneto-rotational instability [32]. The question of centrifugal launching of jets by disks therefore reduces to whether such fine tuned regimes are self-consistently achievable.

This question is explored in [33] by studying separable solutions for the dynamics inside the disk up to its outskirts, any quantity $G(r, z)$ being assumed to be factorable as $G(r, z) = $ $(r/r_0)^s f(z/h(r))$, where $h(r)$ has been found to scale as r. It has been shown that for solutions to exist the following conditions should be fulfilled: (a) The plasma β in the disk should be of order unity (if too large, the field is weak and we are in the magneto-rotational instability regime, if too small, the field lines are too stiff to be sufficiently bent); (b) The effective viscous Reynolds number based on v_r and r should be large, of order r/h, for otherwise the necessary kink in the field at crossing the disk is diffused away; (c) the turbulent magnetic diffusivity should scale as $\alpha_m v_{A0} h$, h being the disk thickness and v_{A0} the Alfvén velocity at its midplane, which also implies $\alpha_m \ll 1$ for $\beta = 1$ conditions, for otherwise the diffusivity would again be too large. Similar independent work [34] reached identical conclusions. These seem to be the necessary conditions for an accretion-ejection machine to self-consistently operate in a thin, ionized, stationary, accretion disk. In such a machine accretion and ejection conspire, the latter removing angular momentum to allow the former to proceed, and the former actually driving the ejection. The potential energy lost during accretion goes into work used to propel the wind. Therefore such systems are rather "silent" in radiative terms.

Both accretion and ejection flow have been included in the picture of a single quadrupolar plasma circulation in [35] where accretion, by which is meant motion towards the central object, and ejection are described on an equal footing. This model assumes parallel velocities and fields and is self similar. The accretion region appears to be rather thick [36], while the ejection flow is not particularly well focused. Note that there is no accretion on the central point mass of this model, the latter effect being viewed as a perturbation to the general quadrupolar circulation.

5.3. Relativistic Jet Propulsion

It is conceivable that rotating MHD winds asymptote to relativistic speeds, for example if the wind source is the environment of a Kerr black hole threaded by an external magnetic field as first suggested in [37]. Solutions for winds in Kerr geometry have been obtained for a slowly rotating hole sitting at the center of an accretion disk of a given inner radius and threaded by an external magnetic field trapped by the disk [7]. The current circulation in this system is dictated by regularity conditions at the inner and outer Alfvén surfaces, which determines both the current and the rate of field rotation, and then also the electric charge density, which is close to the Goldreich-Julian charge density [38], with "sparking gaps" [39] known in pulsar magnetospheres. Such gaps can accelerate high energy particles that would develop γ-ray/pairs cascades, filling the system with a secondary plasma.

At this point the interaction between the photon environment and the pair-plasma becomes a major issue when the plasma has relativistic bulk motion and possibly relativistic thermal energies as well. The directional properties of the ambiant radiation field have strong effects not only on photon transfer but also on the pair-plasma avalanche and dynamics. A relativistic beam of monoenergetic particles is braked when ramming into a soft photon field but a non-relativistic one is accelerated when directionnaly lit. At a certain intermediate Lorentz factor there is equilibrium. A standing bunch of ultra-relativistic particles lit from below suffers an upward bulk force, because the hardest collisions are between the up-going photons and the down-going leptons: this is the Compton rocket effect [40] When such a bunch of relativistic particles also moves with a relativistic bulk motion, the magnitude and sense of the Compton force on it become a matter for quantitative analysis [41]. The effect of the Compton force and of γ-ray radiation transfer has been included in the analysis of the dynamics of such lepton-plasma blobs [42]. It has been shown that in a first phase the bulk motion of the lit relativistic plasma is locked to the condition that there be zero net Compton force, after which the motion has assymptotically reached a rather high bulk Lorentz factor of order

$$\gamma_b = \left(l < \gamma >^2\right)^{1/7} \qquad (30)$$

where $< \gamma >$ is the average "thermal" Lorentz factor, in the blob's rest frame, and l is the source compactness, defined by $l = (\mathcal{L}\sigma_T)/(r_i m_e c^3)$, where $\mathcal{L}$ is the object's luminosity, σ_T the Thomson cross section, r_i the internal radius of the disk anf m_e the electron's mass. Bulk Lorentz factors of order 20 can be reached if the average thermal Lorentz factor is of order 10^3 - 10^4.

Without reheating of the pair plasma, the thermal Lorentz factor would however quickly decline due to Compton cooling of the leptons. The new point that is made here [42,43] is that plasma microinstabilities may keep the pairs super-hot, precisely because of the bulk relativistic streaming. Actually the plasma in the source may be a 2-fluid one [44], with the pair-plasma streaming through a baryonic plasma consisting of material of the MHD wind having found its way to the axis, where the high-energy action is going on. The baryonic component may be dominant or, on the contrary, be a pollution to the lepton plasma. Beam-plasma instabilities can be triggered in this· system and develop into different regimes according to whether the baryonic plasma or the lepton plasma is the most massive one [43,45].

It is important to realize that, although the effect of these unstable waves is to promote friction between the two components, saturation of their relative streaming may well not occur since the intense lepton gas heating causes a large Compton boost on it. As a result, instabilities bring the system to a highly dissipative state without actually quenching the interaction, tremendously heating the pair plasma and further increasing the Compton boost of the bulk flow. The dynamics of the pair shower, assuming $< \gamma >^2$ to be maintained at high values by these instabilities, has been calculated [42] and, as a byproduct, the high energy radiation spectrum of the jet has been obtained. It compares well with observed spectra, in particular by the presence in the MeV range of a Döppler-upshifted pair annihilation spectral break.

Other ideas deserve to be mentioned here, in particular the concept of proton-initiated cascades, where showers would be started by baryonic particles of extremely high energy, namely $\gamma_p = 10^9 - 10^{11}$, generating hard photons between keV and TeV energies by neutral pion decay in the shower and subsequent synchrotron cascade reprocessing [46].

6. NUMERICAL SIMULATIONS

It is not possible in this limited review to do justice to the impressive amount of work that has been carried out numerically to simulate the development and stationary regime of rotating MHD winds. Usually, perfect MHD conditions are imposed and the wind-emitting object enters only in the form of a set of boundary conditions. Relaxation to a stationary state is not granted, especially so when the back reaction of angular momentum extraction by the MHD flow on the source is considered. Early calculations pictured impulsive outflows from systems that were initially violently out of equilibrium [28], but later work considered impulsive accretion/ejection starting from close to equilibrium situations [47] also in general-relativistic gravity [48]. In the latter case a pressure-driven wind appeared to emanate from the region devoid of stable orbit around a Schwarzschild hole.

The approach to a stationary state has been investigated [49] by imposing keplerian rotation at the boundary. The calculation has been run to quasi-stationarity, the wind being in this case pushed by both pressure gradient and B_ϕ coiling. The velocity of the flow eventually exceeded the fast mode speed. Similar calculations have been performed in a regime where the plasma is centrifuged away [50]. Collimation by the pinch force has been observed and a stationary state has been achieved. However, eventhough the rotation speed at the disk surface and the flux distribution on it were fixed, it sometimes happened that no stationary regime was reached and that episodic outflow occured instead [51], which developed into series of schocks. Such regimes have been found when propulsion is by toroidal field coiling. Oscillations in the degree of coiling are well observed in these simulation. They non-linearly steepen into fast MHD shocks. Plasma compression occurs between regions of enhanced toroidal field. A knotty structure is also reported in fully relativistic calculations [52].

7. OTHER QUESTIONS

Stationnary axisymmetric and strongly coiled MHD flows with strong velocity gradient are subject to instabilities. The main question in connection with this is whether the non-linear development of these instabilities results in the destruction of the organized flow or endows it with a turbulent structure with eddies smaller than the jet global scale. This issue has been first considered in HD in the context of the Kelvin-Helmholtz (KH) instability [53,54]. In cylindrical geometry, perturbations come either in the form of surface waves, confined about the interface between differently streaming fluids, or in the form of body modes which propagate inside the jet body. Body waves are typical of supersonic jets. High density in the jet reduces the instability. Light jets do not survive the non-linear development of the KH instability, while even heavy jets are strongly perturbed but conserve coherent directionality.

The addition of strong axial magnetic fields tends to lessen the instability. In simple MHD situations it is found that MHD jets do not develop persistent cat's eye vortices, which are destroyed by Lorentz forces and by reconnection [55]. The density and fields diffuse a long way out of the jet into cocoons, smoothing out the velocity gradient, and isolating the two differentially streaming fluids from eachother by a broad turbulent layer.

Inclusion of electric currents in the stability analysis is beginning [56]. Such currents cause new specific current-driven instabilities, like kink modes, the coupling of which to KH modes needs to be studied. It has been shown [56] that magnetized current-carrying jets moving faster than the fast mode speed exhibit enhanced stiffness, so that the wavelengths most unstable to the KH instability become longer. In the trans-magnetosonic regime, though, the current appears to be destabilizing to the KH modes. The influence on instability properties of the current profile, which is so important in laboratory plasmas, is presently under study.

Questions related to the structure of the jet's head, and the complicated associated shock structure have been dealt with numerically only [57, 58].

REFERENCES

1. P. Mészáros, M. Rees APJ 482 (1997) L29.
2. H. Ardavan MNRAS 189 (1979) 397.
3. Z.Y. Li PHD thesis, U. of Colorado (1993).
4. V. Beskin, V. Par'ev Phys. Uspekhi 36 (1993) 529.
5. S. Nitta, T. Takahashi, A. Tomimatsu Phys. Rev. D 44 (1991) 2295.
6. M. Camenzind AA 156 (1986) 137.
7. V. Beskin Phys. Uspekhi 40 (7) (1997) 659.
8. J. Heyvaerts, C. Norman APJ 347 (1989) 1055.
9. R. Polovin, V. Demutskii Fundamentals of MHD (1990) Consultant Bureau N.Y.
10. S. Bogovalov MNRAS 280 (1996) 39.
11. K. Tsinganos, C. Sauty, G. Surlantzis, E. Trussoni, J. Contopoulos MNRAS 283 (1996) 811.
12. V. Beskin, I. Kuznetsova JETP 86 (1998) 421.
13. T. Chiueh, Z.Y. Li, M. Begelman APJ 377 (1991) 462.
14. R. Blandford, D. Payne MNRAS 199 (1982) 883.
15. Z.Y. Li, T. Chiueh, M. Begelman APJ 394 (1992) 459
16. J. Contopoulos APJ 432 (1994) 508.
17. J. Contopoulos APJ 446 (1995) 67.
18. R. Lovelace, H. Berk, J. Contopoulos APJ 379 (1991) 696
19. J. Contopoulos and R. Lovelace APJ 429 (1994) 139.
20. K. Chan, R. Henriksen APJ 241 (1980) 534.
21. C. Sauty, K. Tsinganos AA 287 (1994) 893.
22. T. Lery, J. Heyvaerts, S. Appl, C. Norman AA 337 (1998) 603.
23. J. Heyvaerts, C. Norman (1999) in prep.
24. S. Appl, M. Camenzind AA 274 (1993) 699.
25. C. Fendt, M. Camenzind, S. Appl AA 300 (1995) 791.
26. V. Beskin, L. Malyshkin Astronomy Letters 22 (1996) 475.
27. J. Contopoulos APJ 450 (1995) 616
28. K. Shibata, Y. Uchida Pub. Astron. Soc. Jap. 38 (1986) 631.
29. N. Shakura, R. Sunyaev AA 24 (1973) 337.
30. S. Lubow, J. Papaloizou, J. Pringle MNRAS 267 (1994) 235.
31. J. Heyvaerts, E. Priest, A. Bardou APJ 473 (1996) 403.
32. S. Balbus, J. Hawley APJ 376 (1991) 214.
33. J. Ferreira, G. Pelletier AA 276 (1993) 625.
34. Z.Y. Li APJ 444 (1995) 848.
35. R. Henriksen, D. Valls-Gabaud MNRAS 266 (1994) 681.
36. J. Fiege, R. Henriksen MNRAS 281 (1996) 1038.
37. R. Blandford, R. Znajek MNRAS 179 (1977) 433.
38. P. Goldreich, W. Julian APJ 157 (1969) 869.
39. A. Cheng, M. Ruderman, P. Sutherland APJ 203 (1976) 209.
40. S. O'Dell APJ 294 (1981) L147.
41. C. Dermer, R. Schlickheiser APJ 416 (1993) 458.
42. A. Markowith, G. Henri, G. Pelletier MNRAS 277 (1995) 681.
43. A. Markowith, G. Pelletier, G. Henri, AA Supp 120 (1996) 563.
44. H. Sol, G. Pelletier, E. Asséo MNRAS 237 (1989) 411.
45. G. Pelletier, A. Markowith APJ 502 (1998) 598.
46. K. Mannheim AA 269 (1993) 67.
47. R. Matsumoto, Y. Uchida, S. Hirose, K. Shibata, M. Hayashi, A. Ferrrari, G. Bodo, C. Norman APJ 461 (1996) 115.
48. S. Koide, K. Shibata, T. Kudoh (1999) Abstracts of 19th Texas Symp. Relat. Astr.
49. G. Ustyugova A. Koldoba M. Romanova, V. Chechetkin, R. Lovelace, APJ 439 (1995) L 39.
50. R. Ouyed, R. Pudritz APJ 482 (1997) 712.
51. R. Ouyed, R. Pudritz APJ 484 (1997) 794.
52. M. Van Putten APJ 467 (1996) L57.
53. A. Ferrari Ann. Rev. AA 36 (1998) 539.
54. M. Birkinshaw in "Beams and Jets in Astrophysics" ed. P.A. Hughes, Cambridge U. Press (1991) 279.
55. K. Min APJ 482 (1997) 733.
56. S. Appl, M. Camenzind AA 256 (1992) 354.
57. M. Norman, L. Smarr, K. Winkler, M. Smith APJ 113 (1982) 285.
58. D. Clarke in "Energy transport in Radio Galaxies and Quasars" ed. H. Röser, K. Meisenheimer, Springer Verlag (1994) 243.

ELSEVIER

Nuclear Physics B (Proc. Suppl.) 80 (2000) 63–77

NUCLEAR PHYSICS B
**PROCEEDINGS
SUPPLEMENTS**

www.elsevier.nl/locate/npe

Gamma-Ray Bursts and Bursters

P. Mészáros [a*]

[a]Dept. of Astronomy & Astrophysics, Pennsylvania State University, University Park, PA 16802, USA

Major advances have been made in the field of gamma-ray bursts in the last two years. The successful discovery of X-ray, optical and radio afterglows, which were predicted by theory, has made possible the identification of host galaxies at cosmological distances. The energy release inferred in these outbursts place them among the most energetic and violent events in the Universe. Current models envisage this to be the outcome of a cataclysmic event leading to a relativistically expanding fireball, in which particles are accelerated at shocks and produce nonthermal radiation. The substantial agreement between observations and the theoretical predictions of the standard fireball shock model provide confirmation of the basic aspects of this scenario. The continued observations show a diversity of behavior, providing valuable constraints for more detailed, post-standard models which incorporate more realistic physical features. Crucial questions being now addressed are the beaming at different energies and its implications for the energetics, the time structure of the afterglow, its dependence on the central engine or progenitor system behavior, and the role of the environment on the evolution of the afterglow.

1. INTRODUCTION

The discovery of gamma-ray bursts (GRB) by the Vela military test-ban treaty satellites was announced in 1973 [45], and was quickly confirmed by Soviet Konus satellite measurements [52]. Then, for 23 years GRB remained essentially just that: brief outbursts of gamma-rays which pierced, for a brief instant, an otherwise pitch-black gamma-ray sky. An intense debate festered for a long time on whether they were objects in our galaxy or at cosmological distances. The first major breakthrough came in 1992 with the launch of the Compton Gamma-Ray Observatory, whose superb results are summarized in a review by Fishman & Meegan [23]. In particular the all-sky survey from the BATSE instrument showed that bursts were isotropically distributed, strongly suggesting either a cosmological or an extended galactic halo distribution, with essentially zero dipole and quadrupole components. The spectra are definitely non-thermal, typically fitted in the MeV range by broken power-laws whose energy per decade νF_ν peak is in the range 50-500 KeV [3], the power law sometimes ex-

tending to GeV energies [38]. GRB appeared to leave no detectable traces at other wavelengths, except in some cases briefly in X-rays [119,13]. The gamma-ray durations range from 10^{-3} s to about 10^3 s, with a roughly bimodal distribution of long bursts of $t_b \gtrsim 2$ s and short bursts of $t_b \lesssim 2$s [48], and substructure sometimes down to milliseconds. The gamma-ray light curves range from smooth, fast-rise and quasi-exponential decay (FREDs), through curves with several peaks, to highly variable curves with many peaks [23,49]. The pulse distribution is complex [90,72], and the time histories can provide clues for the geometry of the emitting regions [21,22].

Theoretically, it was clear from early on that if GRB are cosmological, enormous energies are liberated in a small volume in a very short time, and an $e^\pm - \gamma$ fireball must form [78,32,117], which would expand relativistically. The main difficulty with this was that a smoothly expanding fireball would convert most of its energy into kinetic energy of accelerated baryons (rather than photons), and would produce a quasi-thermal spectrum, while the typical timescales would not explain events much longer than milliseconds. This problem was solved with the introduction of the "fireball shock model" [98,57], based on the re-

*Also Institute for Theoretical Physics, UCSB, and Ctr. for Gravitational Physics & Geometry, Penn State Univ. Supported by NASA NAG-5 2857 and NSF PHY94-07194

alization that shocks are likely to arise, at the latest when the fireball runs into an external medium, which would occur after the fireball is optically thin and would reconvert the kinetic energy into nonthermal radiation. The complicated light curves can be understood in terms of internal shocks [99] in the outflow itself, before it runs into the external medium, caused by velocity variations in the outflow from the source,

The next major breakthrough came in early 1997 when the Italian-Dutch satellite Beppo-SAX succeeded in providing accurate X-ray measurements which, after a delay of 4-6 hours for processing, led to positions [14], allowing follow-ups at optical and other wavelengths, e.g. [127]. This paved the way for the measurement of redshift distances, the identification of candidate host galaxies, and the confirmation that they were indeed at cosmological distances [68,18,50]. The detection of other GRB afterglows followed in rapid succession, sometimes extending to radio [24,25] and over timescales of many months [128], and in a number of cases resulted in the identification of candidate host galaxies, e.g. [108,9,74], etc. The study of afterglows has provided strong confirmation for the generic fireball shock model of GRB. This model in fact led to a correct prediction [61], in advance of the observations, of the quantitative nature of afterglows at wavelengths longer than γ-rays, which were in substantial agreement with the data [129,121,132,101,136].

A major issue raised by the measurement of large redshifts, e.g. [50,51], is that the measured γ-ray fluences imply a total energy of order $10^{54}(\Omega_\gamma/4\pi)$ ergs, where $\Delta\Omega_\gamma$ is the solid angle into which the gamma-rays are beamed. A beamed jet would clearly alleviate the energy requirements, but it is only recently that tentative evidence has been reported for evidence of a jet [51,27,12]. Whether a jet is present or not, such energies are possible [62] in the context of compact mergers involving neutron star-neutron star (NS-NS) or black hole-neutron star (BH-NS) binaries, or in hypernova/collapsar models involving a massive stellar progenitor [81,95]. In both cases, one is led to rely on MHD extraction of the spin energy of a disrupted torus and/or a central fast spinning BH, which can power a relativistic

fireball resulting in the observed radiation.

While it is at present unclear which, if any, of these progenitors is responsible for GRB, or whether perhaps different progenitors represent different subclasses of GRB, there is general agreement that they all would be expected to lead to the generic fireball shock scenario mentioned above. Much of the current effort is dedicated to understanding the progenitors more specifically, and trying to determine what effect, if any, they have on the observable burst and afterglow.

2. Black Hole/Debris systems as generic GRB energy sources

It has become increasingly apparent in the last few years that *most* plausible GRB progenitors suggested so far are expected to lead to a system with a central BH plus a temporary debris torus around it. Scenarios leading to this include, e.g. NS-NS or NS-BH mergers, Helium core - black hole [He/BH] or white dwarf - black hole [WD-BH] mergers, and the wide category labeled as hypernova or collapsars including failed supernova Ib [SNe Ib], single or binary Wolf-Rayet [WR] collapse, etc. [80,137,62,81,95], and accretion-induced collapse [130,96]. An important point is that the overall energetics from these various progenitors do not differ by more than about one order of magnitude [66]. Another possibility is massive black holes ($\sim 10^3 - 10^5 M_\odot$) in the halos of galaxies. Some related models involve a compact binary or a temporarily rotationally stabilized neutron star, perhaps with a superstrong field, e.g. [125,123,130,118], which ultimately also should lead to a BH plus debris torus.

Two large reservoirs of energy are available in these systems: the binding energy of the orbiting debris, and the spin energy of the black hole [62]. The first can provide up to 42% of the rest mass energy of the disk, for a maximally rotating black hole, while the second can provide up to 29% of the rest mass of the black hole itself. The question is how to extract this energy.

One energy extraction mechanisms is the $\nu\bar{\nu} \to e^+e^-$ process [20], which can tap the thermal energy of the torus produced by viscous dissipation. To be efficient, the neutrinos must escape before being advected into the hole; on the other hand, the efficiency of conversion into pairs (which scales with the square of the neutrino density) is low if the neutrino production is too gradual. Typical estimates suggest a fireball of $\lesssim 10^{51}$ erg [106,28,53], except perhaps in the collapsar case where [95] estimate $10^{52.3}$ ergs for optimum parameters. If the fireball is collimated into a solid angle Ω_j then of course the apparent "isotropized" energy would be larger by a factor $(4\pi/\Omega_j)$, but unless Ω_j is $\lesssim 10^{-3} - 10^{-4}$ this would fail to satisfy the apparent isotropized energy of 4×10^{54} ergs deduced for GRB 990123 [51].

An alternative, and more efficient mechanism for tapping the energy of the torus may be through dissipation of magnetic fields generated by the differential rotation in the torus [80,71,62, 43]. Even before the BH forms, a NS-NS merging system might lead to winding up of the fields and dissipation in the last stages before the merger [55,129].

However, a larger energy source is available in the hole itself, especially if formed from a coalescing compact binary, since then it is guaranteed to be rapidly spinning. Being more massive, it could contain more energy than the torus. The energy extractable in principle through MHD coupling to the rotation of the hole by the B-Z (Blandford & Znajek [5]) effect could then be even larger than that contained in the orbiting debris [62,81]. Collectively, any such MHD outflows have been referred to as Poynting jets.

The various progenitors differ only slightly in the mass of the BH and that of the debris torus they produce, but they may differ more markedly in the amount of rotational energy contained in the BH. Strong magnetic fields, of order 10^{15} G, are needed needed to carry away the rotational or gravitational energy in a time scale of tens of seconds [125,123], which may be generated on such timescales by a convective dynamo mechanism, the conditions for which are satisfied in freshly collapsed neutron stars or neutron star tori [19,46]. If the magnetic fields do not thread the BH, then a Poynting outflow can at most carry the gravitational binding energy of

the torus. For a maximally rotating and for a slow-rotating BH this is

$$E_t = \epsilon M_\odot c^2 \begin{cases} 0.42(M_d/M_\odot) \text{ ergs}, & \text{(fast rot.);} \\ 0.06(M_d/M_\odot) \text{ ergs}, & \text{(slow rot.),} \end{cases} \quad (1)$$

where ϵ is the efficiency in converting gravitational into MHD jet energy. The torus or disk mass in a NS-NS merger is[107] $M_d \sim 10^{-1} - 10^{-2} M_\odot$, and for a NS-BH, a He-BH, WD-BH merger or a binary WR collapse it may be estimated at [81,28] $M_d \sim 1 M_\odot$. In the HeWD-BH merger and WR collapse the mass of the disk is uncertain due to lack of calculations on continued accretion from the envelope, so $1 M_\odot$ is just a rough estimate. The maximum torus-based MHD energy extraction is then

$$E_{max,t} \sim \begin{cases} 8 \times 10^{53}\epsilon(M_d/M_\odot) \text{ ergs}, & ; \\ 1.2 \times 10^{53}\epsilon(M_d/M_\odot) \text{ ergs}, & \\ 0.8 \times 10^{53}\epsilon(M_d/0.1M_\odot) \text{ ergs}, & . \end{cases}$$

for the NS-BH, He/WD-BH or collapsar case; the (slow rotating) failed SN Ib case; and NS-NS case, respectively.

If the magnetic fields in the torus thread the BH, the spin energy of the BH can in principle be extracted via the [5] (B-Z) mechanism ([62]). The extractable energy is

$$E_{bh} \sim \epsilon f(a) M_{bh} c^2 , \qquad (2)$$

where ϵ is the MHD efficiency factor, $f(a) = 1 - ([1 + \sqrt{1-a^2}]/2)^{1/2} \leq 0.29$ is the rotational efficiency factor, and $a = Jc/GM^2$ is the rotation parameter, which equals 1 for a maximally rotating black hole. The $f(a)$ rotational factor is is small unless a is close to 1, where it rises sharply to its maximum value $f(1) = 0.29$, so the main requirement is a rapidly rotating black hole, $a \gtrsim 0.5$. For a maximally rotating BH, the extractable energy is therefore

$$E_{max,bh} \sim 0.29\epsilon M_{bh}c^2 \sim 5 \times 10^{53}\epsilon(M_{bh}/M_\odot) \text{ ergs.} (3)$$

Rapid rotation is guaranteed in a NS-NS merger, since the radius (especially for a soft equation of state) is close to that of a black hole and the final orbital spin period is close to the required maximal spin rotation period. The central BH will have a mass [106,107] of about $2.5 M_\odot$, so the NS-NS system can power a jet of up to $E_{NS-NS} \lesssim 1.3 \times 10^{54} \epsilon (M_{bh}/2.5 M_\odot)$ ergs. A maximal rotation rate may also be possible in a He-BH merger, depending on what fraction of the He core gets accreted along the rotation axis as opposed to along the equator [28], and the same should apply to the binary fast-rotating WR scenario, which probably does not differ much in its final details from the He-BH merger. For a fast rotating BH of $2.5 - 3M_\odot$ threaded by the magnetic field, the maximal energy carried out by the jet is then similar or somewhat larger than in the NS-NS case. The scenarios less likely to produce a fast rotating BH are the NS-BH merger (where the rotation parameter could be limited to $a \leq M_{ns}/M_{bh}$, unless the BH is already fast-rotating) and the failed SNe Ib (where the last material to fall in would have maximum angular momentum, but the material that was initially close to the hole has less angular momentum). The electromagnetic energy extraction from the BH in these could be limited by the $f(a)$ factor, but a lower limit would be given by the energy available from the gravitational energy of the disk, in the second line of equation (1).

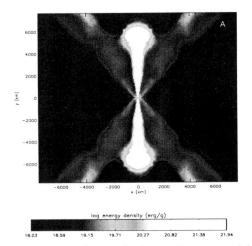

log energy density (erg/g)

Figure 1. Jet formation in a collapsar model leading to black hole from collapse of a fast rotating He core [53]

The total energetics differ thus between the various models at most by a factor 20 for a Poynting (MHD) jet powered by the torus binding energy, whereas for Poynting jets powered by the BH spin energy they differ at most by a factor of a few, depending on the rotation parameter. For instance, allowing for a total efficiency of 50%, a NS-NS merger whose jet is powered by the torus binding energy would require a beaming of the γ-rays by a factor $(4\pi/\Omega_j) \sim 100$, or beaming by a factor ~ 10 if the jet is powered by the B-Z mechanism, to produce the equivalent of an isotropic energy of 4×10^{54} ergs. The beaming requirements of BH-NS and some of the collapsar scenarios are even less constraining, either when tapping the torus or the BH. Thus, even the most extreme energy requirements inferred observationally thus far can be plausibly satisfied by scenarios leading to a BH plus torus system.

The major difference between the various models is expected to be in the *location* where the burst occurs relative to the host galaxy (see §7). They are also likely to differ substantially in the efficiency of producing a directly observable relativistic outflow, as well as in the amount of collimation of the jet they produce. The conditions for the efficient escape of a high-Γ jet are less propitious if the "engine" is surrounded by an extensive envelope. In this case the jet has to "punch through" the envelope, and its ability to do so may be crucially dependent on the level of viscosity achieved in the debris torus (e.g. [53]), higher viscosities leading to more powerful jets (see Figure 1). The simulations, so far, are nonrelativistic and one can only infer that high enough viscosities can lead to jets capable of punching though a massive (several $M_\odot$) envelope. This is facilitated, of course, if the envelope is fast-rotating, as in this case there is a centrifugally induced column density minimum along the spin axis, which might be small enough to allow punch-through to occur. If they do, a very tightly collimated beam may arise. "Cleaner" environments, such as NS-BH or NS-NS merger, or rotational support loss/accretion induced collapse to BH would have much less material to be pushed out of the way by a jet, while their energy is, to order of magnitude, similar to that in massive stellar progenitor cases. In these cases, on the other hand, there is no natural choke to collimate a jet, which might therefore be somewhat wider than in massive progenitor cases.

3. The Fireball Shock Scenario

Irrespective of the details of the progenitor, the resulting fireball is expected to be initially highly optically thick. From causality considerations the initial dimensions must be of order $ct_{var} \lesssim 10^7$ cm, where t_{var} is the variability timescale, and the luminosities must be much higher than a solar Eddington limit. Since most of the spectral energy is observed above 0.5 MeV, the optical depth against $\gamma\gamma \to e^\pm$ is large, and an $e^\pm, \gamma$ fireball is expected. Due to the highly super-Eddington luminosity, this fireball must expand. Since in many bursts one observes a large fraction of the total energy at photon energies $\epsilon_\gamma \gtrsim 1 GeV$, somehow the flow must be able to avoid degrading these photons ($\gamma\gamma \to e^\pm$ would lead, in a stationary or slowly expanding flow, to photons just below 0.511 MeV[34]). In order to avoid this, it seems inescapable that the flow must be expanding with a very high Lorentz factor, since in this case the relative angle at which the photons collide is less than Γ^{-1} and the threshold for the pair production is effectively diminished. The bulk Lorentz factor must be

$$\Gamma \gtrsim 10^2 (\epsilon_{\gamma, 10 \text{GeV}} \epsilon_{t, \text{MeV}})^{1/2} , \qquad (4)$$

in order for photons with energy $\epsilon_\gamma \gtrsim 10$ GeV to escape annihilation against target photons of energy $\epsilon_t \sim 1$ MeV [54,34]. Thus, simply from observations and general physical considerations, a relativistically expanding fireball is expected. ¿From general considerations [58], one can see that an outflow arising from an initial energy E_o imparted to a mass $M_o << E_0/c^2$ within a radius r_l will lead to an expansion. Initially the bulk Lorentz $\Gamma \propto r$, while comoving temperature drops $\propto r^{-1}$; however, Γ cannot increase beyond $\Gamma_{max} \sim \eta \sim E_o/M_o c^2$, which is achieved at a radius $r/r_l \sim \eta$, beyond which the flow continues to coast with $\Gamma \sim \eta \sim$ constant [58].

$$\Gamma \sim \begin{cases} (r/r_l) , & \text{for } r/r_l \lesssim \eta; \\ \eta , & \text{for } r/r_l \gtrsim \eta. \end{cases} \qquad (5)$$

However, the observed γ-ray spectrum observed is generally a broken power law, i.e., highly nonthermal. The optically thick $e^\pm \gamma$ fireball cannot, by itself, produce such a spectrum (it would tend rather to produce a modified blackbody, [78,32]). In addition, the expansion would lead to a conversion of internal energy into kinetic energy of expansion, so even after the fireball becomes optically thin, it would be highly inefficient, most of the energy being in the kinetic energy of the associated protons, rather than in photons.

The most likely way to achieve a nonthermal spectrum in an energetically efficient manner is if the kinetic energy of the flow is re-converted into random energy via shocks, after the flow has become optically thin [98]. This is a plausible scenario, in which two cases can be distinguished. In the first case (a) the expanding fireball runs into an external medium (the ISM, or a pre-ejected stellar wind[98,57,41,109]. The second possibility (b) is that [99,77], even before external shocks occur, internal shocks develop in the relativistic wind itself, faster portions of the flow catching up with the slower portions. This is a completely generic model, which is independent of the specific nature of the progenitor, as long as it delivers the appropriate amount of energy ($\gtrsim 10^{52}$ erg) in a small enough region ($\lesssim 10^7$ cm). This model has been successful in explaining the major observational properties of the gamma-ray emission, and is the main paradigm used for interpreting the GRB observations.

External shocks will occur in an impulsive outflow of total energy E_o in an external medium of average particle density n_o at a radius and on a timescale

$$r_{dec} \sim \quad 10^{17} E_{53}^{1/3} n_o^{-1/3} \eta_2^{-2/3} \text{ cm} ,$$
$$t_{dec} \sim \quad r_{dec}/(c\Gamma^2) \sim 3 \times 10^2 E_{53}^{1/3} n_o^{-1/3} \eta_2^{-8/3} \text{ s} ,$$

where the lab-frame energy of the swept-up external matter ($\Gamma^2 m_p c^2$ per proton) equals the initial energy E_o of the fireball, and $\eta = \Gamma = 10^2 \eta_2$ is the final bulk Lorentz factor of the ejecta. The typical observer-frame dynamic time of the shock (assuming the cooling time is shorter than this) is $t_{dec} \sim r_{dec}/c\Gamma^2 \sim$ seconds, for typical parameters, and $t_b \sim t_{dec}$ would be the burst duration (the impulsive assumption requires that the initial energy input occur in a time shorter than t_{dyn}). Variability on timescales shorter than t_{dec} may occur on the cooling timescale or on the dynamic timescale for inhomogeneities in the external medium, but generally this is not ideal for reproducing highly variable profiles[111]. However, it can reproduce bursts with several peaks[83] and may therefore be applicable to the class of long, smooth bursts.

The same behavior $\Gamma \propto r$ with comoving temperature $\propto r^{-1}$, followed by saturation $\Gamma_{max} \sim \eta$ at the same radius $r/r_l \sim \eta$ occurs in a wind scenario [79], if one assumes that a lab-frame luminosity L_o and mass outflow $\dot{M}_o$ are injected at $r \sim r_l$ and continuously maintained over a time t_w; here $\eta = L_o/\dot{M}_o c^2$. In such wind model, internal shocks will occur at a radius and over a timescale [99]

$$
\begin{aligned}
r_{dis} &\sim \quad ct_{var}\eta^2 \sim 3 \times 10^{14} t_{var}\eta_2^2 \text{ cm,} \\
t_w &\gg \qquad t_{var} \sim r_{dis}/(c\eta^2) \text{ s,}
\end{aligned}
$$

where shells of different energies $\Delta\eta \sim \eta$ initially separated by ct_v (where $t_v \leq t_w$ is the timescale of typical variations in the energy at r_l) catch up with each other. In order for internal shocks to occur above the wind photosphere $r_{ph} \sim \dot{M}\sigma_T/(4\pi m_p c\Gamma^2) = 1.2 \times 10^{14} L_{53}\eta_2^{-3}$ cm, but also at radii greater than the saturation radius (so that most of the energy does not come out in the photospheric quasi-thermal radiation component) one needs to have $7.5 \times 10^1 L_{51}^{1/5} t_{var}^{-1/5} \lesssim \eta 3 \times 10^2 L_{53}^{1/4} t_{var}^{-1/4}$. This type of models have the advantage[99] that they allow an arbitrarily complicated light curve, the shortest variation timescale $t_{var} \gtrsim 10^{-3}$ s being limited only by the dynamic timescale at r_l, where the energy input may be expected to vary chaotically. Such internal shocks have been shown explicitly to reproduce (and be required by) some of the more complicated light curves[111,47,87] (see however [16]).

4. The Simple Standard Afterglow Model

The dynamics of GRB and their afterglows can be understood in a fairly simple manner, independently of any uncertainties about the progenitor systems, using a generalization of the method used to model supernova remnants. The simplest hypothesis is that the afterglow is due to a relativistic expanding blast wave, which decelerates as time goes on [61]. The complex time structure of some bursts suggests that the central trigger may continue for up to 100 seconds, the γ-rays possibly being due to internal shocks. However, at much later times all memory of the initial time structure would be lost: essentially all that matters is how much energy and momentum has been injected; the injection can be regarded as instantaneous in the context of the much longer afterglow. As pointed in the original fireball shock paper [98], the external shock bolometric luminosity builds up and decays as

$$
L \propto \begin{cases} t^2 & \text{rise} \\ t^{-(1+q)} & \text{decay} \end{cases} . \tag{6}
$$

The first line is obtained equating, in the contact discontinuity frame, the kinetic flux $L/4\pi r^2$ to the external ram pressure $\rho_{ext}\Gamma^2$ during the initial phase where $\Gamma \sim$ constant, $r \propto t$, while the second follows from energy conservation $L \propto E/t$ under adiabatic conditions (q takes into account radiative effects or bolometric corrections; the flux per unit frequency rises in the same way, and decays with $q \geq 1$ in equ. (6)). At the deceleration radius (6) the fireball energy and the bulk Lorentz factor decrease by a factor ~ 2 over a timescale $t_{dec} \sim r_{dec}/(c\Gamma^2)$, and thereafter the bulk Lorentz factor decreases as a power law in radius. This is

$$
\Gamma \propto r^{-g} \propto t^{-g/(1+2g)} \ , \ r \propto t^{1/(1+2g)}, \tag{7}
$$

with $g = (3, 3/2)$ for the radiative (adiabatic) regime, in which $\rho r^3 \Gamma \sim$ constant ($\rho r^3 \Gamma^2 \sim$ constant). At late times, a similarity solution [6] solution with $g = 7/2$ may be reached.

The spectrum of radiation is likely to be due to synchrotron radiation, whose peak frequency in the observer frame is $\nu_m \propto \Gamma B'\gamma^2$, and both the comoving field B' and electron Lorentz factor γ are likely to be proportional to Γ [57]. This implies that as Γ decreases, so will ν_m, and the radiation will move to longer wavelengths. This led [76,42] to early discussions, based on the forward blast wave, of the possibility of detecting at late times a radio or optical afterglow of the GRB. A more detailed treatment of the fireball dynamics indicate that approximately equal amounts of energy are radiated by the forward blast wave, moving with $\sim \Gamma$ into the surrounding medium, and by a reverse shock propagating with $\Gamma_r - 1 \sim 1$ back into the ejecta [57]. The electrons are therefore shocked to much higher energies in the forward shock than in the reverse shock, producing a two-step synchrotron spectrum which during the deceleration time t_{dec} peaks in the optical (reverse) and in the γ/X (forward) [59,60]. The predicted fluences in the optical for typical bursts at cosmological distances were $\sim 10^{-7.5}$ erg cm^{-2} s^{-1}, or about a 9th magnitude prompt optical flash [61] of duration comparable to the γ-rays, in agreement with a recent prompt optical detection in GRB 990123 [2]. Detailed calculations and predictions of the time evolution of such a forward and reverse shock afterglow model ([61]) preceded the observations of the first afterglow GRB970228 ([14,127]), which was detected in γ-rays, X-rays and several optical bands, and was followed up for a number of months.

The simplest spherical afterglow model generally concentrates only on the properties of the forward blast wave radiation, for which the flux at a given frequency and the synchrotron peak frequency decay at a rate [61,67]

$$F_\nu \propto t^{[3-2g(1-2\beta)]/(1+2g)} \quad , \quad \nu_m \propto t^{-4g/(1+2g)}, \quad (8)$$

where g is the exponent of Γ (equ. [7]) and β is the photon spectral energy slope. The decay rate of the forward shock F_ν in equ.(8) is typically slower than that of the reverse shock [61], and the reason why the "simplest" model was stripped down to its forward shock component only is that, for the first two years 1997-1998, afterglows were followed in more detail only after the several hours needed by Beppo-SAX to acquire accurate positions, by which time both reverse external shock and internal shock components are expected to have become unobservable. This simple standard model has been remarkably successful at explaining the gross features and light curves of GRB 970228, GRB 970508 (after 2 days; for early rise, see §5) e.g. [136,121,132,101] (see Figure 2).

This simplest afterglow model produces at any given time a three-segment power law spectrum

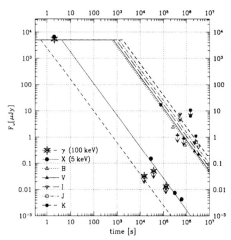

Figure 2. GRB 970228 light-curves compared [136] to the blast wave model predictions of [61]

with two breaks. At low frequencies there is a steeply rising synchrotron self-absorbed spectrum up to a self-absorption break ν_a, followed by a +1/3 energy index spectrum up to the synchrotron break ν_m corresponding to the minimum energy γ_m of the power-law accelerated electrons, and then a $-(p-1)/2$ energy spectrum above this break, for electrons in the adiabatic regime (where γ^{-p} is the electron energy distribution above γ_m). A fourth segment and a third break is expected at energies where the electron cooling time becomes short compared to the expansion time, with a spectral slope $-p/2$ above that. With this third "cooling" break ν_b, first calculated in [65] and more explicitly detailed in [112], one has what has come to be called the simple "standard" model of GRB afterglows. One of the predictions of this model [61] is that the relation between the temporal decay index α, for $g = 3/2$ in $\Gamma \propto r^{-g}$, is related to the photon spectral energy index β through through

$$F_\nu \propto t^\alpha \nu^\beta \quad , \text{with} \quad \alpha = (3/2)\beta . \quad (9)$$

This relationship appears to be valid in many (although not all) cases, especially after the first few days, and is compatible with an electron spectral index $p \sim 2.2 - 2.5$ which is typical of shock acceleration, e.g. [132,112,135], etc. As the remnant expands the photon spectrum moves to lower frequencies, and the flux in a given band decays as a power law in time, whose index can change as breaks move through it. For the simple standard model, snapshot overall spectra have been deduced by extrapolating spectra at different wavebands and times using assumed simple time dependences [133,135]. These can be used to derive rough fits for the different physical parameters of the burst and environment, e.g. the total energy E, the magnetic and electron-proton coupling parameters ϵ_B and ϵ_e and the external density n_o (see Figure 3).

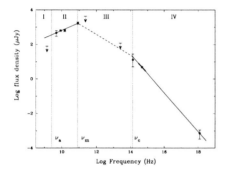

Figure 3. Snapshot spectrum of GRB 970508 at $t = 12$ days and standard afterglow model fit [135]

Since the simple afterglow model has generally proved quite robust, it is worth reviewing the assumptions made in it. The following apply both to the "simple standard" model using forward shocks only, and to the original [61] version including both forward and reverse shocks:
a) A single value of E_o and $\Gamma_o = \eta$ is used,
b) the external medium n_{ext} is homogeneous,
c) the accelerated electron spectral index p, the magnetic field and electron to proton equipartion ratios ε_B and ε_e do not change in time,
d) the expansion is relativistic and the dynamics are given by $\Gamma \propto r^{-3/2}$ (adiabatic),
d) the outflow is spherical (or angle independent inside some jet solid angle Ω_j),
e) the observed radiation is characterized by the scaling relations along the line of sight.

These assumptions, even if correct over some range, clearly would break down after some time. Estimates for the time needed to reach the non-relativistic expansion regime are typically $\lesssim$ month(s) ([129]), or less if there is an initial radiative regime $\Gamma \propto r^{-3}$. However, even when electron radiative times are shorter than the expansion time, it is unclear whether a regime $\Gamma \propto r^{-3}$ should occur, since it would require strong electron-proton coupling [65]. As far as sphericity, the standard model can be straightforwardly generalized to the case where the energy is assumed to be channeled initially into a solid angle $\Omega_j < 4\pi$ [58]. In this case [104,105] a change occurs after Γ drops below $\Omega_j^{-1/2}$, after which the side of the jet becomes observable, and soon thereafter one expects a faster decay of Γ if the jet starts to expands sideways, leading to a decrease in the brightness. A calculation based on the sideways expansion, using the usual scaling laws for a single central line of sight [105] leads then to a steepening of the light curve. Until recently, no evidence for a steepening could be found in afterglows over several months. E.g., in GRB 971214 [97], a snapshot standard model fit and the lack of a break in the late light curve could be, in principle, interpreted as evidence for lack of a jet, leading to an (isotropic) energy estimate of $10^{53.5}$ ergs. While such large energy outputs are possible in *either* NS-NS, NS-BH mergers [62] or in hypernova/collapsar models [81,95] using MHD extraction of the spin energy of a disrupted torus and/or a central fast spinning BH, it is worth stressing that what these snapshot fits constrain is only the *energy per solid angle* [66]. Also, the expectation of a break after some weeks or months (e.g., due to Γ dropping either below a few, or below $\Omega_j^{-1/2}$) is based upon the simple impulsive (angle-independent delta or top-hat function) energy input approximation. The latter is useful, but departures from it would be natural, and certainly not surprising. In fact, as discussed below, tentative evidence for beaming in one obejct has recently been reported [51,27,12], but it is difficulty to ascertain, and could be masked by a number of commonly expected effects.

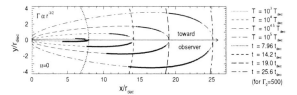

Figure 4. Ring-like equal-arrival time T surfaces of an afterglow, based on [85]

5. "Post-standard" Afterglow Models

In a realistic situation, one could expect any of several fairly natural departures from the simple standard model to occur. The first one is that the emitting region seen by the observer resembles a ring [133,84,110] (see Figure 4). This effect may, in fact, be important in giving rise to the radio scintillation pattern seen in several afterglows, since this requires the emitting source to be of small dimensions, which is aided if the emission is ring-like, e.g. in the example of GRB 970508 [134] (Figure 5).

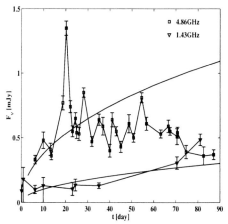

Figure 5. Radio afterglow light-curves of GRB970508 at 4.86GHz and 1.43GHz, compared with the predictions of [134].

One expects afterglows to show a diversity in their decay rates, not only due to different β but

also from the possibility of a non-standard relation between the temporal decay index α and the spectral energy index β, different from equ. (9). The most obvious departure from the simplest standard model occurs if the external medium is inhomogeneous: for instance, for $n_{ext} \propto r^{-d}$, the energy conservation condition is $\Gamma^2 r^{3-d} \sim$ constant, which changes significantly the temporal decay rates [65]. Such a power law dependence is expected if the external medium is a wind, say from an evolved progenitor star as implied in the hypernova scenario (such winds are generally used to fit supernova remnant models). Another obvious non-standard effect, which it is reasonable to expect, is departures from a simple impulsive injection approximation (i.e. from a delta or top hat function with a single value for E_o and Γ_o). An example is if the mass and energy injected during the burst duration t_w (say tens of seconds) obeys $M(>\Gamma) \propto \Gamma^{-s}$, $E(>\Gamma) \propto \Gamma^{1-s}$, i.e. more energy emitted with lower Lorentz factors at later times (but still shorter than the gamma-ray pulse duration). This would drastically change the temporal decay rate and extend the afterglow lifetime in the relativistic regime, providing a late "energy refreshment" to the blast wave on time scales comparable to the afterglow time scale [100]. These two cases lead to a decay rate

$$\Gamma \propto r^{-g} \propto \begin{cases} r^{-(3-d)/2} & ; n_{ext} \propto r^{-d}; \\ r^{-3/(2+s)} & ; E(>\Gamma) \propto \Gamma^{1-s}. \end{cases} \quad (10)$$

Expressions for the temporal decay index $\alpha(\beta, s, d)$ in $F_\nu \propto t^\alpha$ are given by [65,100], which now depend also on s and/or d (and not just on β as in the simple standard relation of equ.(9). The result is that the decay can be flatter (or steeper, depending on s and d) than the simple standard $\alpha = (3/2)\beta$. A third non-standard effect, which is entirely natural, occurs when the energy and/or the bulk Lorentz factor injected are some function of the angle. A simple case is $E_o \propto \theta^{-j}$, $\Gamma_o \propto \theta^{-k}$ within a range of angles; this leads to the outflow at different angles shocking at different radii and its radiation arriving at the observed at different delayed times, and it has a marked effect on the time dependence of the afterglow [65], with $\alpha = \alpha(\beta, j, k)$ flatter or steeper than the standard value, depending on j, k. Thus in general, a

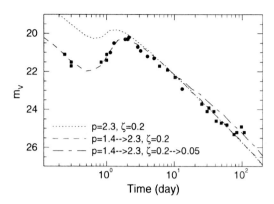

Figure 6. Optical light-curve of GRB 970508, fitted with a non-uniform injection model (a similar fit can be obtained with an off-axis jet plus a weaker isotropic component) [88]

temporal decay index which is a function of more than one parameter

$$F_\nu \propto t^\alpha \nu^\beta \quad \text{, with} \quad \alpha = \alpha(\beta, d, s, j, k, \cdots) \,, \quad (11)$$

is not surprising; what is more remarkable is that, in many cases, the simple standard relation (9) is sufficient to describe the gross overall behavior at late times.

Strong evidence for departures from the simple standard model is provided by, e.g., sharp rises or humps in the light curves followed by a renewed decay, as in GRB 970508 ([89,93]). Detailed time-dependent model fits [88] to the X-ray, optical and radio light curves of GRB 970228 and GRB 970508 show that, in order to explain the humps, a *non-uniform* injection (Figure 6) or an *anisotropic* outflow is required.

These fits indicate that the shock physics may be a function of the shock strength (e.g. the electron index p, injection fraction ζ and/or ϵ_b, ϵ_e change in time), and also indicate that dust absorption is needed to simultaneously fit the X-ray and optical fluxes. The effects of beaming (outflow within a limited range of solid angles) can be significant [86], but are coupled with other effects, and a careful analysis is needed to disentangle them.

One consequence of "post-standard" decay

laws (e.g. from density inhomogeneities, non-uniform injection or anisotropic outflow) is that the transition to a steeper jet regime $\Gamma < \theta_j^{-1} \sim$ few can occur as late as six months to a year after the outburst, depending on details of the energy input. This transition is made more difficult to detect by the fact that, as numerical integration over angles of the ring-like emission [85] show, the transition is very gradual and the effects of sideways expansion effects are not so drastic as inferred [105] from the scaling laws along the central line of sight. This is because even though the flux from the head-on part of the remnant decreases faster, this is more than compensated by the increased emission measure from sweeping up external matter over a larger angle, and by the fact that the extra radiation, arising at larger angles, arrives later and re-fills the steeper light curve. The inference (e.g. [97,105]) that GRB 970508 and a few other bursts were isotropic due to the lack of an observable break is predicated entirely on the validity of the *simplest standard* fireball assumption. Since these assumptions are drastic simplifications, and physically plausible generalizations lead to different conclusions, one can interpret the results of [97,105] as arguments indicating that *post-standard* features are, in fact, necessary in some objects.

6. Prompt multi-wavelength flashes, reverse shocks and jets

Prompt optical, X-ray and GeV flashes from reverse and forward shocks, as well as from internal shocks, have been calculated in theoretical fireball shock models for a number of years [59,60,82,61,113], as have been jets (e.g. [55,58, 60], and in more detail [104,88,86,105]). Thus, while in recent years they were not explicitly part of the "simple standard" model, they are not strictly "post"-standard either, since they generally use the "standard" assumptions, and they have a long history. However, observational evidence for these effects were largely lacking, until the detection of a prompt (within 22 s) optical flash from GRB 990123 with ROTSE by [2], together with X-ray, optical and radio follow-ups citekul99,gal99,fru99,and99,cas99,hjo99. GRB

990123 is so far unique not only for its prompt optical detection, but also by the fact that if it were emitting isotropically, based on its redshift $z = 1.6$ [51,1] its energy would be the largest of any GRB so far, 4×10^{54} ergs. It is, however, also the first (tentative) case in which there is evidence for jet-like emission [51,27,12]. An additional, uncommon feature is that a radio afterglow appeared after only one day, only to disappear the next [30,51].

The prompt optical light curve of GRB 990123 decays initially as $\propto t^{-2.5}$ to $\propto t^{-1.6}$ [2], much steeper than the typical $\propto t^{-1.1}$ of previous optical afterglows detected after several hours. However, after about 10 minutes its decay rate moderates, and appears to join smoothly onto a slower decay rate $\propto t^{-1.1}$ measured with large telescopes [30,51,27,12] after hours and days. The prompt optical flash peaked at 9-th magnitude after 55 s [2], and in fact a 9-th magnitude prompt flash with a steeper decay rate had been predicted more than two years ago [61], from the synchrotron radiation of the reverse shock in GRB afterglows at cosmological redshifts (see also [113]). An optical flash contemporaneous with the γ-ray burst, coming from the reverse shock and with fluence corresponding to that magnitude, had also been predicted earlier [59,60]. An origin of the optical prompt flash in internal shocks [61,67] cannot be ruled out yet, but is less likely since the optical light curve and the γ-rays appear not to correlate well [114,30] (but the early optical light curve has only three points). The subsequent slower decay agrees with the predictions for the forward component of the external shock [61,114,67].

The evidence for a jet is possibly the most exciting, although must still be considered tentative. It is based on an apparent steepening of the light curve after about three days [51,27,12]. This is harder to establish than the decay of the two previous earlier portions of the light curve, since by this time the flux has decreased to a level where the detector noise and the light of the host galaxy become important. However, after correcting for this, the r-band data appears to steepen significantly. (In the K-band, where the noise level is higher, a steepening is not obvi-

ous, but the issue should be settled with further Space Telescope observations). If real, this steepening is probably due to the transition between early relativistic expansion, when the light-cone is narrower than the jet opening, and the late expansion, when the light-cone has become wider than the jet, leading to a drop in the effective flux [104,51,67,105]. A rough estimate leads to a jet opening angle of 3-5 degrees, which would reduce the total energy requirements to about 4×10^{52} ergs. This is about two order of magnitude less than the binding energy of a few solar rest masses, which, even allowing for substantial inefficiencies, is compatible with currently favored scenarios (e.g. [95,53]) based on a stellar collapse or a compact binary merger.

7. Location and Environmental Effects

The location of the afterglow relative to the host galaxy center can provide clues both for the nature of the progenitor and for the external density encountered by the fireball. A hypernova model would be expected to occur inside a galaxy in a high density environment $n_o > 10^3 - 10^5$ cm^{-3}. Most of the detected and well identified afterglows are inside the projected image of the host galaxy [7], and some also show evidence for a dense medium at least in front of the afterglow ([75]). For a number of bursts there are constraints from the lack of a detectable, even faint, host galaxy [116], but at least for Beppo-SAX bursts (which is sensitive only to long bursts $t_b \gtrsim 20$ s) the success rate in finding candidate hosts is high.

In NS-NS mergers one would expect a BH plus debris torus system and roughly the same total energy as in a hypernova model, but the mean distance traveled from birth is of order several Kpc [10], leading to a burst presumably in a less dense environment. The fits of [135] to the observational data on GRB 970508 and GRB 971214 in fact suggest external densities in the range of $n_o = 0.04$-0.4 cm^{-3}, which would be more typical of a tenuous interstellar medium. These could be within the volume of the galaxy, but for NS-NS on average one would expect as many GRB inside as outside. This is based on an estimate of the mean NS-NS merger time of 10^8 years; other estimated merger times (e.g. 10^7 years, [126]) would give a burst much closer to the birth site. BH-NS mergers would also occur in timescales $\lesssim 10^7$ years, and would be expected to give bursts well inside the host galaxy ([10]; see however [28]). In at least one "snapshot" standard afterglow spectral fit for GRB 980329 [103] the deduced external density is $n_o \sim 10^3$ cm^{-3}. In some of the other detected afterglows there is other evidence for a relatively dense gaseous environments, as suggested, e.g. by evidence for dust [102] in GRB970508, the absence of an optical afterglow and presence of strong soft X-ray absorption [33,69] in GRB 970828, the lack an an optical afterglow in the (radio-detected) afterglow ([122]) of GRB980329, and spectral fits to the low energy portion of the X-ray afterglow of several bursts [75]. The latter observations may be suggestive of hypernova models [81,28], involving the collapse of a massive star or its merger with a compact companion.

One important caveat is that all afterglows found so far are based on Beppo-SAX positions, which is sensitive only to long bursts $t_b \gtrsim 20$ s [39]. This is significant, since it appears likely that NS-NS mergers lead [53] to short bursts with $t_b \lesssim 10$ s. To make sure that a population of short GRB afterglows is not being missed will probably need to await results from HETE [35] and from the planned Swift [120] mission, which is designed to accurately locate 300 GRB/yr.

An interesting case is the apparent coincidence of GRB 980425 with the unusual SN Ib/Ic 1998bw [29], which may represent a new class of SN [40,8]. If true, this could imply that some or perhaps all GRB could be associated with SN Ib/Ic [131], differing only in their viewing angles relative to a very narrow jet. Alternatively, the GRB could be (e.g. [138]) a new subclass of GRB with lower energy $E_\gamma \sim 10^{48}(\Omega_j/4\pi)$ erg, only rarely observable, while the great majority of the observed GRB would have the energies $E_\gamma \sim 10^{54}(\Omega_j/4\pi)$ ergs as inferred from high redshift observations. The difficulties are that it would require extreme collimations by factors $10^{-3} - 10^{-4}$, and the statistical association is so far not significant [44].

The environment in which a GRB occurs should also influence the nature of the afterglows

in other ways. The blast wave and reverse shock that give rise to the X-rays, optical, etc occur over timescales proportional to $t_{dec} \propto n_{ext}^{-1/3}$ (equ.[6]) which is longer in lower density environments, so for the same energy the flux is lower, roughly $F_\nu \propto E_o n_{ext}^{1/2}$, contributing also to make afterglows in the intergalactic medium harder to detect. However, in addition to affecting broadband fluxes, one may also expect specific spectral signatures from the external medium imprinted in the X-ray and optical continuum, such as atomic edges and lines [4,91,64]. These may be used both to diagnose the chemical abundances and the ionization state (or local separation from the burst), as well as serving as potential alternative redshift indicators. (In addition, the outflowing ejecta itself may also contribute blue-shifted edge and line features, especially if metal-rich blobs or filaments are entrained in the flow from the disrupted progenitor debris [63], which could serve as diagnostic for the progenitor composition and outflow Lorentz factor). To distinguish between progenitors (§2), an interesting prediction ([64]; see also [31,11]) is that the presence of a measurable Fe K-α X-ray *emission* line could be a diagnostic of a hypernova, since in this case one may expect a massive envelope at a radius comparable to a light-day where $\tau_T \lesssim 1$, capable of reprocessing the X-ray continuum by recombination and fluorescence. Two groups [94,139] have in fact recently reported the possible detection of Fe emission lines in GRB 970508 and GRB 970828.

8. Conclusions

The fireball shock model of gamma-ray bursts has proved quite robust in providing a consistent overall interpretation of the major features of these objects at various frequencies and over timescales ranging from the short initial burst to afterglows extending over many months. The standard internal shock scenario is able to reproduce the properties of the γ-ray light curves, while external shocks involving a forward blast wave and a reverse shock are successful in reproducing the afterglows observed in X-rays, optical and radio. The "simple standard model" of afterglows, involving four spectral slopes and three breaks is quite useful in understanding the 'snapshot' multiwavelength spectra of most afterglows. However, the effects associated with a jet-like outflow and the possible differential beaming at various energies requires further investigations, both theoretical and observational. Caution is required in interpreting the observations on the basis of the simple standard model. For instance, more detailed numerical models, as opposed to the more common analytical scaling law models, show that the contributions of radiation from different angles and the gradual transition between different dynamical and radiative regimes lead to a considerable rounding-off of the spectral shoulders and light-curve slope changes, so that breaks cannot be easily located unless the spectral sampling is dense and continuous, both in frequency and in time. Some of the observed light curves with humps, e.g. in GRB 970508, require 'post-standard' model features (i.e. beyond those assumed in the standard model), such as either non-uniform injection episodes or anisotropic outflows. Time-dependent multiwavelength fits [88] of some bursts also indicate that the parameters characterizing the shock physics change with time. Even without humps or slope changes, a non-standard relation between the spectral and temporal decay slope is observed in several objects, e.g. in GRB 990123 [51]. These are, in our view [67], a strong indication for "post-standard" effects in such bursts.

Much progress has been made in understanding how gamma-rays can arise in fireballs produced by brief events depositing a large amount of energy in a small volume, and in deriving the generic properties of the long wavelength afterglows that follow from this. There still remain a number of mysteries, especially concerning the identity of their progenitors, the nature of the triggering mechanism, the transport of the energy, the time scales involved, and the nature and effects of beaming. However, even if we do not yet understand the details of the gamma-ray burst central engine, it is clear that these phenomena are among the most powerful transients in the Universe, and they could serve as powerful beacons for probing the high redshift ($z > 5$) universe. The modeling of the burst mechanism itself, as well as the resulting outflows and radiation, will continue to be a formidable challenge to theorists and to computational techniques. However, the theoretical understanding appears to be converging, and with dedicated new and planned observational missions under way, the prospects for significant progress are realistic.

I am grateful to Martin Rees for stimulating collaborations, as well as to Alin Panaitescu, Hara Papathanassiou and Ralph Wijers.

REFERENCES

1. Andersen et al. 1999, Science, March 26 issue
2. Akerlof, C., *et al.*, 1999, Nature, in press (astro-ph/9903271)
3. Band, D., *et al.*, 1993, Ap.J., 413, 281
4. Bisnovatyi-Kogan, G & Timokhin, A, 1997, Astr. Rep. 41, 423
5. Blandford, R.D. & Znajek, R.L., 1977, MNRAS, 179, 433
6. Blandford, R.D. & McKee, C., 1976, Phys.Fluids 19, 1130
7. Bloom, J., etal, 1998, A& A Supp.,in press (Procs. Rome Conference on GRB)
8. Bloom, J, *et al.*, 1998, ApJ 506, L105
9. Bloom, J., etal, 1998, ApJ 507, L25
10. Bloom, J., Sigurdsson, S. & Pols, O., 1999, MNRAS in press (astro-ph/9805222)
11. Böttcher, M, *et al.*, 1998, astro-ph/9809156
12. Castro-Tirado et al., 1999, Science, March 26 issue
13. Connors, A & Hueter, G.J. 1998, ApJ 501, 307
14. Costa, E., et al., 1997, Nature, 387, 783
15. Daigne, F & Mochkovitch, R, 1998, MNRAS, 296, 275
16. Dermer, C & Mitman, K, 1998, astro-ph/9809411
17. Djorgovski, S.G. *et al.*, 1997, IAUC 6660
18. Djorgovski, S.G. *et al.*, 1998, ApJ 508, L17
19. Duncan, R & Thompson, C, 1992, ApJ, 392, L9
20. Eichler, D., Livio, M., Piran, T. and Schramm, D.N., 1989, Nature, 340, 126
21. Fenimore, E, Madras, C, and Nayakshin, S, 1996, ApJ, 473, 998
22. Fenimore, E, *et al.*, 1998, ApJ in press (astro-

ph/9802200)

23. Fishman, G. & Meegan, C., 1995, Ann.Rev Astr.Ap.,33, 415

24. Frail, D, et al., 1997, Nature, 389, 261

25. Frail, D, A& A Supp.,in press (Procs. Rome Conference on GRB)

26. Fruchter, A, astro-ph/9810224

27. Fruchter, A. etal, Ap.J. submitted (astro-ph/9902236)

28. Fryer, C & Woosley, S, 1998, ApJ(Lett) subm (astro-ph/9804167

29. Galama, T. et al., 1998, Nature, in press (astro-ph/9806175)

30. Galama, T. et al., 1999, Nature, in press (astro-ph/9903021)

31. Ghisellini, G, et al., 198, astro-ph/9808156

32. Goodman, J., 1986, ApJ, 1986, 308, L47

33. Groot, P. et al., 1997; in Gamma-Ray Bursts, Meegan, C., Preece, R. & Koshut, T., eds., 1997 (AIP: New York), p. 557

34. Harding, A.K. and Baring, M.G., 1994, in Gamma-ray Bursts, ed. G. Fishman, et al., p. 520 (AIP 307, NY)

35. HETE, http://space.mit.edu/HETE/

36. Hjorth, etal, 1999, Science, March 26 issue

37. Horack, J. M., Mallozzi, R. S., & Koshut, T. M. 1996, ApJ, 466, 21

38. Hurley, K., et al., 1994, Nature, 372,652

39. Hurley, K., A& A Supp.,in press (Procs. Rome Conference on GRB)

40. Iwamoto, K, et al., 1998, Nature 395, 672

41. Katz, J., 1994a, ApJ, 422, 248

42. Katz, J., 1994b, ApJ, 432, L107

43. Katz, J.I., 1997, ApJ, 490, 633

44. Kippen, R.M. et al., 1998, ApJ subm (astro-ph/9806364)

45. Klebesadel, R, Strong, I & Olson, R, 1973, ApJ, 182, L85

46. Kluzniak, W & Ruderman, M, 1998, ApJ 508, L113

47. Kobayashi, S, Piran, T & Sari, R, 1998, ApJ, 490, 92

48. Kouveliotou, C., et al., 1993, Ap.J., 413, L101

49. Kouveliotou, C., 1998, review at the APS Spring mtg, Indianapolis, IN

50. Kulkarni, S., et al., 1998, Nature, 393, 35

51. Kulkarni, S., et al., 1999, Nature, in press (astro-ph/9902272)

52. Mazets, EP, Golenetskii, SV & Ilinskii, VN, 1974, JETP Lett. 19, 77

53. Macfadyen, A & Woosley, S, 1999, ApJ n press (astro-ph/9810274)

54. Mészáros , P., 1995, 17th Texas Symp. Relativistic Astrophysics, H. Böhringer et al, N.Y. Acad. Sci., 440

55. Mészáros , P & Rees, M.J., 1992, ApJ, 397, 570

56. Mészáros , P & Rees, MJ, 1992b, MNRAS, 257, 29P

57. Mészáros , P. & Rees, M.J., 1993a, ApJ, 405, 278

58. Mészáros , P., Laguna, P & Rees, M.J., 1993, ApJ, 415, 181

59. Mészáros , P. and Rees, M.J., 1993b, Ap.J., 418, L59

60. Mészáros , P., Rees, M.J. & Papathanassiou, H, 1994, Ap.J., 432, 181

61. Mészáros , P & Rees, M.J., 1997a, ApJ, 476, 232

62. Mészáros , P & Rees, M.J., 1997b, ApJ, 482, L29

63. Mészáros , P & Rees, M.J., 1998a, ApJ, 502, L105

64. Mészáros , P & Rees, M.J., 1998b, MNRAS, 299, L10

65. Mészáros , P, Rees, M.J. & Wijers, R, 1998, Ap.J., 499, 301 (astro-ph/9709273)

66. Mészáros , P, Rees, M.J. & Wijers, R, 1998b, New Astron, in press (astro-ph/9808106)

67. Mészáros , P & Rees, M.J., MNRAS, in press (astro-ph/9902367)

68. Metzger, M et al., 1997, Nature, 387, 878

69. Murakami, T. et al., 1997, in Gamma-Ray Bursts, Meegan, C., Preece, R. & Koshut, T., eds., 1997 (AIP: New York), p. 435

70. Narayan, R., Piran, T. & Shemi, A, 1991, ApJ, 379, L17

71. Narayan, R., Paczyński , B. & Piran, T., 1992, Ap.J., 395, L83

72. Norris, J, 1998, private communication

73. Norris, J. P., et al. 1995, ApJ, 439, 542

74. Odewahn, S, et al., 1998, ApJ, 509, L50

75. Owen, A., et al, 1998, Astron.&Astrophys. in press (astro-ph/9809356),

76. Paczyński , B. & Rhoads, J, 1993, Ap.J., 418, L5

77. Paczyński , B. & Xu, G., 1994, ApJ, 427, 708

78. Paczyński , B., 1986, ApJ, 308, L43

79. Paczyński , B., 1990, Ap.J., 363, 218

80. Pacziński, B., 1991, Acta. Astron., 41, 257

81. Paczyński , B., 1998, ApJ, 494, L45

82. Papapathanassiou, H & Mészáros , P, 1996, ApJ, 471, L91

83. Panaitescu, A & Mészáros , P, 1998a, ApJ, 492, 683

84. Panaitescu, A. & Mészáros , P., 1998b, ApJ, 501, 772

85. Panaitescu, A. & Mészáros , P., 1998, ApJ, 493, L31

86. Panaitescu, A. & Mészáros , P., 1999, ApJ, subm. (astro-ph/9806016)

87. Panaitescu, A. & Mészáros , P., 1998d, ApJ, subm (astro-ph/9810258)

88. Panaitescu, A. Mészáros , P & Rees, MJ, 1998, ApJ, 503, 314

89. ApJ 496 (1998) 311

90. Pendleton, G, et al., 1996, Ap[J, 464, 606

91. Perna, R. & Loeb, A., 1998, ApJ, 503, L135

92. Phinney, E.S., 1991, ApJ, 380, L17

93. Piro, L, et al., 1998, A & A, 331, L41

94. Piro, L, et al., 1998b, A& A Supp.,in press (Procs. Rome Conference on GRB)

95. Popham, R., Woosley, S & Fryer, C., 1998, ApJ subm (astro-ph/9807028)

96. Pugliese, G, Falk, H & Biermann, P, 1999, A&A, in press (astro-ph/9903036)

97. Ramprakash, A et al., 1998, Nature 393, 38

98. Rees, M.J. & Mészáros , P., 1992, MNRAS, 258, P41

99. Rees, M.J. & Mészáros , P., 1994, ApJ, 430, L93

100.Rees, M.J. & Mészáros , P., 1998, ApJ, 496, L1

101.Reichart, D., 1997, ApJ, 485, L57

102.Reichart, D., 1998, ApJ, 495, L99

103.Reichart, D & Lamb, D.Q., 1998, A& A Supp,in press (Procs. Rome Conf. on GRB)

104.Rhoads, J, 1997, Ap.J., 487, L1

105.Rhoads, J, 1999, ApJ subm, astro-ph/9903383

106.Ruffert, M., Janka, H.-T., Takahashi, K., Schaefer, G., 1997, A& A, 319, 122

107.Ruffert, M. & Janka, H.-T., 1998, A&A subm (astro-ph/9804132)

108.Sahu, K., et al., 1997, Nature 387, 476

109.Sari, R. & Piran, T., 1995, ApJ, 455, L143

110.Sari, R, 1998, ApJ, 494, L49

111.Sari, R & Piran, T, 1998, ApJ, 485, 270

112.Sari, R , Piran, T & Narayan, R, 1998, ApJ, 497, L17 (astro-ph/9712005)

113.Sari, R & Piran, T., 1999a, A&A submitted (astro-ph/9901105)

114.Sari, R & Piran, T, 1999b, Ap.J. submitted (astro-ph/9902009)

115.Schaefer, B.E. et al, 1997, ApJ, 489, 693

116.Schaefer, B.E., 1998, ApJ in press (astro-ph/9810424)

117.Shemi, A. and Piran, T., 1990, ApJ, 365, L55

118.Spruit, H, 1999, A&A, in press (astro-ph/98...)

119.Strohmeyer, T, et al., 1998, ApJ, 500, 873

120.Swift, http://swift.gsfc.nasa.gov/

121.Tavani, M., 1997, ApJ, 483, L87

122.Taylor, G.B., et al., 1997, Nature, 389, 263

123.Thompson, C., 1994, MNRAS, 270, 480

124.Totani, T, 1997, ApJ, 486, L71

125.Usov, V.V., 1994, MNRAS, 267, 1035

126.van den Heuvel, E, in *X-ray Binaries & Recycled Pulsars* (ed. E.van den Heuvel and S.Rappaport), Kluwer, 1992, pp 233

127.van Paradijs, J, et al., 1997, Nature,386, 686

128.van Paradijs, J, 1998, invited review at APS Spring mtg, Indianapolis, IN

129.Vietri, M., 1997a, ApJ, 478, L9

130.Vietri, M. & Stella, L, 1998, ApJ, in press

131.Wang, L. & Wheeler, J.C., 1998, ApJ subm (astro-ph/9806212)

132.Waxman, E., 1997, ApJ, 485, L5

133.Waxman, E., 1997b, ApJ, 489, L33

134.Waxman, E., Kulkarni, S & Frail, D, 1998, ApJ, 497, 288

135.Wijers, R.A.M.J. & Galama, T., 1998, ApJ, subm (astro-ph/9805341)

136.Wijers, R.A.M.J., Rees, M.J. & Mészáros , P., 1997, MNRAS, 288, L51

137.Woosley, S., 1993, Ap.J., 405, 273

138.Woosley, S., Eastman, R. & Schmidt, B., 1998, ApJ, subm (astro-ph/9806299)

139.Yoshida, A, et al., 1998, A& A Supp.,in press (Procs. Rome Conference on GRB)

Nuclear Physics B (Proc. Suppl.) 80 (2000) 79–93

NUCLEAR PHYSICS B
PROCEEDINGS
SUPPLEMENTS
www.elsevier.nl/locate/npe

Big Bang Nucleosynthesis

Keith A. Olive[a]

[a]Theoretical Physics Institute, School of Physics and Astronomy, University of Minnesota, Minneapolis MN 55455, USA

A brief review of standard big bang nucleosynthesis theory and the related observations of the light element isotopes is presented. Implications of BBN on chemical evolution and constraints on particle properties will also be discussed.

1. INTRODUCTION

The standard model [1–3] of big bang nucleosynthesis (BBN) is based on the inclusion of an extended nuclear network into a homogeneous and isotropic FRLW cosmology. There is now sufficient data to define the standard model in terms of a single parameter, namely the baryon-to-photon ratio, η. Other factors, such as the uncertainties in reaction rates, and the neutron mean-life can be treated by standard statistical and Monte Carlo techniques [4–6] to make predictions (with specified uncertainties) of the abundances of the light elements, D, ^{3}He, ^{4}He, and ^{7}Li. Even the number of neutrino flavors, N_ν, which has long been treated as a parameter can simply be set (=3) in defining the standard model.

In this review, I will compare the predictions of BBN with the available observational determinations of the light element abundances and test for concordance. I will also discuss the implications of these results on the Galactic chemical evolution of the light elements and on limits to particle properties.

1.1. Historical Aside

It is important to bear in mind that there has always been an intimate connection between BBN and the microwave background as a key test to the standard big bang model. Indeed, it was the formulation of BBN which predicted the existence of the microwave background radiation [7]. The argument is rather simple. BBN requires temperatures greater than 100 keV, which according to the standard model time-temperature relation, $t_s T_{\text{MeV}}^2 = 2.4/\sqrt{N}$, where N is the number of relativistic degrees of freedom at temperature T, corresponds to timescales less than about 200 s. The typical cross section for the first link in the nucleosynthetic chain is

$$\sigma v(p + n \rightarrow D + \gamma) \simeq 5 \times 10^{-20} \text{cm}^3/\text{s} \qquad (1)$$

This implies that it was necessary to achieve a density

$$n \sim \frac{1}{\sigma v t} \sim 10^{17} \text{cm}^{-3} \qquad (2)$$

The density in baryons today is known approximately from the density of visible matter to be $n_{B_o} \sim 10^{-7}$ cm^{-3} and since we know that that the density n scales as $R^{-3} \sim T^3$, the temperature today must be

$$T_o = (n_{B_o}/n)^{1/3} T_{\text{BBN}} \sim 10 \text{K} \qquad (3)$$

thus linking two of the most important tests of the Big Bang theory.

2. THEORY

Conditions for the synthesis of the light elements were attained in the early Universe at temperatures $T \lesssim 1$ MeV. At somewhat higher temperatures, weak interaction rates were in equilibrium. In particular, the processes

$$
\begin{aligned}
n + e^+ &\leftrightarrow p + \bar{\nu}_e \\
n + \nu_e &\leftrightarrow p + e^- \\
n &\leftrightarrow p + e^- + \bar{\nu}_e
\end{aligned}
$$

fix the ratio of number densities of neutrons to protons. At $T \gg 1$ MeV, $(n/p) \simeq 1$.

As the temperature fell and approached the point where the weak interaction rates were no longer fast enough to maintain equilibrium, the neutron to proton ratio was given approximately by the Boltzmann factor, $(n/p) \simeq e^{-\Delta m/T}$, where Δm is the neutron-proton mass difference. The final abundance of ^{4}He is very sensitive to the (n/p) ratio

$$Y_p = \frac{2(n/p)}{[1 + (n/p)]} \approx 0.25 \qquad (4)$$

Freeze out occurs at slightly less than an MeV resulting in $(n/p) \sim 1/6$ at this time.

The nucleosynthesis chain begins with the formation of Deuterium through the process, $p + n \rightarrow$ D $+\gamma$. However, because the large number of photons relative to nucleons, $\eta^{-1} = n_\gamma/n_B \sim 10^{10}$, Deuterium production is delayed past the point where the temperature has fallen below the Deuterium binding energy, $E_B = 2.2$ MeV (the average photon energy in a blackbody is $\bar{E}_\gamma \simeq 2.7T$). This is because there are many photons in the exponential tail of the photon energy distribution with energies $E > E_B$ despite the fact that the temperature or $\bar{E}_\gamma$ is less than E_B. During this delay, the neutron-to-proton ratio drops to $(n/p) \sim 1/7$.

The dominant product of big bang nucleosynthesis is ^{4}He resulting in an abundance of close to 25% by mass. Lesser amounts of the other light elements are produced: D and ^{3}He at the level of about 10^{-5} by number, and ^{7}Li at the level of 10^{-10} by number. The resulting abundances of the light elements are shown in Figure 1, which concentrate on the range in η_{10} between 1 and 10. The curves for the ^{4}He mass fraction, Y, bracket the computed range based on the uncertainty of the neutron mean-life which has been taken as $\tau_n = 887 \pm 2$ s. Uncertainties in the produced ^{7}Li abundances have been adopted from the results in Hata et al. [5]. Uncertainties in D and ^{3}He production are small on the scale of this figure. The boxes correspond to the observed abundances and will be discussed below.

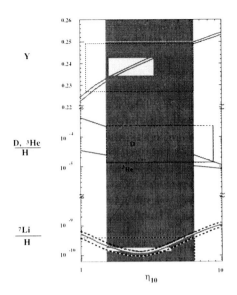

Figure 1. The light element abundances from big bang nucleosynthesis as a function of $\eta_{10} = 10^{10}\eta$.

3. Data

3.1. ^{4}He

The primordial ^{4}He abundance is best determined from observations of HeII $\rightarrow$ HeI recombination lines in extragalactic HII (ionized hydrogen) regions. There is now a good collection of abundance information on the ^{4}He mass fraction, Y, O/H, and N/H in over 70 [8–10] such regions. Since ^{4}He is produced in stars along with heavier elements such as Oxygen, it is then expected that the primordial abundance of ^{4}He can be determined from the intercept of the correlation between Y and O/H, namely $Y_p = Y(O/H \rightarrow 0)$. A detailed analysis of the data including that in [10] found an intercept corresponding to a primordial abundance $Y_p = 0.234 \pm 0.002 \pm 0.005$ [11]. This was updated to include the most recent results of [12] in [13]. The result (which is used in the discussion below) is

$$Y_p = 0.238 \pm 0.002 \pm 0.005 \qquad (5)$$

The first uncertainty is purely statistical and the second uncertainty is an estimate of the systematic uncertainty in the primordial abundance determination [11]. The solid box for ^{4}He in Figure 1 represents the range (at $2\sigma_{\text{stat}}$) from (5). The dashed box extends this by including the systematic uncertainty. A somewhat lower primordial abundance of $Y_p = 0.235 \pm .003$ is found by restricting to the 36 most metal poor regions [13]. These results are consistent with those of a Bayesian analysis [14] based on the 32 points of lowest metallicity. These have been used to calculate a likelihood function for which the peak occurs at $Y_p = 0.238$ and the most likely spread is $w = 0.009$. The 95% CL upper limit to Y_p in this case is 0.245. For further details on this approach see [14].

A global view of the ^{4}He data is shown in Figure 2. What should be absolutely apparent from this figure is the primordial nature of ^{4}He. There are no observations to date which indicate a ^{4}He abundance which is significantly below 23% to 24%. In particular, even in systems in which an element such as Oxygen, which traces stellar activity, is observed at extremely low values (compared with the solar value of O/H $\approx 8.5 \times 10^{-4}$), the ^{4}He abundance is nearly constant. This is far different from all other element abundances (with the exception of ^{7}Li as we will see below). For example, in Figure 3, the N/H vs. O/H correlation is shown. As one can clearly see, the abundance of N/H goes to 0, as O/H goes to 0, indicating a stellar source for Nitrogen.

A more useful plot of the ^{4}He data is shown in Figure 4. Here one sees the correlation of ^{4}He with O/H and the linear regression which leads to primordial abundance given in Eq. (5).

3.2. ^{7}Li

The abundance of ^{7}Li has been determined by observations of over 100 hot, population-II stars, and is found to have a very nearly uniform abundance [15]. For stars with a surface temperature $T > 5500$ K and a metallicity less than about 1/20th solar (so that effects such as stellar convection may not be important), the abundances show little or no dispersion beyond that which is consistent with the errors of individual measurements. Indeed, as detailed in ref. [16,17], much of the work concerning ^{7}Li has to do with the

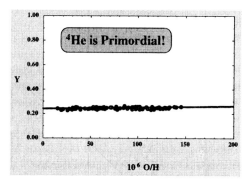

Figure 2. The ^{4}He mass fraction as determined in extragalactic H II regions as a function of O/H.

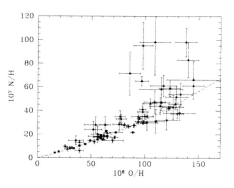

Figure 3. The Nitrogen and Oxygen abundances in the same extragalactic HII regions with observed ^{4}He shown in Figure 2.

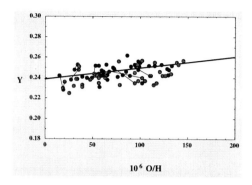

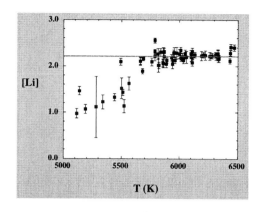

Figure 4. The Helium (Y) and Oxygen (O/H) abundances in extragalactic HII regions, from refs. [8,9] and from ref. [12]. Lines connect the same regions observed by different groups. The regression shown leads to the primordial ^{4}He abundance given in Eq. (5).

Figure 5. The Li abundance in halo stars with [Fe/H] < -1.3, as a function of surface temperature. The dashed line shows the value of the weighted mean of the plateau data.

presence or absence of dispersion and whether or not there is in fact some tiny slope to a [Li] = log ^{7}Li/H + 12 vs. T or [Li] vs. [Fe/H] relationship ([Fe/H] is the log of the Fe/H ratio relative to the solar value).

When the Li data from stars with [Fe/H] < -1.3 is plotted as a function of surface temperature, one sees a plateau emerging for $T > 5500$ K as shown in Figure 5 for the data taken from ref. [16]. As one can see from the figure, at high temperatures, where the convection zone does not go deep below the surface, the Li abundance is uniform. At lower surface temperatures, the surface abundance of Li is depleted as Li is dragged through the hotter interior of the star and destroyed. The lack of dispersion in the plateau region is evidence that this abundance is indeed primordial (or at least very close to it).

I will use the value given in ref. [16] as the best estimate for the mean ^{7}Li abundance and its statistical uncertainty in halo stars

$$\text{Li/H} = (1.6 \pm 0.1) \times 10^{-10} \qquad (6)$$

The small error is is statistical and is due to the large number of stars in which ^{7}Li has been observed. The solid box for ^{7}Li in Figure 1 represents the $2\sigma_{\text{stat}}$ range from (6). There is however

an important source of systematic error due to the possibility that Li has been depleted in these stars, though the lack of dispersion in the Li data limits the amount of depletion. In addition, standard stellar models[18] predict that any depletion of ^{7}Li would be accompanied by a very severe depletion of ^{6}Li. Until recently, ^{6}Li had never been observed in hot pop II stars. The observation[19] of ^{6}Li (which turns out to be consistent with its origin in cosmic-ray nucleosynthesis) is another good indication that ^{7}Li has not been destroyed in these stars [20–22].

Aside from the big bang, Li is produced together with Be and B in accelerated particle interactions such as cosmic ray spallation of C,N,O by protons and α-particles. Li is also produced by $\alpha - \alpha$ fusion. Be and B have been observed in these same pop II stars and in particular there are a dozen or so stars in which both Be and ^{7}Li have been observed. Thus Be (and B though there is still a paucity of data) can be used as a consistency check on primordial Li [23]. Based on the Be abundance found in these stars, one can conclude that no more than 10-20% of the ^{7}Li is due to cosmic ray nucleosynthesis leaving the remainder (an abundance near 10^{-10}) as primordial. A similar conclusion was reached in [17].

The dashed box in Figure 1, accounts for the possibility that as much as half of the primordial ^{7}Li has been destroyed in stars, and that as much as 20% of the observed ^{7}Li may have been produced in cosmic ray collisions rather than in the Big Bang. For ^{7}Li, the uncertainties are clearly dominated by systematic effects.

4. Likelihood Analyses

At this point, having established the primordial abundance of at least two of the light elements, ^{4}He and ^{7}Li, with reasonable certainty, it is possible to test the concordance of BBN theory with observations. Monte Carlo techniques have proven to be a useful form of analysis in this regard [4–6]. Two elements are sufficient for not only constraining the one parameter theory of BBN, but also for testing for consistency [24]. The procedure begins by establishing likelihood functions for the theory and observations. For example, for ^{4}He, the theoretical likelihood function takes the form

$$L_{\text{BBN}}(Y, Y_{\text{BBN}}) = e^{-(Y - Y_{\text{BBN}}(\eta))^2 / 2\sigma_1^2} \qquad (7)$$

where $Y_{\text{BBN}}(\eta)$ is the central value for the ^{4}He mass fraction produced in the big bang as predicted by the theory at a given value of η. σ_1 is the uncertainty in that value derived from the Monte Carlo calculations [5] and is a measure of the theoretical uncertainty in the big bang calculation. Similarly one can write down an expression for the observational likelihood function. Assuming Gaussian errors, the likelihood function for the observations would take a form similar to that in (7).

A total likelihood function for each value of η is derived by convolving the theoretical and observational distributions, which for ^{4}He is given by

$$L^{^4\text{He}}_{\text{total}}(\eta)$$
$$= \int dY \, L_{\text{BBN}}(Y, Y_{\text{BBN}}(\eta)) \, L_{\text{O}}(Y, Y_{\text{O}}) \qquad (8)$$

An analogous calculation is performed [24] for ^{7}Li. The resulting likelihood functions from the observed abundances given in Eqs. (5) and (6) is shown in Figure 6. As one can see there is very good agreement between ^{4}He and ^{7}Li in the range of $\eta_{10} \simeq 1.5 - 5.0$. The double peaked nature of the ^{7}Li likelihood function is due to the presence of a minimum in the predicted lithium abundance. For a given observed value of ^{7}Li, there are two likely values of η.

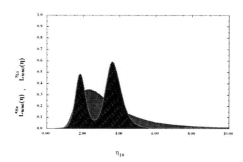

Figure 6. Likelihood distribution for each of ^{4}He and ^{7}Li, shown as a function of η. The one-peak structure of the ^{4}He curve corresponds to the monotonic increase of Y_p with η, while the two peaks for ^{7}Li arise from the minimum in the ^{7}Li abundance prediction.

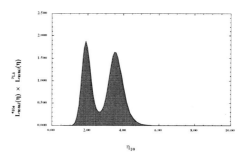

Figure 7. Combined likelihood for simultaneously fitting ^{4}He and ^{7}Li, as a function of η.

The combined likelihood, for fitting both elements simultaneously, is given by the product of the two functions in Figure 6 and is shown in Figure 7. The 95% CL region covers the range $1.55 < \eta_{10} < 4.45$, with the two peaks occurring at $\eta_{10} = 1.9$ and 3.5. This range corresponds to values of Ω_B between

$$0.006 < \Omega_B h^2 < .016 \qquad (9)$$

5. More Data

5.1. D and ^{3}He

Because there are no known astrophysical sites for the production of Deuterium, all observed D is assumed to be primordial. As a result, any firm determination of a Deuterium abundance establishes an upper bound on η which is robust.

Deuterium abundance information is available from several astrophysical environments, each corresponding to a different evolutionary epoch. In the ISM, corresponding to the present epoch, an often quoted measurement for D/H is [25]

$$(\mathrm{D/H})_{\mathrm{ISM}} = 1.60 \pm 0.09^{+0.05}_{-0.10} \times 10^{-5} \qquad (10)$$

This measurement allow us to set the upper limit to $\eta_{10} < 9$ and is shown by the lower right of the solid box in Figure 1. There are however, serious questions regarding homogeneity of this value in the ISM. There may be evidence for considerable dispersion in D/H [26,27] as is the case with ^{3}He [28].

The solar abundance of D/H is inferred from two distinct measurements of ^{3}He. The solar wind measurements of ^{3}He as well as the low temperature components of step-wise heating measurements of ^{3}He in meteorites yield the presolar (D + ^{3}He)/H ratio, as D was efficiently burned to ^{3}He in the Sun's pre-main-sequence phase. These measurements indicate that [29, 30] $((\mathrm{D} + {}^3\mathrm{He})/\mathrm{H})_\odot = (4.1 \pm 0.6 \pm 1.4) \times 10^{-5}$. The high temperature components in meteorites are believed to yield the true solar ^{3}He/H ratio of [29,30] $({}^3\mathrm{He/H})_\odot = (1.5 \pm 0.2 \pm 0.3) \times 10^{-5}$. The difference between these two abundances reveals the presolar D/H ratio, giving,

$$(\mathrm{D/H})_\odot \approx (2.6 \pm 0.6 \pm 1.4) \times 10^{-5} \qquad (11)$$

This value for presolar D/H is consistent with measurements of surface abundances of HD on Jupiter D/H = $2.7 \pm 0.7 \times 10^{-5}$ [31].

Finally, there have been several reported measurements of D/H in high redshift quasar absorption systems. Such measurements are in principle capable of determining the primordial value for D/H and hence η, because of the strong and monotonic dependence of D/H on η. However, at present, detections of D/H using quasar absorption systems do not yield a conclusive value for D/H. As such, it should be cautioned that these values may not turn out to represent the true primordial value and it is very unlikely that both are primordial and indicate an inhomogeneity [32] (a large scale inhomogeneity of the magnitude required to placate all observations is excluded by the isotropy of the microwave background radiation). The first of these measurements [33] indicated a rather high D/H ratio, D/H ≈ 1.9 – 2.5×10^{-4}. Other high D/H ratios were reported in [34]. More recently, a similarly high value of D/H = $2.0 \pm 0.5 \times 10^{-4}$ was reported in a relatively low redshift system (making it less suspect to interloper problems) [35]. This was confirmed in [36] where a 95% CL lower bound to D/H was reported as 8×10^{-5}. However, there are reported low values of D/H in other such systems [37] with values of D/H originally reported as low as $\simeq 2.5 \times 10^{-5}$, significantly lower than the ones quoted above. The abundance in these systems has been revised upwards to about $3.4 \pm 0.3 \times 10^{-5}$ [38]. However, it was also noted [39] that when using mesoturbulent models to account for the velocity field structure in these systems, the abundance may be somewhat higher (3.5 – 5×10^{-5}). This may be quite significant, since at the upper end of this range (5×10^{-5}) all of the element abundances are consistent as will be discussed shortly. I will not enter into the debate as to which if any of these observations may be a better representation of the true primordial D/H ratio. The status of these observations are more fully reviewed in [27]. The upper range of quasar absorber D/H is shown by the dashed box in Figure 1.

There are also several types of ^{3}He measurements. As noted above, meteoritic extractions yield a presolar value for ^{3}He/H. In addition, there are several ISM measurements of ^{3}He in galactic HII regions [28] which show a wide dis-person which may be indicative of pollution or a bias [40] $\left(^3\mathrm{He}/\mathrm{H}\right)_{\mathrm{HII}} \simeq 1 - 5 \times 10^{-5}$. There is also a recent ISM measurement of $^3\mathrm{He}$ [41] with $\left(^3\mathrm{He}/\mathrm{H}\right)_{\mathrm{ISM}} = 2.1^{+.9}_{-.8} \times 10^{-5}$. Finally there are observations of $^3\mathrm{He}$ in planetary nebulae [42] which show a very high $^3\mathrm{He}$ abundance of $^3\mathrm{He}/\mathrm{H}$ $\sim 10^{-3}$. None of these determinations represent the primordial $^3\mathrm{He}$ abundance, and as will be discussed below, their relation to the primordial abundance is heavily dependent on both stellar and chemical evolution.

6. More Analysis

It is interesting to compare the results from the likelihood functions of $^4\mathrm{He}$ and $^7\mathrm{Li}$ with that of D/H. Since D and $^3\mathrm{He}$ are monotonic functions of η, a prediction for η, based on $^4\mathrm{He}$ and $^7\mathrm{Li}$, can be turned into a prediction for D and $^3\mathrm{He}$. The corresponding 95% CL ranges are D/H $= (4.1-25)\times 10^{-5}$ and $^3\mathrm{He}/\mathrm{H} = (1.2-2.6)\times 10^{-5}$. If we did have full confidence in the measured value of D/H in quasar absorption systems, then we could perform the same statistical analysis using $^4\mathrm{He}$, $^7\mathrm{Li}$, and D. To include D/H, one would proceed in much the same way as with the other two light elements. We compute likelihood functions for the BBN predictions as in Eq. (7) and the likelihood function for the observations. These are then convolved as in Eq. (8).

Using D/H $= (2.0\pm 0.5) \times 10^{-4}$ as indicated in the high D/H systems, we can plot the three likelihood functions including $L^{\mathrm{D}}_{\mathrm{total}}(\eta)$ in Figure 8. It is indeed startling how the three peaks, for D, $^4\mathrm{He}$ and $^7\mathrm{Li}$ are in excellent agreement with each other. In Figure 9, the combined distribution is shown. We now have a very clean distribution and prediction for η, $\eta_{10} = 1.8^{+1.6}_{-.3}$ corresponding to $\Omega_B h^2 = .007^{+.005}_{-.001}$. The absence of any overlap with the high-η peak of the $^7\mathrm{Li}$ distribution has considerably lowered the upper limit to η. Overall, the concordance limits in this case are dominated by the Deuterium likelihood function.

If instead, we assume that the low value [38] of D/H $= (3.4 \pm 0.3) \times 10^{-5}$ is the primordial abundance, then we can again compare the likelihood distributions as in Figure 8, now substituting the low D/H value. As one can see from

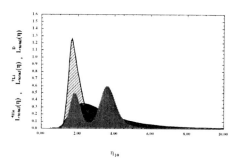

Figure 8. As in Figure 6, with the addition of the likelihood distribution for D/H assuming "high" D/H.

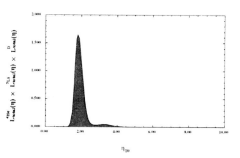

Figure 9. Combined likelihood for simultaneously fitting $^4\mathrm{He}$ and $^7\mathrm{Li}$, and D as a function of η from Figure 8.

Figure 10, there is now hardly any overlap between the D and the ^{7}Li and ^{4}He distributions. The combined distribution shown in Figure 11 is compared with that in Figure 9. In this case, D/H is just compatible (at the 2 σ level) with the other light elements, and the peak of the likelihood function occurs at $\eta_{10} = 4.8^{+0.5}_{-0.6}$. Though one can not use this likelihood analysis to prove the correctness of the high D/H measurements or the incorrectness of the low D/H measurements, the analysis clearly shows the difference in compatibility between the two values of D/H and the observational determinations of ^{4}He and ^{7}Li. To *make* the low D/H measurement compatible, one would have to argue for a shift upwards in ^{4}He to a primordial value of 0.247 (a shift by 0.009) which is not warranted at this time by the data, and a ^{7}Li depletion factor of about 2, which is close to recent upper limits to the amount of depletion [43,21].

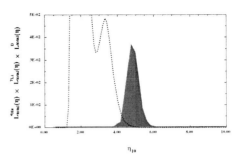

Figure 11. Combined likelihood for simultaneously fitting ^{4}He and ^{7}Li, and low D/H as a function of η. The dashed curve represents the combined distribution shown in Figure 9.

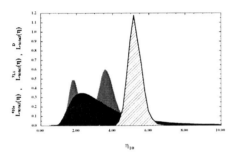

Figure 10. As in Figure 8, with the likelihood distribution for low D/H.

It is important to recall however, that the true uncertainty in the low D/H systems might be somewhat larger. If we allow D/H to be as large as 5×10^{-5}, the peak of the D/H likelihood function shifts down to $\eta_{10} \simeq 4$. In this case, there would be a near perfect overlap with the high η ^{7}Li peak and since the ^{4}He distribution function is very broad, this would be a highly compatible solution.

7. Chemical Evolution

Because we can not directly measure the primordial abundances of any of the light element isotopes, we are required to make some assumptions concerning the evolution of these isotopes. As has been discussed above, ^{4}He is produced in stars along with Oxygen and Nitrogen. ^{7}Li can be destroyed in stars and produced in several (though still uncertain) environments. D is totally destroyed in the star formation process and ^{3}He is both produced and destroyed in stars with fairly uncertain yields. It is therefore preferable, if possible to observe the light element isotopes in a low metallicity environment. Such is the case with ^{4}He and ^{7}Li. These elements are observed in environments which are as low as 1/50th and 1/1000th solar metallicity respectively and we can be fairly assured that the abundance determinations of these isotopes are close to primordial. If the quasar absorption system measurements of D/H stabilize, then this too may be very close to a primordial measurement. Otherwise, to match the solar and present abundances of D and ^{3}He to their primordial values requires a model of galactic chemical evolution.

The main inputs to chemical evolution models are: 1) The initial mass function, $\phi(m)$, indicating the distribution of stellar masses. Typically, a simple power law form for the IMF is chosen,

$\phi(m) \sim m^{-x}$, with $x \simeq -2.7$. This is a fairly good representation of the observed distribution, particularly at larger masses. 2) The star formation rate, ψ. Typical choices for a SFR are $\psi(t) \propto \sigma$ or σ^2 or even a straight exponential $e^{-t/\tau}$. σ is the fraction of mass in gas, $M_{\mathrm{gas}}/M_{\mathrm{tot}}$. 3) The presence of infalling or outflowing gas; and of course 4) the stellar yields. It is the latter, particularly in the case of ^{3}He, that is the cause for so much uncertainty. Chemical evolution models simply set up a series of evolution equations which trace desired quantities.

Deuterium is always a monotonically decreasing function of time in chemical evolution models. The degree to which D is destroyed, is however a model dependent question which depends sensitively on the IMF and SFR. The evolution of ^{3}He is however considerably more complicated. Stellar models predict that substantial amounts of ^{3}He are produced in stars between 1 and 3 $M_\odot$. For example, in the models of Iben and Truran [44] a 1 $M_\odot$ star will yield a ^{3}He abundance which is nearly three times as large as its initial (D+^{3}He) abundance. It should be emphasized that this prediction is in fact consistent with the observation of high ^{3}He/H in planetary nebulae [42].

However, the implementation of standard model ^{3}He yields in chemical evolution models leads to an overproduction of ^{3}He/H particularly at the solar epoch [40,45]. While the overproduction is problematic for any initial value of D/H, it is particularly bad in models with a high primordial D/H. In Scully et al. [46], a dynamically generated supernovae wind model was coupled to models of galactic chemical evolution with the aim of reducing a primordial D/H abundance of 2×10^{-4} to the present ISM value without overproducing heavy elements and remaining consistent with the other observational constraints typically imposed on such models. In Figure 12, the evolution of D/H and ^{3}He/H is shown as a function of time in one of the models with significant Deuterium destruction factors (see ref [46] for details). While such a model can successfully account for the evolution of D/H (and other standard chemical evolution tracers), as one can plainly see, ^{3}He is grossly overproduced (the Deu-

terium data is represented by squares and ^{3}He by circles).

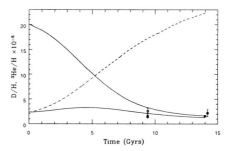

Figure 12. The evolution of D/H and ^{3}He/H with time in units of 10^{-5}. The assumed primordial abundance of D/H is 2×10^{-4}. The dashed ^{3}He curve shows the evolution with standard ^{3}He yields, while the solid curve shows the effect of the reduced yields of [48].

The overproduction of ^{3}He relative to the solar meteoritic value seems to be a generic feature of chemical evolution models when ^{3}He production in low mass stars is included. This result appears to be independent of the chemical evolution model and is directly related to the assumed stellar yields of ^{3}He. It has recently been suggested that at least some low mass stars may indeed be net destroyers of ^{3}He if one includes the effects of extra mixing below the conventional convection zone in low mass stars on the red giant branch [47,48]. The extra mixing does not take place for stars which do not undergo a Helium core flash (i.e. stars $> 1.7 - 2 \, M_\odot$). Thus stars with masses *less than* 1.7 $M_\odot$ are responsible for the ^{3}He destruction. Using the yields of Boothroyd and Malaney [48], it was shown [49] that these reduced ^{3}He yields in low mass stars can account for the relatively low solar and present day ^{3}He/H abundances observed. In fact, in some cases, ^{3}He was underproduced. To account for the ^{3}He evolution and the fact that some low mass stars must be producers of ^{3}He as indicated by the planetary nebulae data, it was suggested that the new yields apply only to a fraction (albeit large) of low mass stars [49,50]. The corresponding evolution [49] of D/H and ^{3}He/H using the reduced yields is shown in Figure 12.

The models of chemical evolution discussed above indicate that it is possible to destroy significant amounts of Deuterium and remain consistent with chemical evolutionary constraints. To do so however, comes with a price. Large Deuterium destruction factors require substantial amounts of stellar processing, which at the same time produce heavy elements. To keep the heavy element abundances in the Galaxy in check, significant Galactic winds enriched in heavy elements must be incorporated. In fact there is some evidence that enriched winds were operative in the early Galaxy. In the X-ray cluster satellites observed by Mushotzky et al. [51] and Loewenstein and Mushotzky [52] the mean Oxygen abundance was found to be roughly half solar. This corresponds to a near solar abundance of heavy elements in the inter-Galactic medium, where apparently little or no star formation has taken place.

If our Galaxy is typical in the Universe, then the models of the type discussed above would indicate that the luminosity density of the Universe at high redshift should also be substantial augmented relative to the present. Recent observations of the luminosity density at high redshift [53] are making it possible for the first time to test models of cosmic chemical evolution. The high redshift observations, are very discriminatory with respect to a given SFR [54]. Models in which the star formation rate is proportional to the gas mass fraction (these are common place in Galactic chemical evolution) have difficulties to fit the multi-color data from $z = 0$ to 1. This includes many of the successful Galactic infall models. In contrast, models with a steeply decreasing SFR are favored. In Figure 13, the predicted luminosity density based on the model with evolution shown in Figure 12 from [46], as compared with the observations (see ref. [54] for details).

While it would be premature to conclude that all models with large Deuterium destruction factors are favored, it does seem that models which do fit the high redshift data destroy significant

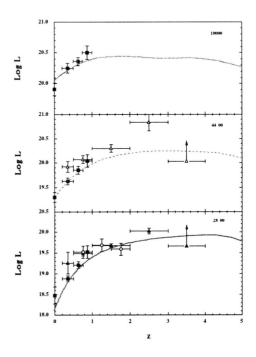

Figure 13. The tricolor luminosity densities (UV, B and IR) at $\lambda = 0.28, 0.44$ and 1.0 μm, in units of $(h/.5)$ WHz^{-1}Mpc^{-3} as a function of redshift for a model shown in 12 which destroys significant amounts of D/H. The data are taken from [53].

amounts of D/H. On the other hand, we can not exclude models which destroy only a small amount of D/H as Galactic models of chemical evolution. In this case, however the evolution of our Galaxy is anomalous with respect to the cosmic average. If the low D/H measurements [37,38] hold up, then it would seem that our Galaxy also has an anomalously high D/H abundance. That is we would predict in this case that the present cosmic abundance of D/H is significantly lower than the observed ISM value. (This conclusion assumes that the ISM D/H abundance is representative of the present epoch. If new results [26] which show values as low as a few $\times 10^{-6}$ are more typical, then low primordial D/H would fit the high redshift data and models with a high degree of D destruction much better.) If the high D/H observations [33–35] hold up, we could conclude that our Galaxy is indeed representative of the cosmic star formation history.

8. Constraints from BBN

Limits on particle physics beyond the standard model are mostly sensitive to the bounds imposed on the ^{4}He abundance. As is well known, the ^{4}He abundance is predominantly determined by the neutron-to-proton ratio just prior to nucleosynthesis and is easily estimated assuming that all neutrons are incorporated into ^{4}He. As discussed earlier, the neutron-to-proton ratio is fixed by its equilibrium value at the freeze-out of the weak interaction rates at a temperature $T_f \sim 1$ MeV modulo the occasional free neutron decay. Furthermore, freeze-out is determined by the competition between the weak interaction rates and the expansion rate of the Universe

$$G_F{}^2 T_f{}^5 \sim \Gamma_{\rm wk}(T_f) = H(T_f) \sim \sqrt{G_N N} T_f{}^2 \quad (12)$$

where N counts the total (equivalent) number of relativistic particle species. The presence of additional neutrino flavors (or any other relativistic species) at the time of nucleosynthesis increases the overall energy density of the Universe and hence the expansion rate leading to a larger value of T_f, (n/p), and ultimately Y_p. Because of the form of Eq. (12) it is clear that just as one can place limits [55] on N, any changes in the weak or

gravitational coupling constants can be similarly constrained (for a recent discussion see ref. [56]).

As discussed above, the limit on N_ν comes about via the change in the expansion rate given by the Hubble parameter,

$$H^2 = \frac{8\pi G}{3}\rho = \frac{8\pi^3 G}{90}[N_{\rm SM} + \frac{7}{8}\Delta N_\nu]T^4 \quad (13)$$

when compared to the weak interaction rates. Here $N_{\rm SM}$ refers to the standard model value for N. At $T \sim 1$ MeV, $N_{\rm SM} = 43/4$. Additional degrees of freedom will lead to an increase in the freeze-out temperature eventually leading to a higher ^{4}He abundance. In fact, one can parameterize the dependence of Y on N_ν by

$$Y = 0.2262 + 0.0131(N_\nu - 3) + 0.0135\ln\eta_{10} \quad (14)$$

in the vicinity of $\eta_{10} \sim 2$. Eq. (14) also shows the weak (log) dependence on η. However, rather than use (14) to obtain a limit, it is preferable to use the likelihood method.

Just as ^{4}He and ^{7}Li were sufficient to determine a value for η, a limit on N_ν can be obtained as well [24,57,58,6]. The likelihood approach utilized above can be extended to include N_ν as a free parameter. Since the light element abundances can be computed as functions of both η and N_ν, the likelihood function can be defined by [57] replacing the quantity $Y_{\rm BBN}(\eta)$ in eq. (7) with $Y_{\rm BBN}(\eta, N_\nu)$ to obtain $L^{^4\rm He}{}_{\rm total}(\eta, N_\nu)$. Again, similar expressions are needed for ^{7}Li and D.

A three-dimensional view of the combined likelihood functions [58] is shown in Figure 14. In this case the high and low η maxima of Figure 7, show up as peaks in the $L - \eta - N_\nu$ space. The likelihood function is labeled L_{47} (and L_{247} when D/H is included). In Figures 15 and 16 the corresponding likelihood functions L_{247} with high and low D/H are shown. Once again one sees an effect of including D/H is to eliminate one of the ^{7}Li peaks. Furthermore, unlike the case discussed in section 6, the likelihood distribution for low D/H is just as strong as that for high D/H, albeit at a lower value of N_ν.

The peaks of the distribution as well as the allowed ranges of η and N_ν are more easily discerned in the contour plots of Figures 17 and 18 which show the 50%, 68% and 95% confidence level contours in L_{47} and L_{247} for high and low D/H as indicated. The crosses show the location of the peaks of the likelihood functions. L_{47} peaks at $N_\nu = 3.2$, $\eta_{10} = 1.85$ and at $N_\nu = 2.6$, $\eta_{10} = 3.6$. The 95% confidence level allows the following ranges in η and N_ν

$$1.7 \le N_\nu \le 4.3 \qquad 1.4 \le \eta_{10} \le 4.9 \qquad (15)$$

Note however that the ranges in η and N_ν are strongly correlated as is evident in Figure 17.

With high D/H, L_{247} peaks at $N_\nu = 3.3$, and also at $\eta_{10} = 1.85$. In this case the 95% contour gives the ranges

$$2.2 \le N_\nu \le 4.4 \qquad 1.4 \le \eta_{10} \le 2.4 \qquad (16)$$

Note that within the 95% CL range, there is also a small area with $\eta_{10} = 3.2 - 3.5$ and $N_\nu = 2.5 - 2.9$.

Similarly, for low D/H, L_{247} peaks at $N_\nu = 2.4$, and $\eta_{10} = 4.55$. The 95% CL upper limit is now $N_\nu < 3.2$, and the range for η is $3.9 < \eta_{10} < 5.4$. It is important to stress that with the increase in the determined value of D/H [38] in the low D/H systems, these abundances are now consistent with the standard model value of $N_\nu = 3$ at the 2 σ level.

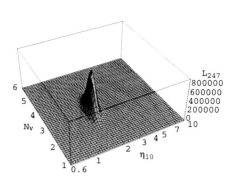

Figure 15. $L_{247}(N_\nu, \eta)$ for observed abundances including high D/H.

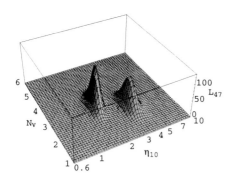

Figure 14. $L_{47}(N_\nu, \eta)$ for observed abundances given by eqs. (5 and 6).

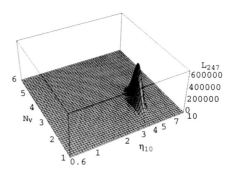

Figure 16. $L_{247}(N_\nu, \eta)$ for observed abundances including low D/H.

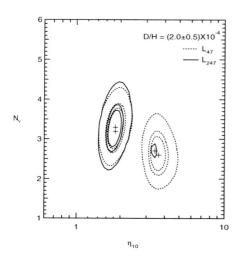

Figure 17. 50%, 68% & 95% C.L. contours of L_{47} and L_{247} where observed abundances are given by eqs. (5 and 6), and high D/H.

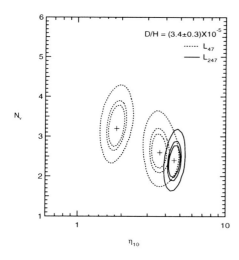

Figure 18. 50%, 68% & 95% C.L. contours of L_{47} and L_{247} where observed abundances are given by eqs. (5 and 6), and low D/H.

9. Summary

To summarize, I would assert that one can conclude that the present data on the abundances of the light element isotopes are consistent with the standard model of big bang nucleosynthesis. Using the isotopes with the best data, ^{4}He and ^{7}Li, it is possible to constrain the theory and obtain a best set of values for the baryon-to-photon ratio of η_{10} and the corresponding value for $\Omega_B h^2$

$$
\begin{aligned}
1.55 &< \eta_{10} < 4.45 \quad 95\%\text{CL} \\
.006 &< \Omega_B h^2 < .016 \quad 95\%\text{CL}
\end{aligned}
\tag{17}
$$

For $0.4 < h < 1$, we have a range $.006 < \Omega_B < .10$. This is a rather low value for the baryon density and would suggest that much of the galactic dark matter is non-baryonic [60]. These predictions are in addition consistent with recent observations of D/H in quasar absorption systems which show a high value. Difficulty remains however, in matching the solar ^{3}He abundance, suggesting a problem with our current understanding of galactic chemical evolution or the stellar evolution of low mass stars as they pertain to ^{3}He.

This work was supported in part by DoE grant DE-FG02-94ER-40823 at the University of Minnesota.

REFERENCES

1. T.P. Walker, G. Steigman, D.N. Schramm, K.A. Olive and K. Kang, *Ap.J.* **376** (1991) 51.
2. S. Sarkar, *Rep. Prog. Phys.* **59** (1996) 1493.
3. K.A. Olive and D.N. Schramm, *Eur.Phys.J.* **C3** (1998) 119; K.A. Olive, astro-ph/9901231.
4. L.M. Krauss and P. Romanelli, *Ap.J.* **358** (1990) 47; L.M. Krauss and P.J. Kernan, *Phys. Lett.* **B347** (1995) 347; M. Smith, L. Kawano, and R.A. Malaney, *Ap.J. Supp.* **85** (1993) 219; P.J. Kernan and L.M. Krauss, *Phys. Rev. Lett.* **72** (1994) 3309.
5. N. Hata, R.J. Scherrer, G. Steigman, D. Thomas, and T.P. Walker, *Ap.J.* **458** (1996) 637.
6. G. Fiorentini, E. Lisi, S. Sarkar, and F.L. Villante, *Phys.Rev.* **D58** (1998) 063506.

7. G. Gamow, *Phys. Rev.* **74** (1948) 505; G. Gamow, *Nature* **162** (1948) 680; R.A. Alpher and R.C. Herman, *Nature* **162** (1948) 774; R.A. Alpher and R.C. Herman, *Phys. Rev.* **75** (1949) 1089.

8. B.E.J. Pagel, E.A. Simonson, R.J. Terlevich and M. Edmunds, *MNRAS* **255** (1992) 325.

9. E. Skillman and R.C. Kennicutt, *Ap.J.* **411** (1993) 655; E. Skillman, R.J. Terlevich, R.C. Kennicutt, D.R. Garnett, and E. Terlevich, *Ap.J.* **431** (1994) 172.

10. Y.I. Izotov, T.X. Thuan, and V.A. Lipovetsky, *Ap.J.* **435** (1994) 647; *Ap.J.S.* **108** (1997) 1.

11. K.A. Olive, E. Skillman, and G. Steigman, *Ap.J.* **483** (1997) 788.

12. Y.I. Izotov, and T.X. Thuan, *Ap.J.* **500** (1998) 188.

13. B.D. Fields and K.A. Olive, *Ap.J.* **506** (1998) 177.

14. C.J. Hogan, K.A. Olive, and S.T. Scully, *Ap.J.* **489** (1997) L119.

15. F. Spite, and M. Spite, *A.A.* **115** (1982) 357; M. Spite, J.P. Maillard, and F. Spite, *A.A.* **141** (1984) 56; F. Spite, and M. Spite, *A.A.* **163** (1986) 140; L.M. Hobbs, and D.K. Duncan, *Ap.J.* **317** (1987) 796; R. Rebolo, P. Molaro, J.E. and Beckman, *A.A.* **192** (1988) 192; M. Spite, F. Spite, R.C. Peterson, and F.H. Chaffee Jr., *A.A.* **172** (1987) L9; R. Rebolo, J.E. Beckman, and P. Molaro, *A.A.* **172** (1987) L17; L.M. Hobbs, and C. Pilachowski, *Ap.J.* **326** (1988) L23; L.M. Hobbs, and J.A. Thorburn, *Ap.J.* **375** (1991) 116; J.A. Thorburn, *Ap.J.* **399** (1992) L83; C.A. Pilachowski, C. Sneden, and J. Booth, *Ap.J.* **407** (1993) 699; L. Hobbs, and J. Thorburn, *Ap.J.* **428** (1994) L25; J.A. Thorburn, and T.C. Beers, *Ap.J.* **404** (1993) L13; F. Spite, and M. Spite, *A.A.* **279** (1993) L9. J.E. Norris, S.G. Ryan, and G.S. Stringfellow, *Ap.J.* **423** (1994) 386; J. Thorburn, *Ap.J.* **421** (1994) 318.

16. P. Molaro, F. Primas, and P. Bonifacio, *A.A.* **295** (1995) L47; P. Bonifacio and P. Molaro, *MNRAS* **285** (1997) 847.

17. S.G. Ryan, J.E. Norris, and T.C. Beers, astro-ph/9903059.

18. C.P. Deliyannis, P. Demarque, and S.D. Kawaler, *Ap.J.Supp.* **73** (1990) 21.

19. V.V. Smith, D.L. Lambert, and P.E. Nissen, *Ap.J.* **408** (1992) 262; *Ap.J.* **506** (1998) 405; L. Hobbs, and J. Thorburn, *Ap.J.* **428** (1994) L25; *Ap.J.* **491** (1997) 772; R. Cayrel, M. Spite, F. Spite, E. Vangioni-Flam, M. Cassé, and J. Audouze, astro-ph/9901205.

20. G. Steigman, B. Fields, K.A. Olive, D.N. Schramm, and T.P. Walker, *Ap.J.* **415** (1993) L35; M. Lemoine, D.N. Schramm, J.W. Truran, and C.J. Copi, *Ap.J.* **478** (1997) 554.

21. M.H. Pinsonneault, T.P. Walker, G. Steigman, and V.K. Naranyanan, *Ap.J.* (1998) submitted.

22. B.D. Fields and K.A. Olive, *New Astronomy*, astro-ph/9811183, in press; E. Vangioni-Flam, M. Cassé, R. Cayrel, J. Audouze, M. Spite, and F. Spite, *New Astronomy*, astro-ph/9811327, in press.

23. T.P. Walker, G. Steigman, D.N. Schramm, K.A. Olive and B. Fields, *Ap.J.* **413** (1993) 562; K.A. Olive, and D.N. Schramm, *Nature* **360** (1993) 439.

24. B.D. Fields and K.A. Olive, *Phys. Lett.* **B368** (1996) 103; B.D. Fields, K. Kainulainen, D. Thomas, and K.A. Olive, *New Astronomy* **1** (1996) 77.

25. J.L. Linsky, et al., *Ap.J.* **402** (1993) 694; J.L. Linsky, et al., *Ap.J.* **451** (1995) 335.

26. E.B. Jenkins, T.M. Tripp, P.R. Wozniak, U.J. Sofia, and G. Sonneborn, astro-ph/9901403.

27. A. Vidal-Madjar, these proceedings

28. D.S. Balser, T.M. Bania, C.J. Brockway, R.T. Rood, and T.L. Wilson, *Ap.J.* **430** (1994) 667; T.M. Bania, D.S. Balser, R.T. Rood, T.L. Wilson, and T.J. Wilson, *Ap.J.S.* **113** (1997) 353.

29. S.T. Scully, M. Cassé, K.A. Olive, D.N. Schramm, J.W. Truran, and E. Vangioni-Flam, *Ap.J.* **462** (1996) 960.

30. J. Geiss, in *Origin and Evolution of the Elements*, eds. N. Prantzos, E. Vangioni-Flam, and M. Cassé (Cambridge: Cambridge University Press, 1993), p. 89.

31. H.B. Niemann, et al. *Science* **272** (1996) 846; P.R. Mahaffy, T.M. Donahue, S.K. Atreya,

T.C. Owen, and H.B. Niemann, *Sp. Sci. Rev.* **84** (1998) 251.

32. C. Copi, K.A. Olive, and D.N. Schramm, *Proc. Nat. Ac. Sci.* **95** (1998) 2758, astro-ph/9606156.

33. R.F. Carswell, M. Rauch, R.J. Weymann, A.J. Cooke, and J.K. Webb, *MNRAS* **268** (1994) L1; A. Songaila, L.L. Cowie, C. Hogan, and M. Rugers, *Nature* **368** (1994) 599.

34. M. Rugers and C.J. Hogan, *A.J.* **111** (1996) 2135; R.F. Carswell, et al. *MNRAS* **278** (1996) 518; E.J. Wampler, et al., *A.A. 316* (1996) 33.

35. J.K. Webb, R.F. Carswell, K.M. Lanzetta, R. Ferlet, M. Lemoine, A. Vidal-Madjar, and D.V. Bowen, *Nature* **388** (1997) 250.

36. D. Tytler, S. Burles, L. Lu, X.-M. Fan, A. Wolfe, and B.D. Savage, astro-ph/9810217.

37. D. Tytler, X.-M. Fan, and S. Burles, *Nature* **381** (1996) 207; S. Burles and D. Tytler, *Ap.J.* **460** (1996) 584.

38. S. Burles and D. Tytler, *Ap.J.* **499** (1998) 699; Ap.J. **507** (1998) 732.

39. S. Levshakov, D. Tytler, and S. Burles, astro-ph/9812114.

40. K.A. Olive, R.T. Rood, D.N. Schramm, J.W. Truran, and E. Vangioni-Flam, *Ap.J.* **444** (1995) 680.

41. G. Gloeckler, and J. Geiss, *Nature* **381** (1996) 210.

42. R.T. Rood, T.M. Bania, and T.L. Wilson, Nature **355** (1992) 618; R.T. Rood, T.M. Bania, T.L. Wilson, and D.S. Balser, 1995, in *the Light Element Abundances, Proceedings of the ESO/EIPC Workshop*, ed. P. Crane, (Berlin:Springer), p. 201; D.S. Balser, T.M. Bania, R.T. Rood, T.L. Wilson, *Ap.J.* **483** (1997) 320.

43. S. Vauclair and C. Charbonnel, *A.A.* **295** (1995) 715.

44. I. Iben, and J.W. Truran, *Ap.J.* **220** (1978) 980.

45. D. Galli, F. Palla, F. Ferrini, and U. Penco, Ap.J. **443** (1995) 536; D. Dearborn, G. Steigman, and M. Tosi, *Ap.J.* **465** (1996) 887.

46. S. Scully, M. Cassé, K.A. Olive, and E. Vangioni-Flam, *Ap.J.* **476** (1997) 521.

47. C. Charbonnel, *A. A.* **282** (1994) 811; C. Charbonnel, *Ap.J.* **453** (1995) L41; C.J. Hogan, *Ap.J.* **441** (1995) L17; G.J. Wasserburg, A.I. Boothroyd, andI.-J. Sackmann, *Ap.J.* **447** (1995) L37; A. Weiss, J. Wagenhuber, and P. Denissenkov, *A.A.* **313** (1996) 581.

48. A.I. Boothroyd, A.I. and R.A. Malaney, astro-ph/9512133.

49. K.A. Olive, D.N. Schramm, S. Scully, and J.W. Truran, *Ap.J.* **479** (1997) 752.

50. D. Galli, L. Stanghellini, M. Tosi, and F. Palla *Ap.J.* **477** (1997) 218.

51. R. Mushotzky, et al., *Ap.J.* **466** (1996) 686

52. M. Loewenstein and R. Mushotzky, *Ap.J.* **466** (1996) 695.

53. S.J. Lilly, O. Le Fevre, F. Hammer, and D. Crampton, *ApJ* **460** (1996) L1; P. Madau, H.C. Ferguson, M.E. Dickenson, M. Giavalisco, C.C. Steidel, and A. Fruchter, *MNRAS* **283** (1996) 1388; A.J. Connolly, A.S. Szalay, M. Dickenson, M.U. SubbaRao, and R.J. Brunner, *ApJ* **486** (1997) L11; M.J. Sawicki, H. Lin, and H.K.C. Yee, *A.J.* **113** (1997) 1.

54. M. Cassé, K.A. Olive, E. Vangioni-Flam, and J. Audouze, *New Astronomy* **3** (1998) 259.

55. G. Steigman, D.N. Schramm, and J. Gunn, *Phys. Lett.* **B66** (1977) 202.

56. B.A. Campbell and K.A. Olive, *Phys. Lett.* **B345** (1995) 429; L. Bergstrom, S. Iguri, and H. Rubinstein. astro-ph/9902157.

57. K.A. Olive and D. Thomas, *Astro. Part. Phys.* **7** (1997) 27.

58. K.A. Olive and D. Thomas, *Astro. Part. Phys.* (in press), hep-ph/9811444.

59. G. Steigman, K.A. Olive, and D.N. Schramm, *Phys. Rev. Lett.* **43** (1979) 239; K.A. Olive, D.N. Schramm, and G. Steigman, *Nucl. Phys.* **B180** (1981) 497.

60. E. Vangioni-Flam and M. Cassé, *Ap.J.* **441** (1995) 471.

ELSEVIER

Nuclear Physics B (Proc. Suppl.) 80 (2000) 95–108

NUCLEAR PHYSICS B
PROCEEDINGS
SUPPLEMENTS

www.elsevier.nl/locate/npe

Searches for Dark Matter

Michel SPIRO, Éric AUBOURG and Nathalie PALANQUE-DELABROUILLE[a]

[a] DAPNIA/SPP, CEA-Saclay, 91191 Gif-sur-Yvette, France

Despite the new results on the estimate of cosmological parameters, the need for dark matter, both baryonic and non baryonic, galactic and intergalactic, is still with us. For baryonic dark matter the remaining possibilities are mostly either intergalactic hot gas or galactic MACHOS (Massive Halo Objects like stellar remnants). For non baryonic dark matter the most likely candidates are the so-called WIMPS, the prototype of which could be the lightest supersymmetric particle. These particles are actively searched for at accelerators, and in our neighbourhood either through direct detection or through their annihilation.

1. Introduction

In this paper we discuss the field of experimental particle astrophysics apart from solar neutrinos which are covered elsewhere. It addresses the issue of the estimate of the cosmological parameters and of the dark matter problem. The high energy cosmic rays are also reviewed. It summarizes the main experimental results obtained up to now. Prospects and new ideas are presented.

2. Estimate of the cosmological parameters

The estimate of the value of Ω, the ratio of the mean energy density in the Universe to the critical energy density, is one of the main issues in modern cosmology. We can measure the components of Ω in various ways:
- from luminous matter (stars), Ω_{lum},
- from the dynamical behavior of stars in spiral galaxies (galactic halos), Ω_{halo}, and on an even greater scale from the dynamical behavior of galaxies in clusters of galaxies ,$\Omega_{clusters}$
- from primordial nucleosynthesis (baryons), Ω_{bar}.

We know from observations that the contributions from dust or gas to Ω are negligible. The estimates for the values of Ω_{lum}, Ω_{halo}, $\Omega_{clusters}$ and Ω_{bar} are shown in figure 3 and compared to the magic value $\Omega = 1$ which is the preferred value for aesthetic and theoretical reasons (to avoid fine tuning in initial conditions, and to agree with inflation theories). From all these values, one can draw two main conclusions:

1) All these estimates are below 1. However, the value $\Omega = 1$ is not excluded, but in order to reach it, it seems unavoidable to invoke intergalactic non baryonic dark matter, such as WIMPs (Weakly Interacting Massive Particles: heavy neutrinos ν_H, or lightest supersymmetric particle LSP...), or light neutrinos such as 10 to 100 eV ν_e or ν_μ or ν_τ, or axions.

2) This is the situation we had before direct estimates of Ω were obtained recently from the measurements of the apparent luminosity of distant SNIa supernovae. SNIa are generally believed to be standard candles. In this case the luminosity distances (distances obtained assuming a $1/d^2$ apparent luminosity dependence) can be determined. Comparing the luminosity distance with the redshift of each SNIa one can measure in principle both the value of Ω_{matter} and Ω_{vacuum}.

Two groups recently collected almost independent sets of distant SNIa [1] and [2]. They get very similar results. The results of the first group are shown in figure 1 and 2. The apparent luminosities of the high-redshift SNIa are on average twenty percent fainter than expected in a $\Omega = 1$ universe with a null cosmological constant ($\Omega_{vacuum} = 0$) where $\Omega = \Omega_{matter} + \Omega_{vacuum}$. Taken at face values, and if one constrains Ω to be 1 because of inflation and also because of the preliminary measurements obtained from inhomogeneities in the microwave background, namely from the location of the first Doppler peak, the best favored values are $\Omega_{matter} = 0.25$

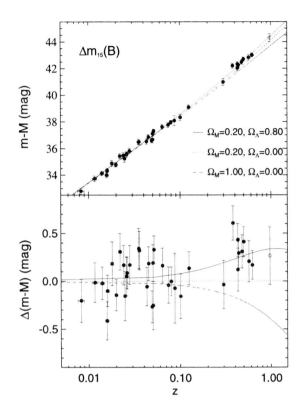

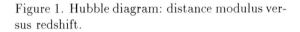

Figure 1. Hubble diagram: distance modulus versus redshift.

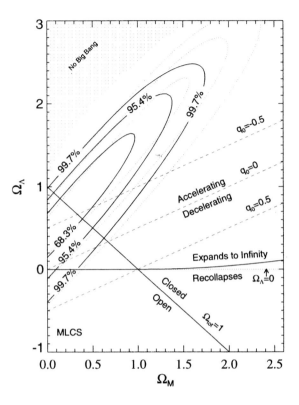

Figure 2. Ω vs Λ contours.

and $\Omega_{vacuum} = 0.75$ in agreement whith the value of $\Omega_{cluster} = 0.2$. A universe closed by ordinary matter, even non baryonic, such that $\Omega_{matter} = 1$ is ruled out at 7σ confidence level. However, before that such a revolutionary result be accepted one has to try to track all the possible sources of systematic errors. This work is not yet over, although no obvious systematic error has been found yet.

3. Evidence for dark matter

The comparison of the allowed range for Ω_{lum}, Ω_{halo} and Ω_{bar} suggests that baryonic dark matter is needed, and that the halos of spiral galaxies, like our own galaxy, could be partly or totally made of MACHOs (Massive Astrophysical Compact Halo Objects) which is almost the only

possibility left for baryonic dark matter: these MACHOs could be either aborted stars (brown dwarfs, planet like objects), or star remnants (white dwarves, neutron stars, black holes)[3].

On the other hand the comparison of the allowed ranges for Ω_{bar} and Ω_{matter} and Ω_{vacuum} implies that the Universe consists primarily of non-baryonic dark matter. Many popular non-baryonic dark matter candidates are in principle detctable. Here we will review efforts to detect either Wimps or axions.

4. Search for MACHO's with the gravitational microlensing technique

It is remarkable that the predicted range for baryonic dark matter and the additional mass needed on the scale of galaxies coincide exactly, making it reasonable to fill the halos of galaxies

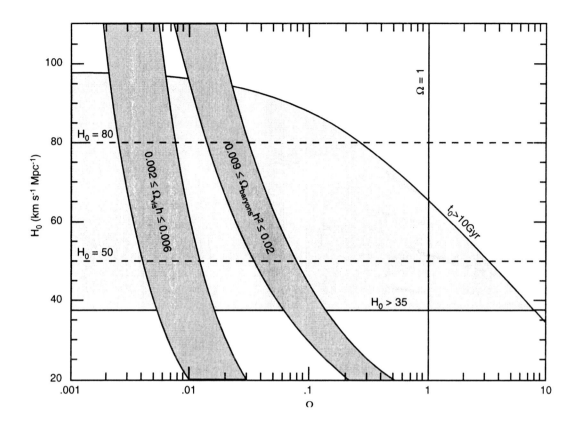

Figure 3. $H_0 - \Omega$ plot, indicating allowed (shaded in gray) and excluded regions. The fraction of visible matter in the Universe, Ω_{vis} is shown, along with the fraction of baryonic matter Ω_B resulting from nucleosynthesis, and the value $\Omega = 1$, theoretically preferred. The two current values for the Hubble constant: $H_0 = 50\,\mathrm{km\,s^{-1}\,Mpc^{-1}}$ and $H_0 = 80\,\mathrm{km\,s^{-1}\,Mpc^{-1}}$ are plotted in dotted lines. Two more limits are indicated: the lower bound $H_0 \geq 35\,\mathrm{km\,s^{-1}\,Mpc^{-1}}$ obtained from white dwarf stars and supernovae, and the lower bound on the age of the Universe $t_0 \geq 10\,\mathrm{Gyr}$ obtained essentially from the age of globular clusters

with baryonic dark matter. This prompted interest in looking for dark compact objects wandering in the halo of the Milky Way, using a new detection technique first proposed by Paczyński[4]: microlensing.

4.1. Gravitational microlensing

According to the principles of general relativity, the light rays from a source star are deflected by the presence of a massive deflector located near the line of sight between the star and the observer. This forms two distorted images of the source, as illustrated in figure 4.

In the particular configuration where all three objects are perfectly aligned, the two images merge into a ring, whose radius is called the Einstein radius R_E

$$R_E = \sqrt{\frac{4GM}{c^2} D_{OS}\, x(1-x)} \qquad (1)$$

where $x = D_{OD}/D_{OS}$ is the ratio of the distance observer-deflector to the distance observer-source and M the mass of the deflector. When probing the dark matter content of the Galactic halo, the source star is typically 60 kpc away (located in one of the Magellanic Clouds) from the observer and the lens typically a solar-mass object or lighter, so the angular separation between the images ($\sim 2R_E/D_{OD}$) is only of the order of the milliarcsecond. Given the limited resolution of optical telescopes with current technology, only the combined light intensity can therefore be recorded. The total magnification, however, is always greater than what the observer would receive from the source in the absence of the lens, which makes the latter detectable. Indeed, if u is the distance of the source to the line of sight in units of the Einstein radius, the amplification is

$$A = \frac{u^2 + 2}{u\sqrt{u^2 + 4}} \qquad (2)$$

As the lens moves in the halo with respect to the line-of-sight, the typical time scale Δt of a microlensing event is given by

$$\Delta t = \frac{R_E}{v_t} \simeq 90\sqrt{M/M_\odot} \qquad (3)$$

where v_t is the transverse velocity of the lens

which to first order can be taken as the rotation velocity of the galactic halo, ie 220 km/s.

The probability that a given star is amplified at a given time is very low, typically $5\,10^{-7}$. Millions of stars thus have to be monitored for years in order to ever be able to detect such a rare event. More details on the general principles of microlensing can be found in [5].

The only targets far enough to probe a large fraction of the galactic halo but close enough to resolve millions of stars are the large and the small Magellanic Clouds (LMC and SMC respectively), observable from the southern hemisphere. Various experiments are involved in the survey of stars located in these two satellite galaxies: EROS 2 (French experiment observing in Chile), MACHO (American experiment observing in Australia) and OGLE 2 (Polish experiment observing in Chile). The following sections present the latest results obtained by these experiments on the contribution of dark compact objects to the mass of the Galactic halo.

Because the mass range to probe extends from planetary objects ($\sim 10^{-7}\ M_\odot$) to stellar dark objects (a few solar masses), the event time-scales that the experiments have to be sensitive to vary from a few hours to a few months (see equation 3). Dedicated experiments and analyses are therefore performed separately to search for either small or large mass deflectors.

4.2. Limits on contribution of small mass objects

To search for planetary mass dark matter in the galactic halo, the EROS 1 CCD experiment, on the one hand, monitored 150 thousand stars in the LMC, with a high efficiency of $\sim 80\%$, thanks to a very good time sampling. On the other hand, the MACHO experiment monitored 8.6 million stars thanks to a large coverage of the LMC, but with only a $\sim 1\%$ efficiency to short time-scale events since their observational strategy was optimized for long time-scale events. Neither of the two experiments found any such event, which allowed them to set quite stringent limits on the maximum contribution of small mass objects to the dark mass of the Galactic halo [6,7].

As explained above, EROS and MACHO have

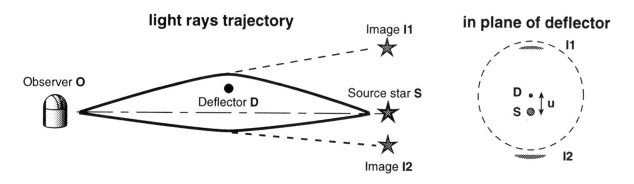

Figure 4. Deflection of light by a massive body D located near the line of sight between the observer O and the source star S. The dotted circle is the Einstein ring.

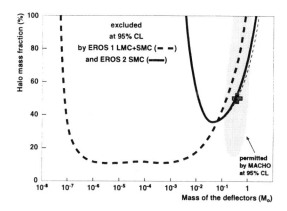

Figure 5. Exclusion diagram showing the 95% CL limits as obtained by EROS 1 (SMC and LMC, dashed curve) and by EROS 2 (SMC, plain curve). Also shown is the 95% CL region compatible with the 6 events detected by the MACHO two year analysis.

chosen very different analysis techniques, and there is little overlap in exposure for the two projects. Combining both sets of data after removal of this small overlap thus yields even stronger limits [8]: not more than $\sim 20\%$ of the halo can be composed of objects in the mass range $[10^{-7}-0.02]\ M_\odot$, at the 95% confidence level. Because we are using δ-function mass distributions

and since the limit is quite flat in the mass interval mentioned above, any mass function that peaks in this range is also excluded at the same confidence level. Moreover, the recent analysis of EROS 2 SMC data [9] yields a strong limit in the $.5\ M_\odot$ range (see figure 5).

4.3. Contribution of high mass objects
4.3.1. Present results toward the LMC

Both the EROS 1 experiment (using photographic plates) and the MACHO experiment (using CCD's) have searched and found long time-scale microlensing candidates. A total of 10 events have been detected, two by EROS [10] and eight by MACHO [11]. although one of them is slightly asymmetric and thus often disregarded as a microlensing candidate and another is an LMC binary event where the lens is most probably located in the LMC itself.

The typical Einstein radius crossing time associated to the LMC events is of the order of 40 days, which implies a surprisingly high most probable mass for the lenses: $\sim 0.5\ M_\odot$. This mass is much larger than the upper limit of $0.08\ M_\odot$ for brown dwarfs, and the lenses could be interpreted as, instead, white dwarfs or black holes. The optical depth implied by the mean duration of the events is compatible with about half that required to account for the dynamical mass of the dark halo. Such an interpretation, however, is not quite accepted among astronomers: by observing younger galaxies where we could detect the light due to a significant white dwarf component in the halo, a limit of 10% has been set on their contribution [12]; furthermore, if indeed half of the dark halos of galaxies consisted of such stellar remnants, we should observe an enrichment of the interstellar medium in Helium, which we do not. There is thus no consensus yet as to the nature of the deflectors causing the observed events.

The 95% CL region allowed by the MACHO experiment due to the detection of 6 events (*ie* disregarding the LMC binary and the asymmetric event) is illustrated in figure 5.

4.3.2. Future data

The interpretation of the present data is ambiguous, and huge error bars remain on both the most probable mass of the deflectors and the halo fraction in compact objects. More statistics is thus required, and several experiments are accumulating data to answer the questions of the presence or not and the nature of dark compact objects in the halo of the Galaxy:

- The EROS 2 experiment is now taking data with a completely redesigned setup and a new strategy. Using a wide field CCD camera (data taken in two colors simultaneously, with in each color a 1 square degree mosaic consisting of eight 2048×2048 CCD's), EROS covered 66 deg^2 on the LMC during the first year of observation (August 1996 - May 1997) and a total of 88 deg^2 the second year. The exposure times and time sampling are adapted to a search for long time-scale events.

- The MACHO experiment is presently analyzing four years of data on 15 deg^2 in the LMC, which means an increase of a factor of 2 in time scale and 1.4 in area. Preliminary results indicate 6 new events with time-scales ranging between 15 and 110 days with an average of about 50 days. This would confirm a high mass for halo deflectors if the lenses are indeed located in the halo of the Galaxy.

- The OGLE 2 experiment uses an upgraded setup and started taking data in summer 1997. They now also cover fields in the Large and the Small Magellanic Clouds (their previous strategy concerned only fields toward the Galactic Center, thus probing disk Dark Matter and not halo Dark Matter).

4.4. Highlights toward the SMC

The Small Magellanic Cloud gives a new line-of-sight through the Milky Way halo and a new population of source stars. The use of various lines-of-sight is very important since a comparison of the event rates is a powerful tool for discriminating between various shapes for the dark halo [13,14]. In addition, this allows for discrimination between various theories for the populations responsible for the LMC lenses [15].

The EROS and MACHO experiments (and more recently the OGLE experiment) thus monitor stars in the SMC to search for microlensing events. EROS recently published the first analysis on SMC data, whose results are presented hereafter.

4.4.1. First analysis of SMC data

The EROS 2 experiment covers the densest 10 square degrees of the Small Magellanic Cloud. On these 10 fields, a total of 5.3 millions light curves were built and subjected to a series of selection criteria and rejection cuts to isolate microlensing candidates [16]. Ten light curves passed all cuts and were inspected individually. Several correspond to physical processes other than microlensing (one of them, for instance, is the light curve of a nova that exploded in the SMC), and only one of the candidates passes this final visual inspection. Its light curve is shown in figure 6. Once corrected for blending — because of the high stellar density of the fields monitored in microlensing surveys, the flux of each reconstructed star generally results from the superposition of the fluxes of many source stars— the event light curve is well fitted by that of a microlensing event with an Einstein radius crossing time of 123 days, a maximum magnification of 2.6 occurring on January 11, 1997 and a $\chi^2/\text{d.o.f.} = 332/217 = 1.5$. The best microlensing fit is for 70% of the monitored flux being amplified and 30% being the contribution of blending unamplified light.

The source star being very bright, the value of

the reduced χ^2 of the fit is surprisingly high. A search for periodicity was therefore performed on the light curve, and a modulation was detected, with an amplitude of 2.5% and a period of 5.2 days. Fitting again the candidate light curve for microlensing allowing for a periodic modulation yields much more satisfactory residuals than before: χ^2/d.o.f. = 199/214. This strongly supports the microlensing interpretation of the observed magnification. The modulation detected was later confirmed by the OGLE experiment, on their own data taken after the event occurred (first points in June 1997). They also confirm the value of the blending coefficient of the source star since their new camera allows the separation of the two components of the blend and thus the individual measure of each of the two fluxes.

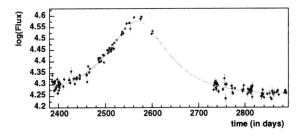

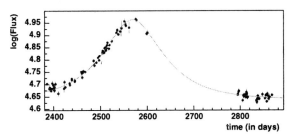

Figure 6. Light curve of microlensing candidate SMC #1, with a microlensing fit including blending superimposed. Time is in days since Jan. 0, 1990 (Julian date 2,447,891.5). Red light curve on top, blue on the bottom.

4.4.2. Estimate of Halo fraction in compact objects and lens mass

The time-scale of the observed event allows one to estimate the fraction of the halo that can be composed of dark compact objects generating microlensing events, independently of their mass. Assuming that the deflector is in the halo of the Milky Way, and considering a standard halo model (ie an isotropic and isothermal spherical halo), the EROS experiment estimated that the detected event is compatible with about 50% of the mass of the halo in dark compact objects. This fraction can vary by as much as a factor of two when considering other halo models (flattened halos for instance, or thinner halos and thicker disks so as to reproduce the rotation curve of the Galaxy but have less mass in the dark halo component).

Using a likelihood analysis also based on the time-scale of the detected event, the most probable mass of the deflector generating the event can be estimated. Under the assumption of a standard halo composed of dark compact objects having a single mass M (ie the mass function is supposed to be a Dirac-function), the most probable mass of the Halo deflector, given with 1σ error bars, is:

$$M = 2.6^{+8.2}_{-2.3} \, \mathrm{M}_\odot \qquad (4)$$

The event has the highest time-scale observed so far, and consequently the highest most probable mass. Only a neutron star or a black hole could be that massive and yet be dark. It is even harder than for the LMC events to explain how the halo of the Galaxy could be filled (even partially) with such heavy dark objects. Other interpretations therefore have to be looked at seriously.

4.4.3. Interpretation as SMC self lensing

The very long time-scale of the observed event suggests that it could show measurable distortions in its light curve due to the motion of the Earth around the Sun (the parallax effect[17]), provided that the Einstein radius projected onto the plane of the Earth is not much larger than the Earth orbital radius. The first detection of parallax in a gravitational microlensing event was observed by [18].

No evidence for distortion due to parallax is detectable on the light curve, implying either a very massive deflector with a very large Einstein radius, or a deflector near the source. The absence of parallax detection implies the following relation between the mass of the deflector and its distance to the observer, at the 95% CL:

$$\frac{M}{M_\odot} \times \frac{x}{1-x} > 0.7 \qquad (5)$$

where, as previously, $x = D_{OD}/D_{OS}$.

If the deflector is in the **halo of the Galaxy** (assuming a standard halo, $x < 0.66$ at the 95% CL) this yields a lower limit on the mass of the deflector: $M > 0.6\ M_\odot$.

If the deflector is located in the **SMC itself**, $1 - x \sim 1/10$ and the mass of the deflector is then $M \sim 0.1 M_\odot$, typical of a brown dwarf or faint star in the SMC.

To validate a possible SMC self lensing interpretation of the first event detected toward this new line-of-sight, it is necessary to check whether the SMC stellar population could provide such an event, in terms of duration and optical depth (probability that at a given time, a given star be magnified).

Various authors have suggested that the SMC is quite elongated along the line-of-sight, with a depth varying from a few kpc to as much as 20 kpc, depending on the region under study. We will approximate the SMC density profile by a prolate ellipsoid:

$$\rho = \frac{\Sigma_0}{2h}\, e^{-|z|/h}\, e^{-r/r_d} \qquad (6)$$

where z is along the line-of-sight and r is transverse to the line-of-sight. The depth h will be a free parameter, allowed to vary between 2.5 and 7.5 kpc. The values of the various parameters of the model are fit to the isophote levels of the cloud (which yields $\Sigma_0 \simeq 400\ M_\odot pc^{-2}$ and $r_d = 0.5$ kpc) and considering a mass-to-light ratio of 3 $M_\odot/L_\odot$ (typical of the values measured in the disk of the Milky Way). For $h = 2.5$, 5.0 or 7.5 kpc, the predicted optical depths are

$\tau = 1.0\ 10^{-7}$, $1.7\ 10^{-7}$ or $1.8\ 10^{-7}$ respectively, to be compared with the experimental optical depth of $3.3\ 10^{-7}$. Considering the very limited statistics we have, the model is consistent with the observations.

Finally, considering a velocity dispersion of 30 km/s in the SMC (and the 123 days time-scale of the event), the mass M of the deflector can be estimated according to an assumed distance between the source star and the lens. If the deflector is 5 kpc (resp. 2.5 kpc) from the source, its mass is $M \sim 0.1\ M_\odot$ (resp. $0.2\ M_\odot$), compatible with the results obtained from the parallax analysis.

An SMC self-lensing interpretation of the first microlensing event detected toward this new line-of-sight is thus quite plausible.

4.4.4. A Binary Lens towards the SMC

After the detection of this first SMC event, the MACHO collaboration alerted the microlensing community of an ongoing microlensing event (IAU circular 6935), which was later identified as a binary source event. In that case, the variation of the amplification is no longer simple: the gravitational potential of the double lens gives rise to cautic lines. When the source star crosses such a caustic, the amplification becomes singular. It is thus possible to resolve the finite size of the star by measuring the duration of the caustic crossing.

The measurements obtained on this microlensing event allowed to predict the date of the second caustic crossing, June 18, 1998. All microlensing collaborations took data this night. Among them, the PLANET collaboration obtained well sampled data at the time of the maximum, and the EROS collaboration equivalent data at the end of the caustic crossing (see figure 7).

Both data sets allowed to put constraints on the duration of the caustic crossing [19,20]. Combining this result with an estimate of the size of the source star, it was then possible to put limits on the proper motion of the lens. The most plausible interpretation for this event is that the lens lies in the SMC itself: only 7% of the halo population has a proper motion compatible with the one measured.

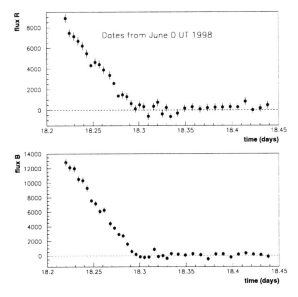

Figure 7. Differential photometry of EROS data taken on 18 June 1998. R data on top, B data on bottom, in ADU.

4.5. Microlensing conclusions

Two targets have been explored so far, in the search for dark matter in the halo of the Milky Way. They are the Large and the Small Magellanic Clouds.

The LMC data collected by the MACHO and the EROS experiments have allowed them to exclude any major contribution to the dark mass of the halo from compact objects in the mass range $10^{-7} \, M_\odot - 0.02 \, M_\odot$. Eight events compatible with microlensing by halo lenses were detected, with an average time-scale of 40 days, which could be interpreted as about 50% of the halo dark matter in the form of $\sim 0.5 M_\odot$ objects. A huge controversy remains as to the nature of these objects.

The SMC data has yielded one event found during the analysis of the first year of data, and one binary event detected online by the MACHO group. The first event has the longest time-scale observed so far: 123 days. If the lens causing the event is assumed to be in the halo of the Galaxy then its most probable mass is $2.6 M_\odot$, with a lower limit of $0.6 M_\odot$ coming from parallax analysis. Such a high mass is very hard to explain. A more plausible explanation is to assume that the lens and the source star are both located in the SMC. For the second event, caustic crossing time indicates clearly that the lens is in the SMC. The mass of the deflectors would then be typical of that of a faint star in the SMC, and the experimental optical depth compatible with a "thick disk" model of the SMC.

In any case, from this single event observed in the direction of the SMC, we can draw an exclusion diagram on the halo MACHO content and compare it with what is obtained in the direction of the LMC (Figure 5). The upper part of the permitted contour of the MACHO collaboration is ruled out.

The status of microlensing experiments and their implications on the galactic structure can be summarized in a few words. With about 100 microlensing events detected toward the Galactic Center, nearly 10 toward the LMC, 1 toward the SMC and none yet toward the Andromeda Galaxy (M31), there is strong evidence for the existence of a bar in the bulge and for lenses residing either in the halo of our Galaxy or in the LMC/SMC themselves. The main question now raised by these results is to determine where the lenses generating the detected events belong. Are they halo objects or intrinsic to either cloud? More statistics is still being accumulated. The MACHO experiment will run until 1999, EROS 2 plans to run until 2002 and OGLE 2 is just starting to take new data. The answer to this problem can then come from at least three possible studies:

- The comparison of the time-scales of the events toward the LMC and the SMC. They are expected to be similar if the deflectors are in the Galactic halo but different (due to the different velocity dispersions) if they are intrinsic to each cloud.

- The analysis of the spatial distribution of LMC events. The events are expected to be distributed evenly over the entire cloud if the lenses belong to the Galactic halo, while they should follow the stellar density of the LMC if they are LMC stars themselves.

- Finally, because the disk of the Andromeda Galaxy is slanted with respect to the line-of-sight, different fractions of its halo will be probed according to which end of the disk is being monitored; this will yield a larger number of events on the far side than on the near side, for microlensing events produced by deflectors in the halo of M31. Several experiments are exploring this line of sight and the first results are expected soon.

5. WIMPs: lightest supersymmetric particle (LSP)

- Big Bang cosmology implies that if the contribution of cold dark matter particles to Ω , *i. e.* , $\Omega > 0.1$, these particles have interactions of the order of the weak scale, i.e. they are WIMPs, and supersymmetry through the lightest supersymmetric particle offers a natural candidate. Indeed one finds that the contribution of WIMPs to Ω depends only on their annihilation cross-section:

$$\Omega = \frac{3 \times 10^{-27}\,\mathrm{cm}^3\,\mathrm{sec}^{-1}}{\sigma_A v} \qquad (7)$$

where v is the velocity at the time of decoupling and is roughly 1/4.

The annihilation cross section can be dimensionally written as α^2/m_χ^2, where α is the fine structure constant. It follows that for Ω around unity $m_\chi = 100 GeV$, the electroweak scale. There is a deep connection between critical cosmological density and the weak scale. The natural range of masses is expected to be

$tens\ of\ GeV < m_\chi < several\ TeV$

- The natural scale of supersymmetry is below a few TeV, to provide a natural explanation of the hierarchy of the grand unification compared to the electroweak unification. Combined with LEP results, we find surprisingly the same allowed mass range as inferred from cosmological arguments, namely between 30 GeV and a few TeV.

5.1. DIRECT and INDIRECT DETECTION of WIMP's

The hypothesis that dark matter particles are gravitationally trapped in the galaxy leads to the conclusion that, like stars, they should have a local Maxwell velocity distribution with a mean spread of 250 km/s. Then, the mean kinetic energy E_r received by a nucleus of mass M_n (in units of GeV) in an elastic collision with a dark matter particle of mass M_x is:

$$E_r = 2\,keV \times M_n M_x^2/(M_x + M_n)^2 \qquad (8)$$

The energy distribution is roughly exponential. The expected event rate for elastic scattering on a given nucleus, assuming that 0.4 GeV/cm^3 is the local density of the halo (needed to account for the flat rotation curve of stars) depends only on the mass and interaction cross section.

Since the annihilation cross section is fixed in standard cosmology (around $10^{-36} cm^2$, the elastic cross section on a nucleon is expected to be of the order of the annihilation cross section times the ratio of the phase space, namely the mass of the nucleon to the square (the reduced mass to the square) divided by the WIMP mass to the square. For a coherent interaction on a heavy nucleus target of mass A one expects an enhancement by A^4 if the neutralino is much heavier than the nucleus (coherence plus reduced mass effects).

These recoil events can be detected in well shielded deep underground devices such as semiconductor diodes, scintillators and cryogenic bolometers. Experimental limits on the cross section of wimps for scattering on germanium[21,22], CaF$_2$[23] and NaI[24] have been published. These limits are shown in figures 8 and 9, translated in terms of single nucleon effective cross section depending on whether we are dealing with pure axial coupling or with coherent N^2 coupling where N is the number of nucleons in the target nucleus. A recent much debated possible evidence for a signal based on the annual modulation (NaI from the DAMA colaboration) is also shown on figure 8.

To compare the sensitivity of the present experiments to the expectations, one can see that one approaches the required quality to detect with

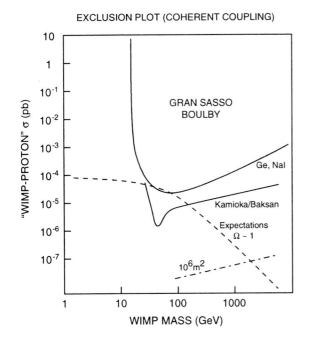

Figure 8. Effective elastic cross-section for coherent coupling.

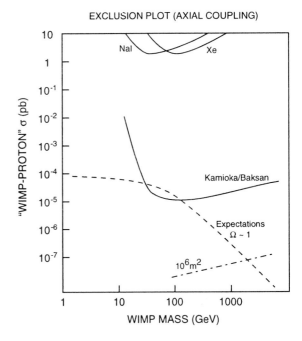

Figure 9. Effective elastic cross-section for axial coupling.

$S/N > 1$ the WIMP interactions if they interact coherently on the target (vector coupling) and have masses in the range of 100 GeV, but one has at least 3 to 4 orders of magnitude too high backgrounds to detect WIMPs with only axial coupling (spin dependent interactions).

Direct searches have no known fundamental limit on their background from radioactive impurities in the detector elements. Considerable progress has been made in lowering this background. Promising techniques to further eliminate the backgrounds include simultaneous detection of phonons and ionization in cryogenic germanium detectors[25], pulse shape analysis in NaI detectors and search for the expected seasonal variation (added velocity due to the movement of the Earth around the Sun) which will be ultimately necessary to confirm any observed signal[26].

Wimps can also be seen indirectly by observing their annihilation products[27]. The most

sensitive indirect technique uses the fact that Wimps can be trapped in the Sun or the Earth. Wimps with galactic orbits that happen to intersect an astronomical body will be trapped if, while traversing the body, they suffer an elastic collision with a nucleus that leaves the wimp with a velocity below the escape velocity[28]. The capture rate is proportional to the elastic sections on the nuclei of the astronomical body. A steady state will eventually be reached when the capture rate is balanced by the annihilation rate. Annihilations in the Sun or the Earth will yield a flux of high energy neutrinos either directly or by decay of annihilation products. The muon neutrinos can be observed in underground detector through their interactions in the rock below the detector yielding upward going muons pointing towards the Sun or the Earth. Presently, the most sensitive limits are those from Kamiokande [29], Baksan [30] and more recently from MACRO (shown at this conference). The flux limits obtained so far on upward going muons can be interpreted in terms of a limit on the effective elastic cross section (figures 8 and 9). These can be then compared with the limits from direct searches.

Indirect detection experiments start to explore the highest part, in terms of cross section, of the neutralino domain, specially through coherent interactions (with Fe in Earth or Oxygen in the Sun). A new generation of neutrino telescopes of the size of $10^6 m^2$ would be perfectly adequate to reach the needed sensitivity for masses from 100 GeV to few TeV, for both coherent and spin dependent type of interactions. This relatively weak dependence on the WIMP mass is because the more massive are the WIMPs the more energetic are the neutrinos, the longer is the range of upwards going muons and the better is the angular resolution towards the sun). The threshold however might limit the sensitivity to low mass WIMPs.

For the future one can note that the indirect searches are limited by the fixed background from atmospheric neutrinos and can therefore expand their limits only in proportion of the square root of the exposure time or detector area. Their are now projects of $1 km^3$ deep underwater or under-ice detectors which would be perfectly adequate for the indirect detection of WIMPs with masses greater than 100 GeV.

We conclude like F. Halzen[31] that the direct method might be superior if the WIMPs interact coherently and their masses are lower to 100 GeV. In all other cases, i.e. for relatively heavy WIMPs and for WIMPs with spin dependent interactions (or incoherent interactions), the indirect method will be competitive or superior, but it depends on the successful deployment of high energy neutrino telescopes of an effective area from 10^4 to $10^6 m^2$ with appropriate low threshold. The energy resolution of the neutrino telescope may be exploited to measure the WIMP mass and suppress the background. A kilometer-size telescope probes WIMP masses to the TeV range, beyond which they are excluded by cosmological considerations.

6. Axions

Axions [32] are hypothetical light scalar particles invented to prevent CP violation in the strong interaction. They would have been produced in the early Universe via both thermal and non-thermal mechanisms and might produce near-critical relic densities if they have masses in the range $m_a \sim 10^{-5}$eV to $\sim 10^{-3}$eV. Axions act as cold dark matter and should be present in the Galactic Halo.

The most popular detection scheme for Galactic axions is based on the expected axion-2 photon coupling [33] which allows the axion to "convert" to a photon of frequency $\nu = m_a c^2/h$ in the presence of a magnetic field. If a microwave cavity is tuned to this frequency, the axions will cause an excess power to be absorbed (compared to neighboring frequencies). If the Halo is dominated by axions, the predicted power is small, about 10^{-21}watts for a cavity of volume $3m^3$ and a magnetic field of 10 T. Since the axion mass is not known, it is necessary to scan over the range of interesting frequencies. Pilot experiments [34,35] have produced limits on the local axion density about a factor 30 above the expected density. Experiments are now in progress to search for axions at the required level of sensitivity [36].

7. Conclusions

As discussed in this paper, two types of dark matter are needed, baryonic and non-baryonic, notwithstanding the recent evidence for vacuum energy density. WIMPs and axions are the main candidates for non-baryonic dark matter. For baryonic dark matter, there are also two basic candidates: intergalactic gas or MACHOs in galactic halos. There are indeed some observations through microlensing effects which tend to support the idea that MACHOs could be the main candidate for baryonic dark matter. However, the high masses inferred from the microlensing events makes this hypothesis very puzzling and the subjecy of baryonic dark matter clearly awaits further obsevations.

It is a pleasure to thank J. Ellis, F. Halzen, J. Rich and D. Vignaud for fruitful discussions and comments.

REFERENCES

1. A.D. Riess et al. *Astronomical Journal*, 116:1009, 1998.
2. R. Pain et al. *this conference*, 1998.
3. B.J. Carr. In J. Tran Tan Vanh, editor, *The early universe and cosmic structures*. Paris: Frontieres, 1990.
4. B. Paczyński. *Astrophysical Journal*, 304:1, 1986.
5. B. Paczyński. *Annu. Rev. Astron. Astrophys.*, 34, 1996.
6. Cecile Renault et al. (eros collaboration). *Astronomy and Astrophysics*, 324:L69, 1997.
7. C. Alcock et al. (macho collaboration). *Astrophysical Journal*, 471:774, 1996.
8. C. Afonso et al. *Astrophysical Journal*, 199:L12, 1998.
9. C. Afonso et al. *Astronomy and Astrophysics*, 1999, in press, astro-ph/9812173.
10. R. Ansari et al. (eros collaboration). *Astronomy and Astrophysics*, 314:94, 1996.
11. C. Alcock et al. (macho collaboration). *Astrophysical Journal*, 490:59, 1997.
12. S. Charlot and J. Silk. *Astrophysical Journal*, 445:124, 1995.
13. P. Sackett and A. Gould. *Astrophysical Journal*, 419:648, 1993.
14. J. Frieman and R. Scoccimarro. *Astrophysical Journal*, 431:L23, 1994.
15. H. Zhao. *submitted to ApJ (astro-ph/9703097)*, 1997.
16. N. Palanque-Delabrouille et al. (eros collaboration). *Astronomy and Astrophysics*, 332:1, 1997.
17. A. Gould. *Astrophysical Journal*, 392:442, 1992.
18. C. Alcock et al. (macho collaboration). *Astrophysical Journal*, 454:L125, 1995.
19. C. Afonso et al. *Astronomy and Astrophysics*, 337:L17, 1998.
20. M.D. Albrow et al. *astro-ph/9807086*, 1998.
21. S.P. Ahlen et al. *Physics Letters*, B195:603, 1987.
22. D. Reusser et al. *Physics Letters*, B255:143, 1991.
23. C. Bacci et al. *Physics Letters*, B293:460, 1992.
24. G. Davies et al. *Physics Letters*, B322:159, 1994.
25. A. Lu et al. In Ed. Frontières, editor, *XXXI Rencontres de Moriond, January 20-27*, 1996.
26. C. Bacci et al. *Astroparticle Physics*, 2:117, 1994.
27. Silk J. *Physical Review Letters*, 55:257, 1985.
28. W.H. Press and D.N. Spergel. *Astrophysical Journal*, 296:679, 1985.
29. M. Mori et al. *Physical Review*, D48:505, 1993.
30. Bolier et al. *Nuclear Physics*, B48:83, 1996.
31. F. Halzen. *Direct and Indirect Detection of WIMPS, preprint*, MAD/PH 95-887, 1995.
32. H. Y. Cheng. *Physics Reports*, 158:1, 1988.
33. P. Sikivie. *Physical Review Letters*, 51:1415, 1983.
34. W. Wuensch et al. *Physical Review*, D40:3153, 1989.
35. C. Hagmann et al. *Physical Review*, D42:1297, 1990.
36. C. Hagmann et al. *Physical Review Letters*, 80:2043, 1998.

ELSEVIER

Nuclear Physics B (Proc. Suppl.) 80 (2000) 109–118

NUCLEAR PHYSICS B
**PROCEEDINGS
SUPPLEMENTS**

www.elsevier.nl/locate/npe

Probing the Pre-Big Bang Universe

G. Veneziano[a]

[a]Theoretical Physics Division, CERN
CH - 1211 Geneva 23

Superstring theory suggests a new cosmology whereby a long inflationary phase *preceeded* a non singular big bang-like event. After discussing how pre-big bang inflation naturally arises from an almost trivial initial state of the Universe, I will describe how present or near-future experiments can provide sensitive probes of how the Universe behaved in the pre-bang era.

1. INTRODUCTION

I would like to begin this talk by asking a very simple question: Did the Universe start "small"? The naive answer is: Yes, of course! However, a serious answer can only be given after defining the two keywords in the question: What do we mean by "start"? and What is "small" relative to? In order to be on the safe side, let us take the "initial" time to be a bit larger than Planck's time, $t_P \sim 10^{-43}$ s. Then, in standard Friedmann–Robertson–Walker (FRW) cosmology, the initial size of the (presently observable) Universe was about 10^{-2} cm. This is of course tiny w.r.t. its present size ($\sim 10^{28}$ cm), yet huge w.r.t. the horizon at that time, i.e. w.r.t. $l_P = ct_P \sim 10^{-33}$ cm. In other words, a few Planck times after the big bang, our observable Universe consisted of $(10^{30})^3 = 10^{90}$ Planckian-size, causally disconnected regions.

More precisely, soon after $t = t_P$, the Universe was characterized by a huge hierarchy between its Hubble radius and inverse temperature on one side, and its spatial-curvature radius and homogeneity scale on the other. The relative factor of (at least) 10^{30} appears as an incredible amount of fine-tuning on the initial state of the Universe, corresponding to a huge asymmetry between time and space derivatives. Was this asymmetry really there? And if so, can it be explained in any more natural way?

It is well known that a generic way to wash out inhomogeneities and spatial curvature consists in introducing, in the history of the Universe, a long period of accelerated expansion, called inflation [1]. This still leaves two alternative solutions: either the Universe was generic at the big bang and became flat and smooth because of a long *post*-bangian inflationary phase; or it was already flat and smooth at the big bang as the result of a long *pre*-bangian inflationary phase.

Assuming, dogmatically, that the Universe (and time itself) started at the big bang, leaves only the first alternative. However, that solution has its own problems, in particular those of fine-tuned initial conditions and inflaton potentials. Besides, it is quite difficult to base standard inflation in the only known candidate theory of quantum gravity, superstring theory. Rather, as we shall argue, superstring theory gives strong hints in favour of the second (pre-big bang) possibility through two of its very basic properties, the first in relation to its short-distance behaviour, the second from its modifications of General Relativity even at large distance.

Before describing these arguments I should stress that pre-big bang inflation has a definite phenomenological advantage over conventional (post-big bang) inflation: Planckian (or string)-scale physics, being no longer washed out by a long subsequent inflationary phase, becomes suddenly accessible to present (or near future) experiments at the millimeter (or Giga-Hz) scale. At the same time, larger-scale experiments (such as those in small-angle CMB anisotropies) will test sub-Plankian physics during the pre-bangian phase.

2. (SUPER)STRING INSPIRATION

2.1. Short Distance

Since the classical (Nambu–Goto) action of a string is proportional to the area A of the surface it sweeps, its quantization must introduce a quantum of length λ_s through:

$$S/\hbar = A/\lambda_s^2 \,. \tag{1}$$

This fundamental length, replacing Planck's constant in quantum string theory [2], plays the role of a minimal observable length, of an ultraviolet cut-off. Thus, in string theory, physical quantities are expected to be bound by appropriate powers of λ_s, e.g.

$$
\begin{aligned}
H^2 &\sim R \sim G\rho < \lambda_s^{-2} \\
&\quad k_B T/\hbar < c\lambda_s^{-1} \\
&\qquad R_{comp} > \lambda_s \,.
\end{aligned}
\tag{2}
$$

In other words, in quantum string theory (QST), relativistic quantum mechanics should solve the singularity problems in much the same way as non-relativistic quantum mechanics solved the singularity problem of the hydrogen atom by putting the electron and the proton a finite distance apart. By the same token, QST gives us a rationale for asking daring questions such as: What was there before the big bang? Certainly, in no other present theory, can such a question be meaningfully asked.

2.2. Large Distance

Even at large distance (low-energy, small curvatures), superstring theory does not automatically give Einstein's General Relativity. Rather, it leads to a scalar-tensor theory of the JBD variety. The new scalar particle/field ϕ, the so-called dilaton, is unavoidable in string theory, and gets reinterpreted as the radius of a new dimension of space in so-called M-theory [3]. By supersymmetry, the dilaton is massless to all orders in perturbation theory, i.e. as long as supersymmetry remains unbroken. This raises the question: Is the dilaton a problem or an opportunity? My answer is that it is possibly both; and while we can try to avoid its potential dangers, we may try to use some of its properties to our advantage ... Let me discuss how.

In string theory ϕ controls the strength of all forces, gravitational and gauge alike. One finds, typically:

$$l_P^2/\lambda_s^2 \sim \alpha_{gauge} \sim e^\phi \,, \tag{3}$$

showing the basic unification of all forces in string theory and the fact that, in our conventions, the weak-coupling region coincides with $\phi \ll -1$. In order not to contradict precision tests of the Equivalence Principle and of the constancy of the gauge and gravitational couplings in the recent past (possibly meaning several million years!) we require [4] the dilaton to have a mass and to be frozen at the bottom of its own potential *today*. This does not exclude, however, the possibility of the dilaton having evolved cosmologically (after all the metric did!) within the weak coupling region where it was practically massless. The amazing (yet simple) observation [5] is that, by so doing, the dilaton may have inflated the Universe!

A simplified argument, which, although not completely accurate, captures the essential physical point, consists in writing the ($k = 0$) Friedmann equation:

$$3H^2 = 8\pi G\rho \,, \tag{4}$$

and in noticing that a growing dilaton (meaning through (3) a growing G) can drive the growth of H even if the energy density of standard matter decreases in an expanding Universe. This new kind of inflation (characterized by growing H and ϕ) has been termed dilaton-driven inflation (DDI). The basic idea of pre-big bang cosmology [5–7] is thus illustrated in Fig. 1: the dilaton started at very large negative values (where it is massless), ran over a potential hill, and finally reached, sometime in our recent past, its final destination at the bottom of its potential ($\phi = \phi_0$). Incidentally, as shown in Fig. 1, the dilaton of string theory can easily roll-up —rather than down— potential hills, as a consequence of its non-standard coupling to gravity.

DDI is not just possible. It exists as a class of (lowest-order) cosmological solutions thanks to the duality symmetries of string cosmology. Under a prototype example of these symmetries, the so-called scale-factor duality [5], a FRW cosmology evolving (at lowest order in derivatives) from

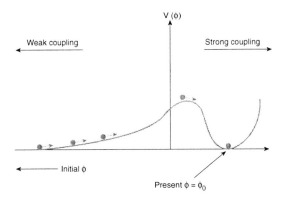

Figure 1.

a singularity in the past is mapped into a DDI cosmology going towards a singularity in the future. Of course, the lowest order approximation breaks down before either singularity is reached. A (stringy) moment away from their respective singularities, these two branches can easily be joined smoothly to give a single non-singular cosmology, at least mathematically. Leaving aside this issue for the moment (see Section 5 for more discussion), let us go back to DDI. Since such a phase is characterized by growing coupling and curvature, it must itself have originated from a regime in which both quantities were very small. We take this as the main lesson/hint to be learned from low-energy string theory by raising it to the level of a new cosmological principle [8] of "Asymptotic Past Triviality".

3. ASYMPTOTIC PAST TRIVIALITY

The concept of Asymptotic Past Triviality (APT) is quite similar to that of "Asymptotic Flatness", familiar from General Relativity [9]. The main differences consist in making only assumptions concerning the asymptotic past (rather than future or space-like infinity) and in the additional presence of the dilaton. It seems physically (and philosophically) satisfactory to identify the beginning with simplicity (see e.g. entropy-related arguments concerning the arrow of time).

What could be simpler than a trivial, empty and flat Universe? Nothing of course! The problem is that such a Universe, besides being uninteresting, is also non-generic. By contrast, asymptotically flat/trivial Universes are initially simple, yet generic in a precise mathematical sense. Their definition involves exactly the right number of arbitrary "integration constants" (here full functions of three variables) to describe a general solution (one with some general, qualitative features, though). This is why, by its very construction, this cosmology cannot be easily dismissed as being fine-tuned.

Can a very rich and complicated Universe, like our own, emerge for such extremely simple initial conditions? This would look much like a miracle. However, as I shall argue below, this is precisely what should be expected, owing to well-known classical and quantum gravitational instabilities.

4. INFLATION AS A CLASSICAL GRAVITATIONAL INSTABILITY

The assumption of APT entitles us to treat the early history of the Universe through the classical field equations of the low-energy (because of the small curvature) tree-level (because of the weak coupling) effective action of string theory. For simplicity, we will illustrate here the simplest case of the gravi-dilaton system already compactified to four space-time dimensions and use pictures (cartoons) rather than formulae.

Even assuming APT, the problem of determining the properties of a generic solution to the field equations is a formidable one. Very luckily, however, we are able to map our problem into one that has been much investigated, both analytically and numerically, in the literature. This is done by going to the so-called "Einstein frame". For our purposes, it simply amounts to the field redefinition

$$g_{\mu\nu} = g_{\mu\nu}^{(E)} e^{\phi - \phi_0} . \tag{5}$$

Our problem is thus reduced to that of studying a massless scalar field minimally coupled to gravity. Such a system has been considered by many authors, in particular by Christodoulou [10], precisely in the regime of interest to us. In line with the APT postulate, in the analogue gravitational collapse problem, one assumes very "weak" initial data with the aim of finding under which conditions gravitational collapse later occurs. Gravitational collapse means that the (Einstein) metric (and the volume of 3-space) shrinks to zero at a space-like singularity. However, typically, the dilaton blows up at that same singularity. Given the relation (5) between the Einstein and the (physical) string metric, we can easily imagine that the latter blows up near the singularity as implied by DDI.

How generically does this happen? In this connection it is crucial to recall the singularity theorems of Hawking and Penrose [11], which state that, under some general assumptions, singularities are inescapable in GR. One can look at the validity of those assumptions in the case at hand and finds that all but one are automatically satisfied. The only condition to be imposed is the existence of a closed trapped surface (a closed surface from where future light cones lie entirely in the region inside the surface). Rigorous results [10] show that this condition cannot be waived: sufficiently weak initial data do not lead to closed trapped surfaces, to collapse, or to singularities. Sufficiently strong initial data do. But where is the border-line? This is not known in general, but precise criteria do exist for particularly symmetric space-times, e.g. for those endowed with spherical symmetry. However, no matter what the general collapse/singularity criterion will eventually turn out to be, we do know that:

- it cannot depend on an over-all additive constant in ϕ;

- it cannot depend on an over-all multiplicative factor in $g_{\mu\nu}$.

This is a simple consequence of the invariance (up to an over-all factor) of the effective action under shifts of the dilaton and rescaling of the metric (these properties depend crucially on the validity of the tree-level low-energy approximation and on the absence of a cosmological constant).

We conclude that, generically, some regions of space will undergo gravitational collapse, will form horizons and singularities therein, but nothing, at the level of our approximations, will be able to fix either the size of the horizon or the value of ϕ at the onset of collapse. When this is translated into the string frame, one is describing, in the region of space-time within the horizon, a period of DDI in which both the initial value of the Hubble parameter and that of ϕ are left arbitrary. These two initial parameters are very important, since they determine the range of validity of our description. In fact, since both curvature and coupling increase during DDI, at some point the low-energy and/or tree-level description is bound to break down. The smaller the initial Hubble parameter (i.e. the larger the initial horizon size) and the smaller the initial coupling, the longer we can follow DDI through the effective action equations and the larger the number of reliable e-folds that we shall gain.

This does answer, in my opinion, the objections raised recently [12] to the PBB scenario according to which it is fine-tuned. The situation here actually resembles that of chaotic inflation [13]. Given some generic (though APT) initial data, we should ask which is the distribution of sizes of the collapsing regions and of couplings therein. Then, only the "tails" of these distributions, i.e. those corresponding to sufficiently large, and sufficiently weakly coupled regions, will produce Universes like ours, the rest will not. The question of how likely a "good" big bang is to take place is not very well posed and can be greatly affected by anthropic considerations.

In conclusion, we may summarize recent progress on the problem of initial conditions by saying that [8]: **Dilaton-driven inflation in string cosmology is as generic as gravitational collapse in General Relativity.** At the same time, having a sufficiently long period of DDI amounts to setting upper limits on two arbitrary moduli of the classical solutions.

Figure 2 gives a $(2+1)$-dimensional sketch of a possible PBB Universe: an original "sea" of dilatonic and gravity waves leads to collapsing regions of different initial size, possibly to a scale-invariant distribution of them. Each one of these collapses is reinterpreted, in the string frame, as the process by which a baby Universe is born after a period of PBB inflationary "pregnancy", with the size of each baby Universe determined by the duration of its pregnancy, i.e. by the initial size of the corresponding collapsing region. Regions initially larger than 10^{-13} cm can generate Universes like ours, smaller ones cannot.

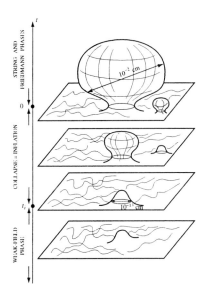

Figure 2.

A basic difference between the large numbers needed in (non-inflationary) FRW cosmology and the large numbers needed in PBB cosmology should be stressed at this point. In the former, the ratio of two classical scales, e.g. of total curvature to its spatial component, which is expected to be $O(1)$, has to be taken as large as 10^{60}. In the latter, the above ratio is initially $O(1)$ in the collapsing/inflating region, and ends up being very large in that region thanks to DDI. However, the common order of magnitude of these two classical quantities is a free parameter, and is taken to be much larger than a classically irrelevant quantum scale.

We can visualize analogies and differences between standard and pre-big bang inflation by comparing Figs. 3a and 3b. In these, we sketch the evolution of the Hubble radius and of a fixed comoving scale (here the one corresponding to the part of the Universe presently observable to us) as a function of time in the two scenarios. The common feature is that the fixed comoving scale was "inside the horizon" for some time during inflation, and possibly very deeply inside at its onset. Also, in both cases, the Hubble radius at the beginning of inflation had to be large in Planck units and the scale of homogeneity had to be at least as large. The difference between the two scenarios is just in the behaviour of the Hubble radius during inflation: increasing in standard inflation (a), decreasing in string cosmology (b). This is what makes PBB's "wine glass" more elegant, and stable! Thus, while standard inflation is still facing the initial-singularity question and needs a non-adiabatic phenomenon to reheat the Universe (a kind of small bang), PBB cosmology faces the singularity problem later, combining it with the exit and heating problems (discussed in Sections V and VIB, respectively).

5. THE EXIT PROBLEM/ CONJECTURE

We have argued that, generically, DDI, when studied at lowest order in derivatives and coupling, evolves towards a singularity of the big bang type. Similarly, at the same level of approximation, the non-inflationary solutions emerge from a singularity. Matching these two branches in a smooth, non-singular way has become known as the (graceful) exit problem in string cosmology [14]. It is, undoubtedly, the most important theoretical problem facing the whole PBB scenario.

There has been quite some progress recently on the exit problem. However, for lack of space, I shall refer the reader to the literature [14] for details. Generically speaking, toy examples have shown that DDI can flow, thanks to higher-curvature corrections, into a de-Sitter-like phase, i.e. into a phase of constant H (curvature) and constant $\dot{\phi}$. This phase is expected to last until loop corrections become important (see next sec-

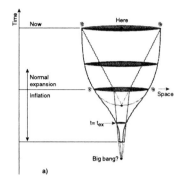

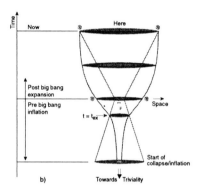

Figure 3.

tion) and give rise to a transition to a radiation-dominated phase. If these toy models serve as an indication, the full exit problem can only be achieved at large coupling and curvature, a situation that should be described by the newly invented M-theory [3].

It was recently pointed out [15] that the reverse order of events is also possible. The coupling may become large *before* the curvature does. In this case, at least for some time, the low-energy limit of M-theory should be adequate: this limit is known [3] to give $D = 11$ supergravity and is therefore amenable to reliable study. It is likely, though not yet clear, that, also in this case, strong curvatures will have to be reached before the exit can be completed. In the following, we will assume that:

- the big bang singularity is avoided thanks

to the softness of string theory;

- full exit to radiation occurs at strong coupling and curvature, according to a criterion given in Section 62.

6. OBSERVABLE RELICS AND HEATING THE PRE-BANG UNIVERSE

6.1. PBB relics

Since there are already several review papers on this subject (e.g. [16]), I will limit myself to mentioning the most recent developments, after recalling the basic physical mechanism underlying particle production in cosmology [17]. A cosmological (i.e. time-dependent) background coupled to a given type of (small) inhomogeneous perturbation Ψ enters the effective low-energy action in the form:

$$I = \frac{1}{2} \int d\eta \, d^3x \, S(\eta) \left[\Psi'^2 - (\nabla \Psi)^2 \right]. \qquad (6)$$

Here η is the conformal-time coordinate, and a prime denotes $\partial/\partial\eta$. The function $S(\eta)$ (sometimes called the "pump" field) is, for any given Ψ, a given function of the scale factor $a(\eta)$, and of other scalar fields (four-dimensional dilaton $\phi(\eta)$, moduli $b_i(\eta)$, etc.), which may appear non-trivially in the background.

While it is clear that a constant pump field S can be reabsorbed in a rescaling of Ψ, and is thus ineffective, a time-dependent S couples non-trivially to the fluctuation and leads to the production of pairs of quanta (with equal and opposite momenta). One can easily determine the pump fields for each one of the most interesting perturbations. The result is:

Gravity waves, dilaton	:	$S = a^2 e^{-\phi}$
Heterotic gauge bosons	:	$S = e^{-\phi}$
Kalb – Ramond, axions	:	$S = a^{-2} e^{-\phi}$. (7)

A distinctive property of string cosmology is that the dilaton ϕ appears in some very specific way in the pump fields. The consequences of this are very interesting:

- For gravitational waves and dilatons, the effect of ϕ is to slow down the behaviour of

a (both a and ϕ grow in the pre-big bang phase). This is the reason why those spectra are quite steep [18] and give small contributions at large scales. The reverse is also true: at short scales, the expected yield in a stochastic background of gravitational waves is much larger than in standard inflationary cosmology. Values of Ω_{GW} in the range of 10^{-6}–10^{-7} are possible in some regions of parameter space, which, according to some estimates of sensitivities [19], could be inside detection capabilities in the near future.

- For gauge bosons there is no amplification of vacuum fluctuations in standard cosmology, since a conformally flat metric (of the type forced upon by inflation) decouples from the electromagnetic (EM) field precisely in $D = 3 + 1$ dimensions. As a very general remark, apart from pathological solutions, the only background field which, through its cosmological variation, can amplify EM (more generally gauge-field) quantum fluctuations is the effective gauge coupling itself [20]. By its very nature, in the pre-big bang scenario the effective gauge coupling inflates together with space during the PBB phase. It is thus automatic that any efficient PBB inflation brings together a huge variation of the effective gauge coupling and thus a very large amplification of the primordial EM fluctuations [21–23]. This can possibly provide the long-sought origin for the primordial seeds of the observed galactic magnetic fields. Notice, however, that, unlike GW, EM perturbations interact quite considerably with the hot plasma of the early (post-big bang) Universe. Thus, converting the primordial seeds into those that may have existed at the proto-galaxy formation epoch is by no means a trivial exercise. Work is in progress to try to adapt existing codes [24] to the evolution of our primordial seeds.

- Finally, for Kalb–Ramond fields and axions, a and ϕ work in the same direction and spectra can be large even at large scales

[25]. An interesting fact is that, unlike the GW spectrum, that of axions is very sensitive to the cosmological behaviour of internal dimensions during the DDI epoch. On one side, this makes the model less predictive. On the other, it tells us that axions represent a window over the multidimensional cosmology expected generically from string theories, which must live in more that four dimensions. Curiously enough, the axion spectrum becomes exactly HZ (i.e. scale-invariant) when all the nine spatial dimensions of superstring theory evolve in a rather symmetric way [22]. In situations near this particularly symmetric one, axions are able to provide a new mechanism for generating large-scale CMB anisotropy and LSS.

A recent calculation [26] of the effect gives, for massless axions,

$$l(l + 1)C_l \sim O(1) \left(\frac{H_{max}}{M_P} \right)^4$$
$$(\eta_0 k_{max})^{-2\alpha} \frac{\Gamma(l + \alpha)}{\Gamma(l - \alpha)} , \qquad (8)$$

where C_l are the usual coefficients of the multipole expansion of $\Delta T / T$

$$\langle \Delta T / T(\vec{n}) \ \ \Delta T / T(\vec{n}') \rangle =$$
$$\sum_l (2l + 1) C_l P_l(\cos \theta) , \qquad (9)$$

and the parameters H_{max}, k_{max}, α are defined by the primordial axion energy spectrum in critical units as:

$$\Omega_{ax}(k) = \left(\frac{H_{max}}{M_P} \right)^2 (k/k_{max})^\alpha . \qquad (10)$$

In string theory, as repeatedly mentioned, we expect $H_{max}/M_P \sim M_s/M_P \sim 1/10$ and $\eta_0 k_{max} \sim 10^{30}$, while the exponent α depends on the explicit PBB background with the above-mentioned HZ case corresponding to $\alpha = 0$. The standard tilt parameter $n = n_s$ (s for scalar) is given by $n = 1 + 2\alpha$ and is found, by COBE [27], to lie between 0.9 and 1.5, corresponding to $0 < \alpha < 5$ (a negative α leads to some theoretical problems). With these inputs we can see that the correct normalization ($C_2 \sim 10^{-10}$) is reached for $\alpha \sim 0.1$, which is just in the middle of the allowed range. In other words, unlike in standard inflation, we cannot predict the tilt, but when this is given, we can predict (again unlike in standard inflation) the normalization. Work is now in progress to compute the C_l in the acoustic peak region. The peak structure is expected to be very sensitive to α which, in turn, is basically fixed by the ·overall normalization. Hence an almost parameter-free prediction for the acoustic peaks is expected. Furthermore, our model, being of the isocurvature type, bears some resemblance to the one recently advocated by Peebles [28] and, like his, is expected to contain some calculable amount of non-Gaussianity. This too is being calculated and will be checked by the future satellite measurements (MAP, PLANCK).

- Many other perturbations, which arise in generic compactifications of superstrings, have also been studied, and lead to interesting spectra. For lack of time, I will refer to the existing literature [22,23].

6.2. Heat and entropy as a quantum gravitational instability

Before closing this section, I wish to recall how one sees the very origin of the hot big bang in this scenario. One can easily estimate the total energy stored in the quantum fluctuations, which were amplified by the pre-big bang backgrounds. The result is, roughly,

$$\rho_{quantum} \sim N_{eff} \, H_{max}^4 \, , \qquad (11)$$

where N_{eff} is the effective number of species that are amplified and H_{max} is the maximal curvature scale reached around $t = 0$. We have already argued that $H_{max} \sim M_s = \lambda_s^{-1}$, and we know that, in heterotic string theory, N_{eff} is in the hundreds. Yet this rather huge energy density is

very far from critical, as long as the dilaton is still in the weak-coupling region, justifying our neglect of back-reaction effects. It is very tempting to assume [22] that, precisely when the dilaton reaches a value such that $\rho_{quantum}$ is critical, the Universe will enter the radiation-dominated phase. This PBBB (PBB bootstrap) constraint gives, typically:

$$e^{\phi_{exit}} \sim 1/N_{eff} \ , \qquad (12)$$

i.e. a value for the dilaton close to its present value.

The entropy in these quantum fluctuations can also be estimated following some general results [30]. The result for the density of entropy S is, as expected

$$S \sim N_{eff} H_{max}^3 \, . \qquad (13)$$

It is easy to check that, at the assumed time of exit given by (12), this entropy saturates a recently proposed bound [30]. This also turns out to be a physically acceptable value for the entropy of the Universe just after the big bang: a large entropy on the one hand (about 10^{90}); a small entropy for the total mass and size of the observable Universe on the other, as often pointed out by Penrose [31]. Thus, PBB cosmology neatly explains why the Universe, at the big bang, *looks* so fine-tuned (without being so) and provides a natural arrow of time in the direction of higher entropy. The entropy bound of [30] has also interesting implications for the exit problem discussed in Section 5.

7. CONCLUSIONS

- Pre-big bang (PBB) cosmology is a "top–down" rather than a "bottom–up" approach to cosmology. This should not be forgotten when testing its predictions.

- It does not need to invent an inflaton, or to fine-tune its potential; inflation is "natural" thanks to the duality symmetries of string cosmology.

- It makes use of a classical gravitational instability to inflate the Universe, and of a quantum instability to warm it up.

- The problem of initial conditions "decouples" from the singularity problem; it is classical, scale-free, and unambiguously defined. Issues of fine tuning can be addressed and, I believe, answered.

- The spectrum of large-scale perturbations has become more promising through the invisible axion of string theory, while the possibility of explaining the seeds of galactic magnetic fields remains a unique prediction of the model.

- The main conceptual (technical?) problem remains that of providing a fully convincing mechanism for (and a detailed description of) the pre-to-post-big bang transition. It is very likely that such a mechanism will involve both high curvatures and large coupling and should therefore be discussed in the (yet to be fully constructed) M-theory [3]. New ideas borrowed from such theory and from D-branes [32,15] could help in this respect.

- Once/if this problem will be solved, predictions will become more precise and robust, but, even now, with some mild assumptions, several tests are (or will soon become) possible, e.g.

  - some non-Gaussianity in $\Delta T/T$ correlations is expected, and calculable.

  - the axion-seed mechanism should lead to a characteristic acoustic-peak structure, which is being calculated;

  - it should be possible to convert the predicted seed magnetic fields into observables by using some reliable code for their late evolution;

  - a characteristic spectrum of stochastic gravitational waves is expected to surround us and could be large enough to be measurable within a decade or so.

REFERENCES

1. E. W. Kolb and M. S. Turner, *The Early Universe* (Addison-Wesley, Redwood City, CA, 1990); A.D. Linde, *Particle Physics and Inflationary Cosmology* (Harwood, New York, 1990).

2. G. Veneziano, Europhys. Lett. **2** (1986) 133; *The Challenging Questions*, Erice, 1989, ed. A. Zichichi (Plenum Press, New York, 1990), p. 199.

3. See, e.g., E. Witten, Nucl. Phys. **B443** (1995) 85; P. Horawa and E. Witten, Nucl. Phys. **B460** (1996) 506.

4. T.R. Taylor and G. Veneziano, Phys. Lett. **B213** (1988) 459.

5. G. Veneziano, Phys. Lett. **B265** (1991) 287.

6. M. Gasperini and G. Veneziano, Astropart. Phys. **1** (1993) 317, Mod. Phys. Lett. **A8** (1993) 3701, Phys. Rev. **D50** (1994) 2519.

7. An updated collection of papers on the PBB scenario is available at http://www.to.infn.it/~gasperin/.

8. A. Buonanno, T. Damour and G. Veneziano, Nucl. Phys. **B543** (1999) 275; see also, G. Veneziano, Phys. Lett. **B406** (1997) 297; A. Buonanno, K.A. Meissner, C. Ungarelli and G. Veneziano, Phys. Rev. **D57** (1998) 2543, and references therein.

9. R. Penrose, *Structure of space-time*, in *Battelle Rencontres*, ed. C. Dewitt and J.A. Wheeler, Benjamin, New York, 1968.

10. D. Christodoulou, Commun. Pure Appl. Math. **56** (1993) 1131, and references therein.

11. R. Penrose, Phys. Rev. Lett. **14** (1965) 57; S. W. Hawking and R. Penrose, Proc. Roy. Soc. Lond. **A314** (1970) 529.

12. M. Turner and E. Weinberg, Phys. Rev. **D56** (1997) 4604; N. Kaloper, A. Linde and R. Bousso, Phys. Rev.**D59** (1999) 043508.

13. A. Linde, Phys. Lett. **129B** (1983) 177.

14. R. Brustein and G. Veneziano, Phys. Lett. **B329** (1994) 429; N. Kaloper, R. Madden and K.A. Olive, Nucl. Phys. **B452** (1995) 677, Phys. Lett. **B371** (1996) 34; R. Easther, K. Maeda and D. Wands, Phys. Rev. **D53** (1996) 4247; M. Gasperini, M. Maggiore and G. Veneziano, Nucl. Phys. **B494** (1997) 315; R. Brustein and R. Madden, Phys. Lett. **B410** (1997) 110, Phys. Rev. **D57** (1998) 712.

15. M. Maggiore and A. Riotto, *D-branes and*

Cosmology, hep-th/9811089;
see also T. Banks, W. Fishler and L. Motl, *Duality versus Singularities*, hep-th/9811194.

16. G. Veneziano, in *String Gravity and Physics at the Planck Energy Scale*, Erice, 1995, eds. N. Sanchez and A. Zichichi (Kluver Academic Publishers, Boston, 1996), p. 285; M. Gasperini, ibid., p. 305.

17. See, e.g., V. F. Mukhanov, A. H. Feldman and R. H. Brandenberger, Phys. Rep. **215** (1992) 203.

18. R. Brustein, M. Gasperini, M. Giovannini and G. Veneziano, Phys. Lett. **B361** (1995) 45; R. Brustein et al., Phys. Rev. **D51** (1995) 6744.

19. P. Astone et al., Phys. Lett. **B385** (1996) 421; B. Allen and R. Brustein, Phys. Rev. **D55** (1997) 970.

20. B. Ratra, Astrophys. J. Lett. **391** (1992) L1.

21. M. Gasperini, M. Giovannini and G. Veneziano, Phys. Rev. Lett. **75** (1995) 3796; D. Lemoine and M. Lemoine, Phys. Rev. **D52** (1995) 1955.

22. A. Buonanno, K. A. Meissner, C. Ungarelli and G. Veneziano, JHEP **1** (1998) 4;

23. R. Brustein and M. Hadad, Phys. Rev. **D57** (1998) 725.

24. R. M. Kulsrud, R. Cen, J. P. Ostriker and D. Ryu, Ap. J. **480** (1997) 481.

25. E.J. Copeland, R. Easther and D. Wands, Phys. Rev. D56 (1997) 874; E.J. Copeland, J.E. Lidsey and D. Wands, Nucl. Phys. **B506** (1997) 407.

26. R. Durrer, M. Gasperini, M. Sakellariadou and G. Veneziano, Phys. Lett. **B436** (1998) 66; Phys. Rev. **D59** (1999) 043511; M. Gasperini and G. Veneziano, Phys. Rev. **D59** (1999) 043503.

27. G. F. Smoot et al., Ap. J. **396** (1992) L1; C. L. Bennet et al., Ap. J. **430** (1994) 423.

28. P. J. E. Peebles, *An isocurvature CDM cosmogony. I and II*, astro-ph/9805194, and astro-ph/9805212.

29. M. Gasperini and M. Giovannini, Phys. Lett. **B301** (1993) 334; Class. Quant. Grav. **10** (1993) L133; R. Brandenberger, V. Mukhanov and T. Prokopec, Phys. Rev. Lett. **69** (1992) 3606; Phys. Rev. **D48** (1993) 2443.

30. G. Veneziano, *Pre-bangian origin of our entropy and time arrow*, hep-th/9902126; R. Easther and D. A. Lowe, *Holography, cosmology and the 2nd law of thermodynamics*, hep-th/9902088; D. Bak and S.-J. Rey, *Cosmic Holography*, hep-th/9902173.

31. see, e.g., R. Penrose, *The Emperor's new mind*, (Oxford University Press, New York, 1989), Chapter 7.

32. A. Lukas, B.A. Ovrut and D. Waldram, Phys. Lett. **B393** (1997) 65; Nucl. Phys. **B495** (1997) 365; F. Larsen and F. Wilczek, Phys. Rev. **D55** (1997) 4591; N. Kaloper, Phys. Rev. **D55** (1997) 3394; H. Lu, S. Mukherji and C.N. Pope, Phys. Rev. **D55** (1997) 7926; R. Poppe and S. Schwager, Phys. Lett. **B393** (1997) 51; A. Lukas and B. A. Ovrut, Phys. Lett. **B437** (1998) 291; N. Kaloper, I. Kogan and K. A. Olive, Phys. Rev. **D57** (1998) 7340.

ELSEVIER

Nuclear Physics B (Proc. Suppl.) 80 (2000) 119–132

NUCLEAR PHYSICS B
**PROCEEDINGS
SUPPLEMENTS**

www.elsevier.nl/locate/npe

Cosmic Deuterium

Alfred Vidal-Madjar[a]

[a]Institut d'Astrophysique de Paris, CNRS, 98 bis boulevard Arago, F-75014 Paris, France

The knowledge of the primordial deuterium to hydrogen ratio provides one of the most reliable tests of the early Universe nucleosynthesis models and a direct estimate of the cosmic baryon density. Evaluations have been traditionally made using D/H estimations in the interstellar medium, extrapolated backwards in time with the use of galactic evolution models. Direct primordial D/H measurements have been carried out only recently in the direction of quasars. These measurements of deuterium abundances along with observations made in the solar system and in the interstellar medium are presented.

New results that indicate spatial variations of the deuterium abundance in the interstellar medium at the level of $\sim 50\%$ over scales possibly as small as ~ 10 pc, may question our global vision of deuterium evolution until the causes for the origin of these variations are understood. With a conservative point of view, observations thus suggest that the primordial D/H value should be within the range $1. \times 10^{-5}$–$3. \times 10^{-4}$. leading to a relatively low baryon content Universe.

Since the actual evolution of deuterium from primordial nucleosynthesis to now is not known in details, more observations, hopefully to be made with the Hubble Space Telescope, FUSE the Far Ultraviolet Spectroscopic Explorer (launched in 1999). or from the ground with the largest telescopes (Keck, VLT, ...), should reveal the evolution of that key element, and better constrain its primordial abundance.

1. Introduction

Starting with Alpher, Bethe & Gamov (1948), and until the late sixties, in the frame of Big-Bang nucleosynthesis (BBN) models, the primordial origin of ^{4}He seemed quite plausible, but the site of formation of the other light elements remained slightly mysterious. Truran & Cameron (1971) then argued that in the absence of post-Big-Bang production, deuterium is slowly destroyed during galactic evolution, as it is entirely burned to ^{3}He in stars; in particular they estimated a destruction factor ~ 2. Reeves, Audouze, Fowler & Schramm (1973) thus argued that deuterium, if solely produced in the Big-Bang, would be a monitor of stellar formation and then showed that a baryonic density $\Omega_b = 0.016 \pm 0.005 h^{-2}$ (with $H_0 = 100 h$ km/s/Mpc) could explain the primordial abundance of ^{2}D, ^{3}He, ^{4}He, and possibly some ^{7}Li.

These ideas have been strengthened since then, and hardly, if at all, modified.

Following that global description, Epstein, Lattimer & Schramm (1976) showed that no deuterium should be produced in significant quantities in astrophysical sites other than the Big-Bang. Hence, measured abundances of deuterium would provide lower limits to the primordial abundance and consequently, an upper limit to the cosmic baryon density. It has been long recognized (see *e.g.*, Schramm. 1998; Schramm & Turner 1998) that the primordial abundance of deuterium represents the most sensitive probe of the baryonic density Ω_b .

Deuterium had only been detected in ocean water, at a level D/H$\sim 10^{-4}$ until the late sixties. In the early seventies, Black (1971) and Geiss & Reeves (1972) performed the first indirect measurement of the abundance of deuterium representative of the presolar nebula using combined solar wind and meteorite ^{3}He measurements. Shortly after, Cesarsky et al. (1973) attempted a detection via the radio observation of the 21 cm and 92 cm lines of both H I and D I, and Rogerson & York (1973) successfully measured for the first time the abundance of deuterium in the interstellar medium from H I and D I Lyman absorption lines. These efforts were followed by numerous other studies.

Measurements of the deuterium to hydrogen

ratio in moderate to high redshift absorbers toward quasars have been obtained for the first time in the last few years. These clouds are very metal-deficient, so that their deuterium content should not have been affected by astration of gas or, equivalently, the deuterium abundance measured should be close to primordial. This is in contrast with the presolar nebula and interstellar medium measurements, whose deuterium abundances are modified from the primordial abundance by chemical evolution.

Now three samples of deuterium abundances (measured by number in comparison with hydrogen) are at our disposal, each representative of a given epoch: BBN [primordial abundance $(D/H)_{QSO}^{BBN}$] as observed in high redshift material in front of quasars (QSO), 4.5 Gyrs past [pre-solar abundance $(D/H)_{pre\odot}^{4.5Gy}$] within the solar system and present epoch (Now) [interstellar medium (ISM) abundance $(D/H)_{ISM}^{Now}$]. Note that the inference of a primordial, pre-solar or interstellar D/H ratio from a measurement rests on the assumption of efficient mixing of the material probed by the observations. This seems to be questionned today.

In effect, our goal is ultimately both to know the primordial D/H ratio and to understand the evolution of its abundance with time, in order to constrain the overall amount of star formation. Interstellar (ISM) measurements do not always agree with each other and few observers argue, on the basis of very recent data, that at least part of the scatter is real ; in other words, there exist some unknown processes that affect the D/H ratio in the ISM by $\sim 30 - 50\%$ in some cases, over possibly very small scales. Few plausible explanations will be mentionned. In the case of the presolar nebula abundances, there also exists scatter, but at the present time, it is not clear whether it arises from chemical fractionation of deuterium and hydrogen in molecules, or from some other cause.

For quasar absorption systems (QSO), the situation is not yet clear, although two remarkable measurements of Burles & Tytler (1998a,b,c,d) agree to a common value $(D/H)_{QSO}= 3.4 \pm 0.3 \times 10^{-5}$. As we have learned in the case of the

ISM, the picture may very well change when much more observations will be available. For that reason I prefer to remain very cautious and keep the primordial D/H evaluation issue, open.

I will thus briefly discuss in this review the current determinations of deuterium abundances and focus on the latest results from interstellar measurements. The QSO absorbers are discussed in Section 2, presolar nebula measurements in Section 3 and ISM measurements in Section 4. Section 5 lists possible causes of spatial variations of the $(D/H)_{ISM}$ ratio and Section 6, their consequences. Finally Section 7 summarizes the conclusions and underlines future investigations.

2. Primordial abundance [$(D/H)_{QSO}^{BBN}$]

Adams (1976) first suggested that measurements of the D/H ratio in metal-deficient absorbers on lines of sight to distant quasars offer direct access to the primordial abundance of deuterium. Although of fundamental importance with respect to Big-Bang nucleosynthesis, this measurement is particularly difficult to achieve (Laurent & Vidal-Madjar 1981; Webb et al. 1991).

In the Lyman series of ground state absorption by atomic H I and D I the absorption of deuterium appears 82 km s^{-1} bluewards (shorter wavelengths) of the corresponding H I absorption. In realistic situations, there is only a limited range of b-values (the physical parameter that roughly defines the width of the absorption line and which is related to the temperature and the turbulent velocity) and column densities, for which the absorption due to D I can be well separated from that of H I. Typically, for a single absorber and H I column densities N(H I)$\sim$ 10^{18} cm^{-2}, one would like the H I b-value to range around ~ 15 km s^{-1}, corresponding to temperatures $\sim 10^4$ K (Laurent & Vidal-Madjar 1981; Webb et al. 1991; Jenkins 1996). Such b-values are typical of diffuse ISM clouds, but quite atypical of quasar absorbers, in which the broadening parameter takes values above $\simeq 15 - 20$ km s^{-1}. Good candidate absorbers are thus certainly difficult to find.

Moreover, one rarely observes a single ab-

A. Vidal-Madjar/Nuclear Physics B (Proc. Suppl.) 80 (2000) 119–132

121

sorber. In particular, the Lyman α forest is present at high redshifts $z \geq 2$, with a large density of lines per unit redshift, so that one has to disentangle the H I and D I from the numerous neighbouring weak lines of H I. In particular, one always runs the risk of confusion between a D I line and a weak H I line, at a redshift such that the line falls at the expected position of the D I line; such H I lines are called "interlopers". As a consequence, measuring the $(D/H)_{QSO}$ ratio is a matter of statistics. Burles & Tytler (1998c,d) have estimated that about one out of thirty quasars could offer a suitable candidate.

York et al. (1983) obtained the first upper limit on $(D/H)_{QSO}$ toward Mrk509 at a redshift $z_{abs} = 0.03$ using IUE data. Several years ago, Carswell et al. (1994) and Songaila et al. (1994) reported detections of deuterium absorption toward QSO0014+813, $(D/H)_{QSO} \simeq 25 \times 10^{-5}$ at $z_{abs} = 3.32$, using respectively the Kitt Peak and W.M. Keck telescopes. These authors were cautious in pointing out the possibility that the deuterium feature could actually be due to an H I interloper. A revised analysis of the Keck data by Rugers & Hogan (1996) gave $(D/H)_{QSO} \sim 19 \pm 5 \times 10^{-5}$ apparently confirming the first estimation.

In the subsequent years, the situation has rapidly evolved. The aim here is not to review all of these developments, and I refer the reader to recent existing reviews (Hogan 1997; Vidal-Madjar, Ferlet & Lemoine 1998; Burles & Tytler 1998c,d). As of today, there are three strong claims for a detection of D I, namely $(D/H)_{QSO} = 3.3 \pm 0.3 \times 10^{-5}$ at $z_{abs} = 3.57$ toward QSO1937-1009 (Burles & Tytler 1998a), $(D/H)_{QSO} = 4.0 \pm 0.7 \times 10^{-5}$ at $z_{abs} = 2.50$ toward QSO1009+2956 (Burles & Tytler 1998b), and $(D/H)_{QSO} = 25. \pm 10. \times 10^{-5}$ at $z_{abs} = 0.701$ toward QSO1718+4807 (Webb et al. 1997). We therefore have today two low and one high values of the $(D/H)_{QSO}^{BBN}$ ratio.

These evaluations deserve few additional comments:

- QSO0014+873

  From new observations of QSO0014+813, Burles, Kirkman & Tytler (1999) have demonstrated the presence of an H I inter-

loper in the absorption line that had been identified as D I, so that consequently, no $(D/H)_{QSO}$ ratio could be measured with confidence in this system. However, Hogan (1998) maintains that there is evidence for a high deuterium abundance in this system and that the probability and amount of contamination should be small, basing his arguments on statistical studies of correlations of absorbers on scales $\sim 80\,\mathrm{km\,s^{-1}}$. Songaila (1998) reports a similar finding, from statistical arguments, although based on a relatively small number of lines of sight, and derives $(D/H)_{QSO} \geq 5 \times 10^{-5}$.

- QSO1937-1009

  Songaila (1998) also claims that the estimate of the H I column density of Burles & Tytler toward QSO1937-1009 is incorrect and finds for this system $(D/H)_{QSO} \geq 5 \times 10^{-5}$. However, Burles & Tytler (1998e) contradict again this last point.

- QSO1009+2956

  Tytler et al. (1999) using Monte-Carlo simulations of H I cloud distribution on the line of sight, could also check that the low $(D/H)_{QSO}$ ratios toward QSO1009+2956 as in the case of QSO1937-1009, held.

- QSO1718+4807

  One should note also that Tytler et al. (1999) have reanalyzed the HST data of QSO1718+4807 together with IUE and Keck spectra and concluded that, for a single absorber, $(D/H)_{QSO} = 8 - 57 \times 10^{-5}$. However, they find that if a second H I absorber is allowed for on this line of sight, then the $(D/H)_{QSO}$ ratio becomes an upper limit, $(D/H)_{QSO} \leq 50 \times 10^{-5}$. The result toward QSO1718+4807 is thus not yet conclusive since the HST dataset contains only Lyman α and an associated Si III line and it would be extremely valuable to have data on the whole Lyman series of this absorber.

Finally, note also that Levshakov (1998, for a review) suggests that correlations in turbulent velocity on large spatial scales could seriously affect determinations of the $(D/H)_{QSO}$ ratio. This author, and collaborators, claim that the above high and low measurements of the deuterium abundance are consistent with a single value $D/H \simeq 3.5 - 5.2 \times 10^{-5}$ (see also Levshakov, Tytler & Burles 1999).

Highly confident conclusions at this time are still out of reach, although a trend toward $(D/H)_{QSO}^{BBN} \sim 3.5 \times 10^{-5}$ seems to be emerging as indicated by the recent results of Burles & Tytler (1998a,b). A need for further measurements of the $(D/H)_{QSO}$ ratio is however mandatory as they could change our understanding of the situation. This should be clear from our forthcoming discussion of the measurements of the $(D/H)_{ISM}^{Now}$ ratio.

3. Pre-solar abundance $[(D/H)_{pre\odot}^{4.5Gy}]$

Deuterium beeing burned into ^{3}He in the sun, by measuring the ^{3}He abundance in the solar wind, Geiss & Reeves (1972) determined the abundance in the protosolar nebula and hence found $(D/H)_{pre\odot} \simeq 2.5 \pm 1.0 \times 10^{-5}$. This result was historically the first evaluation of the deuterium abundance of astrophysical significance. It was confirmed by Gautier & Morel (1997) who showed $(D/H)_{pre\odot} = 3.01 \pm 0.17 \times 10^{-5}$. These determinations of $(D/H)_{pre\odot}$ are indirect and linked to the solar $(^4He/^3He)$ ratio and its evolution since the formation of the solar system.

Whereas in cometary water deuterium is enriched by a factor of at least 10 relative to the protosolar ratio (e.g. Bockelée-Morvan et al. 1998; Meier et al. 1998), the giant planets Jupiter and Saturn are considered to be undisturbed deuterium reservoirs, free from production or loss processes (see the recent review by Encrenaz 1999). Thus they should reflect the abundance of their light elements at the time of the formation of the solar system 4.5 Gyrs ago (Owen et al. 1986). The first measurements of the $(D/H)_{pre\odot}$ ratio in the Jovian atmosphere have been performed through methane and its deuterated counterpart CH_3D, yielding $(D/H)_{pre\odot} = 5.1 \pm 2.2 \times 10^{-5}$ (Beer & Taylor 1973). Other molecules such as HD and H_2, yield lower values: $(D/H)_{pre\odot} = 1. - 2.9 \times 10^{-5}$ (Smith et al. 1989).

Recently, new measurements of the $(D/H)_{pre\odot}$ ratio using very different methods were carried out. Two are based on the first results of the far infrared ISO observations of the HD molecule in Jupiter (Encrenaz et al. 1996) and Saturn (Griffin et al. 1996), and lead respectively to $(D/H)_{pre\odot} = 2.2 \pm 0.5 \times 10^{-5}$ and $(D/H)_{pre\odot} = 2.3_{-0.8}^{+1.2} \times 10^{-5}$. Note that the Encrenaz et al. (1996) value was updated to the more reliable value $(D/H)_{pre\odot} = 1.8_{-0.5}^{+1.1} \times 10^{-5}$ by Lellouch et al. (1997). Another is based on the direct observation with HST-GHRS of both H I and D I Lyman α emission at the limb of Jupiter for the first time (Ben Jaffel et al. 1994; 1997) yielding $(D/H)_{pre\odot} = 5.9 \pm 1.4 \times 10^{-5}$. The third one is an in situ measurement with a mass spectrometer onboard the Galileo probe (Niemann et al. 1996) yielding $(D/H)_{pre\odot} = 5.0 \pm 2.0 \times 10^{-5}$ [however this last value has been revised recently toward the lower part of the range, i.e. $(D/H)_{pre\odot} = 2.7 \pm 0.6 \times 10^{-5}$ (Mahaffy et al. 1998)].

It is surprising that measurements that probe almost the same atmospheric region of Jupiter ($\sim$ 1 bar level) lead to a such a large scatter in the D/H ratio. Indeed, the atmospheric composition at that level is the key parameter in the ISO data analysis, the H and D Lyman α spectra modeling and the Galileo mass spectrometer measurements.

It is likely that the differences between these values are due to systematic effect associated with models, such as the CH_4 mixing ratio (Lecluse et al. 1996), the effect of aerosols, the effect of eddy diffusion, or in the case of the mass spectrometer data, instrumental uncertainties. Additional investigations and observations including HST-STIS and FUSE observations will help to resolve this issue.

4. Interstellar abundance $[(D/H)_{ISM}^{Now}]$

The first measurement of the interstellar D/H ratio was reported by Rogerson & York (1973), from *Copernicus* observations of the line of sight to β Cen, giving $(D/H)_{ISM} = 1.4 \pm 0.2 \times 10^{-5}$. In the subsequent years, many other measurements of the interstellar deuterium abundances were carried out from *Copernicus* and IUE observations of the Lyman series of atomic D I and H I (for a review, see *e.g.* Vidal-Madjar, Ferlet & Lemoine 1998). Because absorption by the Lyman series takes place in the far-UV, these measurements require satellite-borne instruments, and the latest observations have been performed using HST and the *Interstellar Medium Absorption Profile Spectrograph* (IMAPS), which afford higher spectral resolution.

In order to measure $(D/H)_{ISM}$, one can also observe deuterated molecules such as HD, DCN, *etc*, and form the ratio of the deuterated molecule column density to its non-deuterated counterpart (H_2, HCN, *etc*). More than twenty different deuterated species have been identified in the ISM, with abundances relative to the non-deuterated counterpart ranging from 10^{-2} to 10^{-6}. Conversely, this means that fractionation effects are important. As a consequence, this method cannot currently provide a precise estimate of the true interstellar D/H ratio. Rather, this method is used in conjunction with estimates of the $(D/H)_{ISM}$ ratio to gather information on the chemistry of the ISM.

Another way to derive the $(D/H)_{ISM}$ ratio comes through radio observations of the hyperfine line of D I at 92 cm. The detection of this line is extremely difficult, but it would allow one to probe more distant interstellar media than the local medium discussed below. However, because a large column density of D is necessary to provide even a weak spin-flip transition, these observations aim at molecular complexes. As a result, the upper limit derived toward Cas A (Heiles et al. 1993) $(D/H)_{ISM} \leq 2.1 \times 10^{-6}$ may as well result from a large differential fraction of D and H being in molecular form in these clouds, as from the fact that one expects the D/H ratio to be lower closer to the galactic center (since D is destroyed by stellar processing). The most recent result is the low significance detection of interstellar D I 92 cm emission performed by Chengalur et al. (1997) toward the galactic anticenter, giving $(D/H)_{ISM} = 3.9 \pm 1.0 \times 10^{-5}$.

Therefore, the most reliable estimate of $(D/H)_{ISM}$ remains the observation of the atomic transitions of D and H of the Lyman series in the far-UV. The relatively low resolution of the *Copernicus* spectra (~ 15 km s^{-1}) usually left the velocity structure unresolved, which could lead to significant errors. These uncertainties were reduced when HST and IMAPS echelle observations provided resolving powers high enough (3.5 to 4 km s^{-1}) to unveil the velocity structure.

Either the Lyman α lines emissions from cool stars or the continua from hot stars have been used as background sources. Whereas cool stars can be selected in the solar vicinity, luminous hot stars are located further away, with distances $\gtrsim 100$ pc. Therefore, the line of sight to hot stars generally comprises more absorbing components than cool stars. However, for cool stars, the modeling of the stellar flux is usually much more difficult than for hot stars. Moreover, lines of species such as N I and O I that lie close to Lyman α cannot be observed, as the flux drops to zero on either side of Lyman α. Hence, in the case of cool stars, the line of sight velocity structure in H I typically has to be traced with Fe II and Mg II ions and this is usually not a good approximation. In contrast, N I and O I were shown to be good tracers of H I in the ISM (Ferlet 1981; York et al. 1983; Meyer, Cardelli & Sofia 1997; Meyer, Jura & Cardelli 1998; Sofia & Jenkins 1998) and hot stars are particularly interesting targets in that respect.

In any case, both types of background sources have offered some remarkable results. In the direction to the cool star Capella, Linsky et al. (1993; 1995) have obtained, using HST: $(D/H)_{ISM} = 1.60 \pm 0.09^{+0.05}_{-0.10} \times 10^{-5}$. On this line of sight, only one absorbing component was detected, the Local Interstellar Cloud (LIC), in which the solar system is embedded (Lallement & Bertin 1992, Ferlet 1999). Several more cool stars have been observed with HST, all compatible with the Capella evaluation (Linsky et al. 1995: Procyon; Linsky & Wood 1996: α Cen A, α Cen B; Piskunov et al. 1997: HR 1099, 31 Com, β Cet, β Cas; Dring et al. 1997: β Cas, α Tri, ϵ Eri, σ Gem, β Gem, 31 Com). The most precise of these measurements has been obtained toward HR 1099 by Piskunov et al. (1997): $(D/H)_{ISM} = 1.46 \pm 0.09 \times 10^{-5}$. None of the other results is accurate enough to place any new constraints on the Linsky et al. (1993; 1995) evaluation.

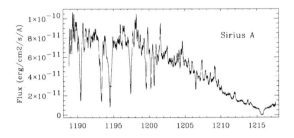

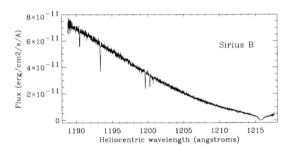

Figure 1. **The Sirius A (top) and B (bottom) Lyman α lines.** These two spectral line profiles underline the interrest of using white dwarfs as background stars to evaluate D/H. In effect in the Sirius A spectrum (top) it is not clear that any interstellar line is even present. In the Sirius B spectrum (bottom) on the contrary they show up obviously (the NI triplet around 1200 Å or the two SiII lines close two 1190 Å). In both spectra the interstellar hydrogen (HI) absorption at 1215.66 Å is clearly seen with the weak deuterium (DI) feature at 1215.33 Å (from Hébrard et al. 1999).

Recently, new observations by HST and IMAPS have become available. HST observations of white dwarfs instead of hot or cool stars can be used to circumvent most of the afore-mentionned difficulties. White dwarfs can be chosen near the Sun and they can also be chosen in the high tem-

perature range, so as to provide a smooth stellar profile at Lyman α. At the same time, the N I triplet at 1200 Å as well as the O I line at 1302 Å are available. Such observations have now been conducted using HST toward three white dwarfs: G191-B2B (Lemoine et al. 1996; Vidal-Madjar et al. 1998), Hz 43 (Landsman et al. 1996) and Sirius B (Hébrard et al. 1999).

Toward G191-B2B, Vidal-Madjar et al. (1998) detected three absorbing clouds using HST-GHRS 3.5 km s^{-1} spectral resolution data. Assuming that all three absorbing components shared the same (D/H)$_{ISM}$ ratio, they measured at Lyman α (D/H)$_{ISM}$ = $1.12 \pm 0.08 \times 10^{-5}$. There is a clear discrepancy between this ratio and that observed toward Capella by Linsky et al. (1993; 1995). As it turns out, one of the three absorbers seen toward G191-B2B is the LIC, also seen toward Capella. Moreover, the angular separation of both targets is 7°. One should thus expect to see the same (D/H)$_{ISM}$ ratio in both LIC line of sights. When this constraint is included in the three-component fit, Vidal-Madjar et al. (1998) find that the average (D/H)$_{ISM}$ in the other two absorbers is $\sim 0.9 \pm 0.1 \times 10^{-5}$. Finally, it is important to note that Vidal-Madjar et al. (1998) re-analyzed the dataset of Linsky et al. (1993; 1995) toward Capella, using the same method of analysis as toward G191-B2B and confirmed the previous estimate. Therefore, the conclusion is that the (D/H)$_{ISM}$ ratio varies by at least $\sim 30\%$ within the local interstellar medium, either from cloud to cloud, and/or within the LIC.

Using HST-GHRS observations, Hébrard et al. (1999) detected two interstellar clouds toward Sirius A and its white dwarf companion Sirius B (see Fig. 1), one of them being identified as the LIC, in agreement with previous HST observation of Sirius A by Lallement et al. (1994). As in the case of G191-B2B, the interstellar structure of this sightline, which is assumed to be the same toward the two stars (separated by less than 4 arcsec at the time of the observation), is constrained by high spectral resolution data of species such as O I, N I, Si II or C II. Whereas the deuterium Lyman α line is well detected in the LIC with an abundance in agreement with the one of Linsky et al. (1993, 1995), no significant D I line is de-

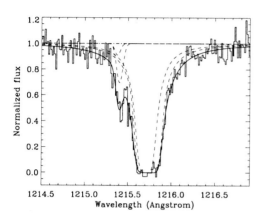

Figure 2. **The reconstructed Sirius A-B Lyman α line.** To avoid the contamination by the stellar wind of Sirius A the blue side of the line is taken from the Sirius B profile, while the red side is taken from the Sirius A line to avoid the perturbation of the red side of the Sirius B line by its photospheric shape distorted by its important gravitational redshift. From this composit spectrum it is possible to simulate the profile by two interstellar components which contributions are shown by the dashed lines. The strongest and redward component is the LIC while the other one should be somewhere between the LIC boundary and Sirius. It's deuterium content seems to be extremely low (see text, from Hébrard et al. 1999).

tected in the other cloud. However, the Lyman α lines toward Sirius A and Sirius B are not simple. Indeed an excess of absorption is seen in the blue wing of the Sirius A Lyman α line and interpreted as the wind from Sirius A. In its white dwarf companion, an excess in absorption is seen in the red wing and interpreted as the core of the Sirius B photospheric Lyman α line. A composite Lyman α profile could nonetheless be constructed (see Fig. 2) and the $(D/H)_{ISM}$ measured in the second cloud is $(D/H)_{ISM} = 0.5^{+1.1}_{-0.5} \times 10^{-5}$ (90% confidence level). The rather large error bar stems primarily from the fact that only medium resolution data were available for the Lyman α region.

Finally, IMAPS on the space shuttle ORFEUS-SPAS II mission was used by Jenkins et al. (1999) to observe at high spectral resolution (4 $km\,s^{-1}$) the Lyman δ and Lyman ϵ lines toward δ Ori. These data allowed an accurate measurement of the D I column density. Together with a new and accurate measurement of the H I column density from Lyman α spectra of δ Ori in the IUE archive, Jenkins et al. (1999) found the value $(D/H)_{ISM} = 0.74^{+0.19}_{-0.13} \times 10^{-5}$, at a 90% confidence level (c.l.), which confirms the *Copernicus* result obtained by Laurent et al. (1979). Compared to Capella (Linsky et al. 1993; 1995) and HR 1099 (Piskunov et al. 1997), this value is very low. This suggests that variations by ~50% are possible in the local interstellar medium.

Using the same analysis techniques as Jenkins et al. (1999), IMAPS was also used combined with IUE archive toward two other stars, γ^2 Vel and ζ Pup to yield the first results $(D/H)_{ISM} = 2.1^{+0.36}_{-0.30} \times 10^{-5}$ and $(D/H)_{ISM} = 1.6^{+0.28}_{-0.23} \times 10^{-5}$, respectively (Jenkins et al. 1998; Sonneborn et al. 1999). The value for γ^2 Vel is marginally inconsistent with the lower value toward Capella, and this disparity may be substantiated further when the error estimates become more refined. We also note that the γ^2 Vel result confirms previous estimates of York & Rogerson (1976), while the ζ Pup result is only in marginal agreement with the Vidal-Madjar et al. (1977) evaluation, both made with *Copernicus*.

5. Interstellar D/H variations

There is now firm evidence for variations of the $(D/H)_{ISM}$ ratio, able to reach $\sim 50\%$, over scales as small as ~ 10 pc. This fact had already been suggested by early *Copernicus* and IUE data, although it was not known whether this was due to the inadequacy of the data and the complexity of the problem, or to real physical effects. The dispersion of all published $(D/H)_{ISM}$ ratios, ranging from 0.5×10^{-5} to 4×10^{-5}, was thus not universally accepted as real (Mc Cullough 1992). Even if some of this scatter may be accounted for by systematic errors, as I have argued above, but I believe that at least part of it is real.

Actually, one should recall that time variations of the $(D/H)_{ISM}$ ratio have already been reported toward ϵ Per (Gry et al. 1983). They were interpreted as the ejection of high velocity hydrogen atoms from the star, which would contaminate the deuterium feature. Such an effect can only mimic an enhancement the D/H ratio, and it is thus worth noting that in at least five cases, the $(D/H)_{ISM}$ ratio was found to be really low: $0.9 \pm 0.1 \times 10^{-5}$ in two components toward G191-B2B (Vidal-Madjar et al. 1998, see Section 4); $0.8 \pm 0.2 \times 10^{-5}$ toward λ Sco (York 1983); $0.5 \pm 0.3 \times 10^{-5}$ toward θ Car (Allen et al. 1992); $0.7 \pm 0.2 \times 10^{-5}$ and $0.65 \pm 0.3 \times 10^{-5}$ toward δ and ϵ Ori (Laurent et al. 1979), recently confirmed in the case of δ Ori by Jenkins et al. (1999): $(D/H)_{ISM} = 0.74^{+0.19}_{-0.13} \times 10^{-5}$ (90% c.l.). Two other lines of sight seem to give low values for the D/H ratio, albeit with larger error bars: $0.5^{+1.1}_{-0.5} \times 10^{-5}$ (90% c.l.) in one of the two components toward Sirius (Hébrard et al. 1999, see Section 4), $0.8^{+0.7}_{-0.4} \times 10^{-5}$ toward BD+28 4211 (Gölz et al. 1998). All the above authors discussed possible systematics but concluded that none of the identified ones could explain such low values of $(D/H)_{ISM}$.

Plausible causes of variations were discussed in detail by Lemoine et al. (1999). They are summarized below under two sets, the first one more related to possible explanations for small scales variations and the second set linked to larger scales galactic or cosmologic fluctuations:

Small scale variations

- Molecular fractionation effects (Watson 1973).

- Vidal-Madjar et al. (1978), and Bruston et al. (1981) have suggested that the anisotropic flux in the solar neighborhood, combined with a differential effect of radiation pressure on H I and D I atoms, would result in the spatial segregation of D I vs. H I.

- Jura (1982) has suggested that the adsorption of D I and H I onto dust grains could be selective.

- Copi, Schramm & Turner (1995), and Copi (1997) have devised a stochastic approach to chemical evolution, in which they compute the evolution of a particular region of space in Monte-Carlo fashion.

- Along similar lines of thought, Lemoine et al. (1999) propose that stellar ejecta like planetary nebula or cool giant winds could introduce inhomogeneities in the $(D/H)_{ISM}$ ratio.

- Recently, Mullan & Linsky (1999) have argued that production of deuterium in stellar flares, by radiative capture of a proton by a free neutron, can produce a non-negligible source of non-primordial deuterium in the ISM, and possibly explain the observed variations.

Large scale variations

- First, Epstein, Lattimer & Schramm (1976) showed that no realistic astrophysical system could produce deuterium by *nucleosynthesis* or *spallation* mechanisms, without, in the latter case, overproducing Li.

- Photodisintegration of ^{4}He can lead to production of deuterium, as exemplified by Boyd, Ferland & Schramm (1989) and Jedamzik & Fuller (1997). The production should be however negligible.

- Cassé & Vangioni-Flam (1998; 1999) have argued that blazars could however actually influence absorbers in a significant way if the absorber is a blob of matter expelled by the central engine. They also argue that creation as well as destruction of deuterium can occur, depending on the γ-ray spectrum.

- Jedamzik & Fuller (1995, 1997) have pointed out that primordial isocurvature baryon fluctuations on mass scales $\leq 10^5 - 10^6 \, M_\odot$ could produce variations by a factor 10 on these scales, and variations of order unity on galactic mass scales $\sim 10^{10} - 10^{12} \, M_\odot$. However, this attractive and original scenario would not apply to $(D/H)_{ISM}$ ratios on very small spatial scales $\sim 10 \, pc$.

The above mechanisms do not agree as to whether D should be enhanced or depleted with respect to H, if any of them operates. Therefore, one cannot conclude which one of the observed interstellar abundances, if any, is more representative of a cosmic abundance that would result solely from Big-Bang production followed by star formation.

6. Discussion

Taking for granted that variations in the $(D/H)_{ISM}$ ratio exist, do we expect to see a similar effect in QSO absorption line systems, and if yes, how would it affect the estimate of the primordial abundance of deuterium?

No one knows the answer to this question, mainly because the nature and the physical environment of absorption systems at high redshift may be very different from that of interstellar clouds in the solar neighborhood, and because the origin of the variation of the $(D/H)_{ISM}$ ratio is not understood for the moment.

The $(D/H)_{ISM}$ ratio is measured in interstellar clouds that typically show: $N(H\,I) \sim 10^{18} \, cm^{-2}$, $n_H \sim 0.1 \, cm^{-3}$, ionization $n_{HI} \sim n_{HII}$, $T \sim 10^4 \, K$, size $L \sim 1 \, pc$, and mass $M \sim 10^{-2} \, M_\odot$. Although the Lyman limit systems have a similar column densities, their physical characteristics may be very different. One opinion is that these

systems are associated with extended gaseous haloes, as one often finds a galaxy at the redshift of the absorber with an impact parameter $R \sim 30 h^{-1} \, kpc$ (Bergeron & Boissé 1991; Steidel 1993).

However, it is not known whether this absorption is continuous and extends on scales $\sim R$, or whether the absorption is due to discrete clouds sufficiently clustered on the scale $\sim R$ to produce absorption with probability ≈ 1. In particular, York et al. (1986) and Yanny & York (1992) have suggested that QSO Lyman α absorption (not necessarily Lyman limit systems) occur in clustered dwarf galaxies undergoing merging. In this case, one expects the absorbers to be much like galactic clouds, in particular, of small spatial extent. It is also usually believed that the QSO UV background is responsible for the ionization properties of these absorbers, in which case one typically derives low densitites $n_H \sim 10^{-3} \, cm^{-3}$, which translates into large masses $\sim 10^8 \, M_\odot$ (plus or minus a few orders of magnitude) if the clouds have a large radius $\sim 30 \, kpc$. However, there are other models for the ionization of these Lyman limit systems. For instance, Viegas & Friaça (1995) have proposed a model where Lyman limit systems originate in galactic haloes, have sizes $\sim$ few kpc, hydrogen densities $n_H \sim 10^{-1}$, and the ionization results from the surrounding hot gas. Lyman limit systems are shrouded in mystery.

Despite these large uncertainties, one can establish a few interesting points. First, the depletion of deuterium by contamination of low mass stellar ejecta has been ruled out by Jedamzik & Fuller (1997). Indeed, the QSO absorbers where D has been detected have been shown to be very metal-poor. The metallicity inferred, typically $[C/H] \leq -2.0$, implies that no more than 1% of the gas been cycled through stars. Note that Timmes et al. (1997) suggest that the incomplete mixing and the smallness of the QSO beam could introduce non-negligible variations in $(D/H)_{QSO}$ ratios.

Differential radiation pressure could affect measured $(D/H)_{QSO}$ ratios if Lyman limit systems are discrete clouds, and their radius is not too large. Indeed, the primary requirement of the model of Vidal-Madjar et al. (1978) and Bruston et al. (1981), is that the radiation flux be anisotropic, and for maximum efficiency, that the line of sight cross the absorber perpendicularly to the direction of the radiation flux. As it turns out, QSO absorbers chosen for measurements of D/H ratios fulfil these criteria. In effect, these sytems are selected for D/H studies if their line of sight is as trivial as possible. This means that the absorber has to be isolated, which, in geometrical terms means, for a spherical distribution, that it has to lie on the boundary. If the radiation flux arises from the central part of the spherical system, it is anisotropic on a cloud at the boundary, and moreover, the line of sight is effectively perpendicular to the flux impinging on the cloud; otherwise, one would expect multiple absorbers on the line of sight. Following Bruston et al. (1981), the diffusion velocity of deuterium atoms is: $v_D \sim 1\,\mathrm{pc.Myr}^{-1}\Phi_{-6}n_{-1}^{-1}T_4^{-1/2}$, for Φ_{-6} in units of 10^{-6} photons/cm^2/s/Hz, n_{-1} total density in units of 0.1 atoms.cm^{-3}, and T_4 in units of 10^4 K.

In the case of the York *et al.* (1986) model, one expects a radiation flux corresponding to $\sim 10^2 - 10^4$ O stars, impinging on a cloud located ~ 1kpc from the center of the dwarf galaxy. This gives a diffusion velocity $v_D \sim 0.02 - 2\,\mathrm{pc/Myr}$. Diffusion, hence segregation, can thus occur over scales of a ~ 1 pc, a typical cloud size, as the relevant timescale is the crossing time ~ 20 Myr for a cloud circulation velocity ~ 50 km/s. However, for the model developed by Viegas & Friaça (1995), the typical diffusion distance for deuterium atoms is $\sim 1 - 10$ pc, for a flux $\Phi_{-6} \sim 0.01 - 0.1$, sustained on a star formation timescale $\sim 10^8$ yrs. This distance is small compared to the modeled cloud size $\sim$ few kpc, and one thus does not expect segregation of deuterium.

Although the above numbers are very qualitative, mainly because of the uncertainties inherent to our knowledge of Lyman limit systems, one cannot rule out an effect of anisotropic radiation. It is actually interesting that the criteria according to which Lyman limit systems are chosen for D/H studies, coincide with those for a maximum effect of radiation pressure. In moderately ion-

ized and small sized ($\sim 1 - 10$ pc) regions, the deuterium abundance could be depleted by a factor 2.

Finally, whatever the right value of the primordial deuterium abundance, that is, either low $\sim 3.5 \times 10^{-5}$, or high $\sim 10^{-4}$, there is satisfying agreement with both Big-Bang nucleosynthesis and the predictions of other light elements abundances, and with chemical evolution and the interstellar abundances of deuterium. See, *e.g.*, Schramm & Turner (1998) for a discussion of the agreement of a low (D/H)$_{QSO}$ with primordial ^{4}He determinations (and statistical errors), and primordial ^{7}Li, and its cosmological implications. High deuterium abundances are known to provide very good agreement with BBN predictions for ^{4}He and ^{7}Li. Although they predict significant astration of deuterium: (D/H)$_{QSO}$/(D/H)$_{ISM}\sim 5 - 10$, it is also known there exist viable chemical evolution models able to account for such a large destruction (*e.g.* Vangioni-Flam & Cassé 1995; Timmes et al. (1997); Scully et al. 1997).

7. Conclusion

The different D/H evaluations reviewed or presented here are shown in Fig. 3 (from Lemoine et al. 1999), as a function of (approximate) time. This figure seems to reveal a trend of decreasing deuterium abundance with time, predicted as early as 1971 (Truran & Cameron 1971). However, if one looks more closely at Fig. 3, there are discrepancies between different evaluations of deuterium abundances at similar cosmic time, which, as I have argued in the case of ISM measurements, cannot be always accounted for in terms of measurements systematics.

Nevertheless, the trend indicated in Fig. 3 seems to show that we are converging toward a reasonable (at least understandable) picture of the cosmic history of deuterium.

FUSE, scheduled for launch in mid 1999, will sharpen this picture, and fill in the gaps to construct a curve of evolution of the abundance of deuterium *vs* time/metallicity. The FUSE Science Team intends to conduct a comprehensive study of the deuterium abundance in the Galaxy through Lyman series absorption of D I between

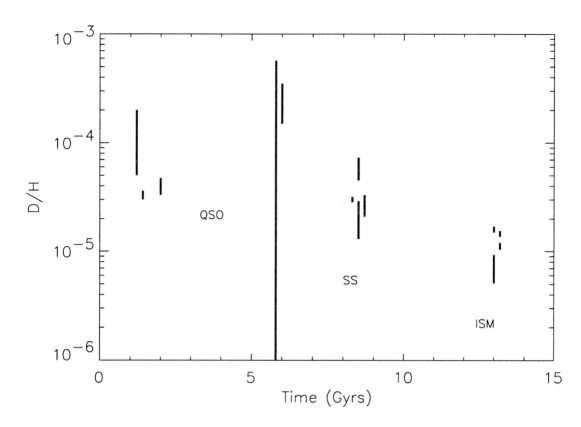

Figure 3. **Deuterium abundance measurements.** The different D/H evaluations are shown as a function of time (for $\Omega_0 = 1$, $q_0 = 0.5$, $H_0 = 50$ km/s/Mpc). The primordial measurements plotted (QSO) are from Burles & Tytler (1998a,b) and Songaila (1997) [high redshifts] and from Webb et al. (1997) and Tytler et al. (1999) [moderate redshift]. Pre-solar values plotted (SS) are from Gautier & Morel (1997) [solar wind] and Lellouch et al. (1997), Ben Jaffel et al. (1997) and Mahaffy et al. (1998) [Jupiter]. Interstellar values (ISM) plotted are the ones from Linsky et al. (1995) [Capella], Piskunov et al. (1997) [HR 1099], Vidal-Madar et al. (1998) [G191-B2B] and Jenkins et al. (1999) [δ Ori].

912 and 1187 Å. Access to a suite of lines in the series provides much stronger constraints on $N(D\,\textsc{i})$ and $N(H\,\textsc{i})$ than single line (*i.e.*, Lyman α) observations alone. The bandpass also contains a large number of lines of $O\,\textsc{i}$, $N\,\textsc{i}$, and $Fe\,\textsc{ii}$ that can be used to trace the metallicity and dust content of the absorbers studied.

The primary goal of the FUSE D/H program is to link the destruction of deuterium to the physical and chemical properties of the interstellar gas. This objective is critical to successful galactic chemical evolution models since astration of deuterium, metal production, and mixing/recycling of the ISM are key ingredients in the models. FUSE observations of D/H in environments with different chemical histories will help to reveal the effectiveness of astration and its dependence upon environmental factors (*e.g.* metallicity, star-formation). A study of regional variations may reveal evidence that supports the proposal about the differential effect of radiation pressure. Finally, D/H measurements in regions of low metallicity will be particularly important benchmarks for relating the high redshift D/H values to present epoch values.

FUSE will be capable of observing deuterium in distant gas clouds beyond the solar neighborhood clouds explored by *Copernicus*, HST, and IMAPS. Therefore, it should be possible to search for large scale variations in D/H related to global star formation and metal gradients, as well as small scale variations in selected regions due to incomplete mixing of the interstellar gas or deuterium decrements in the ejecta of stars.

FUSE, together with HST-STIS and IMAPS in space and the new large telescopes like the Keck, the VLT and several others from the ground, should give access to more precise D/H evaluations and greatly clarify the problem of the chemical evolution of deuterium, and hence much better constrain our understanding of the primordial D/H value.

References :

Adams, T. F.: 1976, A&A 50, 461

Allen, M. M., Jenkins, E. B., & Snow, T. P.: 1992, ApJS 83, 261

Alpher, R. A., Bethe, H. A., & Gamov, G.: 1948, Phys. Rev. 73, 803

Beer, R., & Taylor, F.: 1973, ApJ 179, 309

Ben Jaffel, L., et al.: 1994, Bull. Am. Astron. Soc. 26, 1100

Ben Jaffel, L., et al.: 1997, *The Scientific return of GHRS*

Bergeron, J., Boissé, P.: 1991, AA 243, 344

Black, D. C.: 1971, Nature Physic Sci. 234, 148

Bockelée-Morvan, D., et al.: 1998, Icarus 133, 147

Boyd, R.N., Ferland, G.J., Schramm, D.N.: 1989, ApJ 336, L1

Bruston, P., Audouze, J., Vidal-Madjar, A., & Laurent, C.: 1981, ApJ 243, 161

Burbidge, G. & Hoyle, F.: 1998, ApJ 509, in press

Burles, S., Kirkman, D., Tytler, D.: 1999, ApJ 519, in press

Burles, S., & Tytler, D.: 1998a, ApJ 499, 699

Burles, S., & Tytler, D.: 1998b, ApJ 507, 732

Burles, S., & Tytler, D.: 1998c, , Proceedings of the Second Oak Ridge Symposium on Atomic & Nuclear Astrophysics, Eds. A. Mezzacappa, `astro-ph/9803071`

Burles, S., Tytler, D.: 1998d, Sp. Sc. Rev. 84, 65

Burles, S., Tytler, D.: 1998e, AJ 114, 1330

Carswell, R. F., Rauch, M., Weymann, R. J., Cooke, A. J., & Webb, J. K.: 1994, MNRAS 268, L1

Cassé, M., & Vangioni-Flam, E.: 1998, in Structure and Evolution of the Intergalactic Medium from QSO Absorption Line Systems, Eds. Petitjean, P., & Charlot, S., 331

Cassé, M., Vangioni-Flam, E.: 1999, in 3rd Integral Workshop (ESA), Taormina Sept.1998, in press

Cesarsky, D.A., Moffet, A.T., & Pasachoff, J.M.: 1973, ApJ 180, L1

Chengalur, J. N., Braun, R., & Burton, W. B. : 1997, A&A 318, L35

Copi, C. J., Schramm, D. N., & Turner, M. S.: 1995, ApJ 455, L95

Copi, C.J: 1997, ApJ 487, 704

Cox, D.P., Reynolds, R.J.: 1987, ARAA 25, 303

Dring, A.R., Linsky, J., Murthy, J., Henry, R. C., Moos, W., Vidal-Madjar, A., Audouze, J., & Landsman, W.: 1997, ApJ 488, 760

Encrenaz, Th., et al. : 1996, A&A 315, L397

Encrenaz, Th.: 1999, in International Conference

on 3K Cosmology, F. Melchiori, edt. AIP Publ., in press

Epstein, R. I., Lattimer, J. M., & Schramm, D.N.: 1976, Nature 263, 198

Ferlet, R.: 1981, A&A 98, L1

Ferlet, R.: 1999, Astron. Astrophys. Rev., astro-ph/9902258, in press

Fuller, G. M., & Shu, X.: 1997, ApJ 487, L25

Gautier, D., & Morel, P.:1997, A&A 323, L9

Geiss, J., & Reeves, H.: 1972, A&A 18, 126

Gölz, M., et al.: 1998, Proc. IAU Colloq. 166, The Local Bubble and Beyond, Eds. Breitschwerdt, D., Freyberg, M. J., Trumper, J., 75

Gnedin, N. Y., Ostriker, J. P.: 1992, ApJ 400, 1

Grevesse, N., Noels, A.: 1993, in Origin and evolution of the elements, eds M. Cassé, N. Prantzos, E. Vangioni-Flam (CUP), p15

Griffin, M. J., et al. : 1996, A&A 315, L389

Gry, C., Laurent, C., & Vidal-Madjar, A.: 1983, A&A 124, 99

Hébrard, G., Mallouris, C., Vidal-Madjar, A., et al.: 1999, to be submitted to A&A

Heiles, C., McCullough, P., & Glassgold, A.: 1993, ApJS 89, 271

Hogan, C. J.: 1997, Proceedings of the ISSI workshop, Primordial Nuclei and their Galactic Evolution, astro-ph/9712031

Hogan, C. J.: 1998, Space Sci. Rev. 84, 127

Jedamzik, K., Fuller, G.M.: 1995, ApJ 452, 33

Jedamzik, K., Fuller, G.M.: 1997, ApJ 483, 560

Jenkins, E.B.: 1996, in Cosmic Abundances, eds S.S. Holt, G. Sonneborn (ASP Conf. Series, 99), p90

Jenkins, E. B., et al.: 1998, in ESO workshop – Chemical Evolution from Zero to High Redshift, Oct. 14-16

Jenkins, E. B., Tripp, T. M., Wozniak, P. R., Sofia, U. J., & Sonneborn, G.: 1999, submitted to ApJ, preprint astro-ph/9901403

Jura, M. A.: 1982, *Four Years of IUE Research*, NASA CP 2238, 54

Kingsburgh, R.L., Barlow, M.J.: 1994, MNRAS 271, 257

Lallement, R., & Bertin, P.: 1992, A&A 266, 479

Lallement, R., Bertin, P., Ferlet, R., Vidal-Madjar, A., & Bertaux, J.L.: 1994, A&A 286, 898

Landsman, W., Sofia, U. J., & Bergeron, P.: 1996, Science with the Hubble Space Telescope - II, STScI, 454

Laurent, C., & Vidal-Madjar, A.: 1981, Haute résolution spectrale en astrophysique, Deuxième Colloque National Français du Télescope Spatial, Orsay, 145

Laurent, C., Vidal-Madjar, A., & York, D. G.: 1979, ApJ 229, 923

Lecluse, C., Robert, F., Gautier, D., & Guiraud, M.: 1996, Plan. Space Sci. 44, 1579

Lellouch, E., et al.: 1997, Proceedings of the ISO workshop at Vispa, Oct. 97, ESA-SP 419, 131

Lemoine, M., Vidal-Madjar, A., Bertin, P., Ferlet, R., Gry, C., Lallement, R.: 1996, A&A 308, 601

Lemoine, M., et al.: 1999, New Astronomy, in press, preprint astro-ph/9903043

Levshakov, S.A.: 1998, to appear in the Proceedings of the Xth Rencontres de Blois, astro-ph/9808295

Levshakov, S.A., Tytler, D., Burles, S.: 1999, AJ submitted, preprint astro-ph/9812114

Linsky, J. L., Brown, A., Gayley, K., Diplas, A., Savage, B. D., Ayres, T. R., Landsman, W., Shore, S. W., Heap, S. R.: 1993, ApJ 402, 694

Linsky, J. L., Diplas, A., Wood, B. E., Brown, A., Ayres, T. R., Savage, B. D.: 1995, ApJ 451, 335

Linsky, J., & Wood, B. E.: 1996, ApJ 463, 254

McCullough, P.R.: 1992, ApJ 390, 213

Mahaffy, P. R., Donahue, T. M., Atreya, S. K., Owen, T. C., & Nieman, H. B.: 1998, Space Sci. Rev. 84, 251

Meier, R., et al.: 1998, Science 279, 842

Meyer, D.M., Cardelli, J.A., Sofia, U.J.: 1997, ApJ 490, L103

Meyer, D.M., Jura, M., Cardelli, J.A.: 1998, ApJ 493, 222

Mullan, D.J., Linsky, J.L.: 1999, ApJ 511, 502

Niemann, H. B., et al.: 1996, Science 272, 846

Owen, T., Lutz, B., & De Bergh, C.: 1986, Nature 320, 244

Piskunov, N., Wood, B. E., Linsky, J.L., Dempsey, R. C., & Ayres, T. R.: 1997, ApJ 474, 315

Pottasch, S.R.: 1983, Planetary Nebulae (D. Reidel, Astrophysics and Space Science Library)

Reeves, H., Audouze, J., Fowler, W. A., & Schramm, D. N.: 1973, ApJ 179, 909

Rogerson, J., & York, D.: 1973, ApJ 186, L95

Routly, P.M., Spitzer, L.: 1952, ApJ 115, 227

Rugers, M., Hogan, C.J.: 1996, ApJ 459, L1

Schramm, D.N.: 1998, Sp. Sc. Rev. 84, 3

Schramm, D.N., Turner, M.S.: 1998, Rev. Mod. Phys. 70, 303

Scully, S. T., Cassé, M., Olive. K. A., & Vangioni-Flam, E.: 1997, ApJ 476, 521

Smith, WM. H., Schemp, W. V., & Baines, K. H.: 1989, ApJ 336, 967

Sofia U.J., Jenkins, E.B.: 1998, ApJ 499, 951

Songaila, A., Cowie, L. L., Hogan, C. J., & Rugers, M.: 1994, Nature 368, 599

Songaila, A.: 1998, in Structure and evolution of the IGM from QSO absorption lines, eds. P. Petitjean & S. Charlot (Frontières, Paris)

Sonneborn, G., et al.: 1999, ApJL, in preparation

Steidel, C.C.: 1993, in The evolution of galaxies and their environment, eds. J. Shull, H.A.Thronson (Kluwer, Dordrecht), p263

Truran, J.W., Cameron, A.G.W.: 1971, Ap. Sp. Sci., 14, 179

Timmes, F.X., Truran, J.W., Lauroesch, J.T., York, D.G.: 1997, ApJ 476, 464

Tytler, D., Burles, S., Lu, L., Fan, X.-M., Wolfe, A., & Savage, B.: 1999, AJ 117, 63

Vallerga, J.V., Vedder, P.W., Carig, N., Welsh, B.Y.: 1993, ApJ 411, 729

Vangioni-Flam, E., & Cassé, M.: 1995, ApJ 441, 471

Vidal-Madjar, A., Laurent, C., Bonnet, R.M., & York, D.G.: 1977, ApJ 211, 91

Vidal-Madjar, A., Laurent, C., Bruston, P., & Audouze, J.: 1978, ApJ 223, 589

Vidal-Madjar, A., Ferlet, R., Lemoine, M.: 1998, Sp. Sc. Rev. 84, 297

Vidal-Madjar, A., Lemoine, M., Ferlet, R., Hébrard, G., Koester, D., Audouze, J., Cassé, M., Vangioni-Flam, E., & Webb, J.: 1998, A&A 338, 694

Viegas, S.M., Friaça, A.C.S.: 1995, MNRAS 272, L35

Watson, W. D.: 1973, ApJ 182, L73

Webb, J.K., Carswell, R. F., Irwin, M. J., Penston, M. V.: 1991, MNRAS 250, 657

Webb, J. K., Carswell, R. F., Lanzetta, K. M., Ferlet, R., Lemoine, M., Vidal-Madjar, A., & Bowen, D. V.: 1997, Nature 388, 250

Yanny, B., York, D.G.: 1992, ApJ 391, 569

York, D.G., Rogerson, J.B. Jr: 1976, ApJ 203, 378

York, D.G.: 1983, ApJ 264, 17

York, D. G., Spitzer, L., Bohlin, R. C., Hill, J., Jenkins, E. B., Savage, B. D., Snow, T. P.: 1983, ApJ 266, L55

York, D. G., Ratcliff, S., Blades, J. C., Wu, C. C., Cowie, L. L., & Morton, D. C.: 1984, ApJ 276, 92

York, D.G., Dopita, M., Green, R., Bechtold, J.: 1986, ApJ 311, 610

Highlight talks

ELSEVIER

Nuclear Physics B (Proc. Suppl.) 80 (2000) 135–142

NUCLEAR PHYSICS B
**PROCEEDINGS
SUPPLEMENTS**

www.elsevier.nl/locate/npe

Hypernovae, Collapsars, and Gamma-Ray Bursts

D. H. Hartmann [a] and A. I. MacFadyen [b]

[a]Department of Physics and Astronomy, Clemson University, Clemson, SC 29634

[b]Astronomy Department, UC Santa Cruz, Santa Cruz, CA 95064

Cosmic gamma-ray bursts now appropriately hold the distinction of being the "largest explosions in the universe". Their afterglows are often brighter than supernovae, thus often referred to as "hypernovae". Their kinetic energies may also be greater, or at least highly collimated, and require a new source of energy. Recent photometric and spectroscopic observations of the afterglow emission have provided a major breakthrough in our understanding of these powerful explosions. The data place at least some bursts at large distances and in association with faint host galaxies. But what is (or are) the underlying cause(s) of these violent events? The answer to this question remains uncertain, but several theoretical arguments point towards the creation of hyperaccreting black holes with accretion rates from 10^{-4} to 10 solar masses per second, whose accretion disks produce narrow jets of relativistically expanding plasma. We review the basic concepts of one of these models, the "collapsar model".

Discovered more than three decades ago (Klebesadel *et al.* 1973) gamma-ray bursts (GRBs) remain mysterious. However, distance estimates for a subset of bursts, all of the long, comparatively soft variety, have been obtained in recent years, placing GRBs firmly into the realm of cosmology. Since the launch of the Compton Observatory in 1991, data from BATSE provided growing evidence for a cosmological distance scale. This followed from the isotropy of their angular distribution (Figure 1) combined with a deficit of faint bursts relative to expectations based on a uniform distribution in Euclidean space. For several years it was debated whether or not the data ruled out the possibility of a very extended Galactic halo distribution of GRBs (this "great debate" is described in several articles in Vol. 107 of the Proceedings of the Astronomical Society of the Pacific). The debate ended when distances were finally established through the discovery of fading afterglow emission (see Hartmann 1999 for a recent review).

1. GRBs in the Beppo-SAX Era

With the launch of the Italian-Dutch Beppo-SAX X-ray satellite (Piro *et al.* 1998) the study of GRBs underwent a major transition. SAX has so far established the existence of about a dozen

1892 BATSE Gamma-Ray Bursts

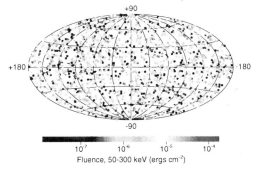

Figure 1. Distribution on the sky of 1892 GRBs observed by BATSE. The map shows burst locations (without indicating position uncertainties) in Galactic coordinates. There are no preferred directions, either of Galactic or extra-galactic significance. The color code indicates total burst fluence. GRBs are distributed isotropically, independent of their brightness, duration, spectrum, or any other characteristic. The BATSE 4B catalog (Paciesas *et al.* 1999) is available at astro/ph-9903205.

fading X-ray counterparts persisting at detectable levels for days. GRB970228 was the first for which an afterglow was detected. The arc-minute accuracy of the location allowed ground-based telescopes to successfully search for similarly fading counterparts in the optical (van Paradijs et al. 1997). Further studies with HST and Keck found extended emission region around the source, but its nature could not be established. Then history repeated itself with GRB970508, except that this time there was no evidence for an extended source. However, spectroscopy with Keck II found a set of absorption lines in the spectrum of the OT that suggested a burst redshift of z = 0.835 (Metzger et al. 1997). After those breakthroughs, the study of GRBs became an international multi-wavelengths effort, involving a very large and growing group of astronomers from the radio to the X-ray domain.

Skipping over many exciting details, we highlight a few bursts and what was learned from them. GRB971214 broke the redshift record: z = 3.43 (Kulkarni et al. 1998) which implies a total burst energy of over 10^{54} ergs (assuming isotropic emission) - clearly a challenge to theorists. GRB980425 surprised us with a potential association with a peculiar Type Ic supernova (SN1998bw), located in a nearby galaxy at z = 0.008 (Galama et al. 1998). Models that explain the observations (Woosley, Eastman, & Schmidt 1998; Iwamoto et al. 1998) require a very large kinetic energy, in excess of 10^{52} ergs if the explosion was isotropic, perhaps less if it was not (Höflich, Wheeler, & Wang 1998). High velocities are required to explain the spectrum, and about 0.5 $M_\odot$ of ^{56}Ni is needed to power the light curve. GRB980425 was likewise unusual. Assuming isotropic emission, a total energy of only 10^{48} ergs was released in about 20 s duration. This is much less than the canonical 10^{52} ergs required for bursts at z ~ 1-2 (see below).

Several more bursts were detected at redshifts of order unity and in essentially all cases was the candidate host galaxy faint (R ~ 22 − 26). In some cases estimates of the star formation rate in these hosts was possible from emission lines or the UV continuum, suggesting typical values of a few $M_\odot$ yr^{-1}. The recent event GRB990510

was found to be beyond z = 1.62, and observations with the VLT led to the discovery of polarization (1.7 %) in the optical band (Covino et al. 1999). GRB970508 was the first burst with an established radio afterglow (Frail et al. 1997), but there are now three other events with delayed radio emission. The radio lightcurves undergo a transition from a phase of rapid oscillations to a smooth decline. This behavior can be explained with scintillation as the radio waves propagate through the local ISM of the Milky Way (Goodman 1997).

Perhaps the most spectacular jem in the jewelry box is GRB990123, at z ~ 1.6. Figure 2 shows the HST image of the afterglow and an associated group of galaxies (Fruchter et al. 1999 . This burst established the existence of bright, truly simultaneous optical emission. The ROTSE experiment observed a peak brightness of V ~ 9 (Akerlof et al. 1999), almost visible to the naked eye ! And the story goes on as SAX continues to operate and as the HETE2 satellite is readied for launch later this year. After that, AGILE, GLAST, INTEGRAL, SWIFT, and other international efforts are underway to catch more bursts and their afterglows.

We now know that at least some GRBs are at cosmological distances, in fact the measured redshifts suggest that GRBs compete in distance with quasars and high redshift galaxies. It is possible that GRBs trace the cosmic star formation history, which would make GRBs a new and valuable tool of observational cosmology. So far, the OTs are surprisingly close to the hosts and most likely not associated with the centers of active galaxies. The high redshifts imply large energies, comparable in some cases to a solar mass rest energy. This is beyond the reach of most theoretical models, and geometric beaming into a small solid angle is often invoked as a way out of this energy dilemma. But even if beaming is significant, the explosive release of more than 10^{51} ergs of energy (electromagnetic or in particle form) into a small volume inevitably leads to an opaque region (a "fireball") which will rapidly expand to relativistic velocities. The Lorentz factors achieved in this expansion depend on the amount of baryon loading. The escape of hard photons from the burst

Figure 2. The HST image of the optical afterglow of GRB990123, obtained February 8/9, 1999 (Fruchter *et al.* 1999). The OT is the bright point source at the center of the image. North is up, East to the left. The extended objects seen in this image are probably a group of proto-galactic fragments in the process of merging with the dominant irregular galaxy of magnitude V = 24.2.

(EGRET aboard the Compton Observatory has established emission up to $\sim$ 20 GeV) requires bulk Lorentz factors greater than $\Gamma \sim 100$. The resulting relativistic beaming of the photons reduces the optical depth in various ways, which is needed to let the high energy photons escape.

The basic picture is then a relativistically expanding shell (or shells) of electrons, positrons, and baryons: the fireball scenario (Meszaros and Rees 1997; Piran 1999). In this fireball scenario the GRB emission is believed to result from shocks in the expanding fireball, where shock amplified magnetic fields allow radiative cooling via electron synchrotron radiation. The shocks could be due to internal energy dissipation from colliding shells with different initial Lorentz factors, or due to external shocks occurring when the shell(s) run into an external medium. Based on the ob-

served rapid fluctuations in the light curves of GRBs and the relatively smooth afterglow emission it is now often assumed that the GRB is due to internal shocks, while the low energy afterglows originate from the interaction of the burst ejecta with the surrounding medium (ISM or pre-burst ejecta/winds from the progenitor). One can think of the burst afterglows as the relativistic analog of supernova remnants. The observations and underlying physics of these afterglows is described in detail by P. Meszaros (review presented at this symposium, see astro/ph-9904038) and by T. Piran (1999). We now turn our attention to theoretical ideas aimed at explaining the underlying mechanisms causing GRBs at cosmological distances at a rate of one event per million years per host.

2. The central engine(s)

The first demands one makes upon theoretical models for cosmological GRBs are that they provide the necessary event rate and energy. The observed fluences and redshifts imply a total energy of order

$$E_{\text{tot}} \sim 10^{54} f_\Omega \text{ ergs} \qquad (1)$$

where f_Ω is the geometric beaming fraction. While models with strong jets alleviate the energy requirements, they correspondingly increase the required burst frequency. With a burst rate of 1/day from sampling to redshifts of a few one derives a specific rate of roughly 1 GRB per host per million years. This rate is consistent with estimates of merger rates for binary systems of compact stars (NS/NS, NS/BH, NS/WD, see Fryer *et al.* (1999) and Brown *et al.* 1999 for recent discussions). However, a beaming factor of ~ 0.01 increases this to one event every 10,000 years. While this rate is probably too high for mergers, collapsar models based on a small fraction of single, rapidly rotating massive stars can produce such high rates, requiring only $\sim$ 1 % of a galaxy's supernova rate. The optical afterglow of GRB990510 displayed a significant steepening of the decay about 10 days after the burst, which provides strong support for jet geometry (e.g., Harrison *et al.* 1999) reducing the isotropic energy of $\sim 3 \times 10^{53}$ ergs by a factor 300. As a canonical GRB energy scale we may thus use 10^{52} ergs.

This much energy is available in principle in several kinds of model, but it is often very hard to efficiently extract the energy. And last but not least, the question of how to convert the energy to the observed γ-ray spectra and low-energy afterglows remains a challenge to theorists. In the past years a generic GRB model has emerged as the most likely configuration providing the central engines: a stellar mass Black Hole surrounded by a transient Accretion Disk (BHAD). There are several progenitor systems that eventually converge on this configuration (Fryer *et al.* 1999). The BHAD systems produced can vary substantially, which is perhaps useful to explain different GRB classes distinguished by duration, spectral hardness, or other parameters.

Mergers of compact binaries containing Neutron Stars (NS) and Black Holes (BH) occur after a delay of order 10^8 years, so that any substantial systemic velocity imparted by the supernovae creating the NS/BH would allow a substantial fraction of the GRBs from these systems to move away from their host galaxies. The present observational situation argues in favor of a close association between afterglows and host galaxies, arguing against this class of mergers as the source of most GRBs. However, the low density environment of GRBs separated widely from their hosts might not lead to bright afterglows resulting in a strong selection effect. On the other hand, a close association with the host is expected in models that trace star formation without substantial delays, such as the formation of a BH inside a rotating massive star (the "collapsar model" (Woosley 1993; Hartmann & Woosley 1995; Popham *et al.* 1998; Paczynski 1998; MacFadyen & Woosley 1999).

NS/NS mergers are expected to generate accretion disks with rather low mass ($0.01 - 0.1$ $M_\odot$), while collapsars are endowed with massive disks (few $M_\odot$). Janka & Ruffert (1996) have shown that neutrino anti-neutrino annihilations are not capable of extracting a sufficient fraction of the binding energy ($\sim 10^{53}$ ergs). But there are two other reservoirs. First, a large amount of energy is stored in the binding energy of the

rotating debris disk. This resource can provide up to 10^{50} erg if the energy extraction from the disk is restricted to occurring by neutrinos. MHD processes can, with even greater uncertainty, extract up to two or even three orders of magnitude more (see the discussion in P. Meszaros' review in these proceedings). Furthermore, the Blandford-Znajek (1977) process can be invoked to extract the potentially large amount of energy stored in the rotation of the black hole. Here the black hole is coupled to the accretion disk by magnetic fields that thread the event horizon and are anchored in the disk. To extract the energy efficiently and rapidly magnetic field strengths of order 10^{15} Gauss are required. The maxium fraction of $M_{BH}c^2$ that can be extracted is 29%, and can be reached in principle for a maximally rotating BH, i.e., one for which the rotation parameter is maximal

$$a = Jc/GM^2 = 1. \qquad (2)$$

Livio *et al.* (1999) argued that the electromagnetic output from the inner disk regions is expected (in most cases) to dominate over that from the hole, and that therefore the black hole spin is irrelevant to the power output from the system. However, Lee, Wijers, and Brown (1999) reanalised the strength of the Blandford-Znajek process and find that efficiencies of 9% can in fact be realised, and that, contrary to the findings of Livio *et al.*, the process is not necessarily dominated by the disk. While we are far from a well developed theory for the central BHAD-engine of a GRB, it seems as if an energy requirement of $E_{grb} \sim 10^{52}$ ergs will be the least of our worries on the way tward a theory of the GRB mechanism(s).

3. Collapsars

The term "hypernova" is an observational term, not a theory. In our opinion, the often used phrase "hypernova model" is undefined. We prefer to use "hypernova" to describe the observational manifestation of a supernova-like outburst in a massive star that has a total of over 10^{52} ergs of isotropic equivalent energy (IEE) or an optial brightness for a sustained period of over 10^{44}

erg/s. The mechanism is not specified. If the hypernova is powered by a specific mechanism, such as a collapsar, a super-magnetic neutron star, etc, one should refer to the model by this name (collapsar model, pulsar model, etc). There could be many different models that explain a variety of events that all lead to a hypernova. Perhaps it is too early to fix the terminology, but we suggest the use of "hypernova model" to be discontinued.

The collapsar model (Woosley 1993; MacFadyen & Woosley 1999) refers to the explosion of a massive star, possibly with the production of a high energy transient, by the formation of a hyperaccreting black hole and a jet in its middle. The jet may be powered by neutrinos or by MHD processes. By hyperaccreting we mean accretion rates from 10^{-4} to 10 $M_\odot$ s^{-1}.

One particular variant of this model, the one discussed most extensively for producing common GRBs, is based on the formation of a powerful jet lasting over 10 s in the helium stripped core of a star whose main sequence mass was over 35 $M_\odot$. At the end of its life, such a star may collapse directly to a black hole. In the unlikely case that the star does not rotate, it simply disappears with no energetic outburst. However, if the collapsing star is endowed with angular momentum of j $\gtrsim 3 \times 10^{16}$ cm^2 s^{-1} ($M_{bh}/M_\odot$), acquired either through the merger of a binary system or from single star evolution, then an accretion disk forms around the newly born black hole. MHD turbulence in the disk provides a source of viscosity, parameterized as $\nu = \alpha c_s H$ where H is the disk scale height. The disk is therefor capable of transporting angular momentum outward and allowing gas to feed the black hole as the star continues to collapse onto the disk. MacFadyen & Woosley (1999) have simulated such a collapse of the 14 $M_\odot$ helium core of a 35 $M_\odot$ star and the ensuing disk accretion into the hole. The hydrogen envelope of the star is assumed to have been lost through a wind or transferred to a binary companion.

The collapse proceeds through several stages: first, the star collapses at all angles until centrifugal forces slow the equatorial infall and a disk forms; second, accretion continues along the rotation axis until the polar density declines suffi-

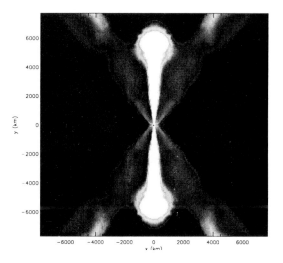

Figure 3. Energy density in the jet and surrounding area 0.824s after its initiation (from MacFadyen & Woosley 1999). The jet has now moved 7,000 km. The jet remains highly focused with an opening angle (half-width) of about 10 degrees. The red regions at polar angles of 35 degrees are the plumes formed earlier by dissipation in the disk.

ciently to allow a channel for energy escape; third, energy generated in the accretion disk or tapped from the spin energy of the black hole is transported to the polar region by MHD processes or neutrinos; fourth, the energy escapes along the pole in a jet; fifth, the GRB and associated lower energy transients are produced after the jet escapes the star and interacts with itself (through internal shocks) and with the external medium (the pre-explosion stellar wind and/or the ISM).

For the particular 14 $M_\odot$ model studied by Mac-Fadyen & Woosley which had mostly constant specific angular momentum of $10^{17} cm^2 s^{-1}$, and viscosity of $\alpha = 0.1$ accretion through the disk stayed near $0.1 M_\odot s^{-1}$ for the twenty seconds simulated. The hole mass grew from 2 $M_\odot$ to 4 $M_\odot$ and the normalized Kerr parameter (eq. 2), measuring the black hole spin, grew from .5 to .98. Accretion at this rate corresponds to an inflow of $\dot{M}_{acc}c^2 \approx 1.8 \times 10^{53} ergs^{-1}$ of rest mass energy through the disk. The gravitational potential of a black hole is an efficient engine for converting a significant fraction of accreted rest mass into potentially useful forms of energy for powering an explosion. Well known numbers for the binding energy which must be released as gas is lowered through the disk to the innermost stable circular orbits around non-rotating and maximally rotating black holes are 6% and 42%. In viscous accretion disks some of the gravitational binding energy is converted to heat as gas spiraling in nearly circular orbits rubs against neighboring gas. While some of the energy released may simply be carried with the accreting gas into the hole, in most cases a significant fraction of the accreted rest mass energy can be expected to escape the disk. As discussed above, in addition to the radiation of disk binding energy, the rotation of the black hole may itself be a source of energy for the burst. While the details of the mechanisms capable of extracting energy from the black hole - disk system using MHD processes are still being worked out, it is in principle possible to extract 6 − 42 % of of the rest mass energy accreted through the disk and perhaps an additional ∼ 29% of the mass of a spinning black hole.

For the case of neutrino mediated transport it is possible to calculate the efficiency of energy

extraction directly. Since the disk calculation of MacFadyen & Woosley (1999) match the semi-analytic, relativistic, disk models of Popham, Woosley & Fryer (1998) in the regions where they overlap, the disk temperature and neutrino luminosity in the hot inner disk from the analytic models can be used to calculate the total neutrino luminosity and neutrino annihilation rate for the accretion rate, hole mass and Kerr parameter calculated in the 2D simulations. During the 20 seconds of evolution calculated, 3.3×10^{53} erg of neutrino energy was emitted and 1.2×10^{52} ergs were deposited by neutrino annihilation in the polar region above the hole. This represents 9% of the accreted rest mass radiated as neutrinos and 0.3% converted to a pair fireball by neutrino annihilation as neutrinos emitted from one side of the disk collide with anti-neutrinos from the opposite side and annihilate. These numbers represent the "optimistic" case described by MacFadyen & Woosley. It is worth emphasizing that even this optimistic case is conservative compared to the energy potentially available from MHD processes in the disk or extraction of the rotational energy of the black hole.

The response of the collapsed star plus accretion torus to an energy input consistent with the calculated accretion rate is of interest. Simulated thermal energy deposition of $10^{51} ergs^{-1}$ ($\sim 0.005 \dot{M}_{acc}c^2$) near the pole of the accretion disk results in beamed expansion along the rotation axis - jets beamed to $\sim 1\%$ of the sky (Figure 3). While the gas initially expands isotropically the density structure of the accretion disk and evacuated polar axis naturally channels the expanding gas into a jet with opening half angle of ≈ 10 degrees. This particular model could therefor achieve an isotropic energy equivalent of 8 × $10^{53} f_\gamma$ erg, where f_γ is the efficiency of converting explosion energy into gamma rays.

It is worth noting a major difference between collapsars and merging compact objects. The collapse of a helium star occurs over much longer time scales due to their extended size and relatively low density. Collapsars can power a GRB jet for hundreds of seconds as the collapsing stellar envelope continues to feed the accretion disk and power the jet. Short time scale variability in the jet energy can come both from variations in the disk accretion rate due to hydrodynamical instabilities and from the interaction of the jet with the outer layers of the star as it pushes through. Merging compact objects, on the other hand, will have a time scale of only tens of milliseconds if neutrino losses are important (Janka & Ruffert 1996) and last tens of seconds only if the disk viscosity is very low.

The collapsar mechanism may also operate in stars which have extended envelopes, either because of lower mass loss (due perhaps to low metallicity), or due to accretion from a binary companion. In this case the observed event in unlikely to be a classical GRB since the time for the jet to break out of the stellar surface (1,000 - 100,000 s) exceeds the engine duration time. The engine turns off because of starvation due to the decreasing accretion rate either because the envelope density is low or because shocks from the jet sweep around the star and eject the stellar envelope at all polar angles. A very powerful asymmetric explosion will result in this case with possible soft X-ray transient or weak GRB due to shock acceleration along the steep density gradient at the stellar surface. SN1998bw and 1997cy may be examples of such a "smothered collapsar". Here we note growing evidence for a "supernova component" in afterglow lightcurves. Due to the extreme brightness of the GRB afterglow the light from an associated supernova can be very hard to detect. If the afterglow decay is particularly steep and if the lightcurve is carefully monitored for a sufficiently long time, the supernova signal could be separable from the GRB afterglow. In addition to the more direct evidence from the pair GRB980425 - SN1998bw there is now additional support for a supernova light component from GRB970228 (Reichart 1999), GRB980326 (Bloom *et al.* 1999), GRB980519 (Chevalier & Li 1999), and perhaps even more bursts.

4. Conclusion

Just as an accreting supermassive black hole can give rise to a variety of observational phenomena - blazars, BL-Lac objects, Seyferts, quasars, and radio galaxies - depending on the environment in which the black hole finds itself and the viewing angle - a hyperaccreting stellar mass black hole can also be created in various ways and display various attributes. In a massive helium star, the prompt formation of a black hole might make a classical GRB of typical duration 20 s. In a massive red supergiant star though, the result might be an asymmetric supernova with anomalously large energy (SN 1997cy) and little or no accompanying GRB. In a helium star in which the jet energy source is lost before the jet has had time to break through the surface one may get an intermediate case, a Type Ib supernova with a weak GRB. In a massive helium star in which the black hole forms over a longer period (owing to fallback in an otherwise successful supernova), a long GRB could be made.

Hyperaccreting holes can also be set up by the merger of compact objects and give rise to short GRBs - or they could be set up by the merger of white dwarfs or helium stars with black holes and give long GRBs. Indeed, what we have traditionally called GRBs may just be the leading edge of a collection of high energy phenomena whose diversity has yet to be fully sampled.

Acknowledgements: Part of this work has been supported by NASA (NAG5-2843 and MIT SC A292701) and the NSF (AST-97-31569). We acknowledge many helpful conversations with Chris Fryer, Bob Popham, and Stan Woosley. We thank Andi Fruchter and the HST GRB Team for providing the afterglow image of GRB990123, and the BATSE Team for providing public access to up-to-date GRB data, such as the angular distribution shown in Figure 1. A portion of this work was carried out at the Max Planck Institute (MPI) for Astrophysics and the MPI for Extraterrestrial Physics (Garching, FRG), and we gratefully acknowledge the support and hospitality of these two institutions.

REFERENCES

1. Akerlof, C., et al. 1999, Nature 398, 400
2. Bloom, J. S., *et al.* 1999, astro-ph/9905301
3. Blandford, R. D. & Znajek, R. L. 1977, MNRAS, 179, 433
4. Brown, G. E., Wijers, R. A. M. J., Lee, C.-H.,

& Bethe, H. A. 1999, ApJ, in press

5. Chevalier, R. A., & Li, Z.-Y. 1999, astro-ph/9904417

6. Covino, S. et al. 1999, IAUC7172

7. Frail, D., et al. 1997, IAU6662

8. Fruchter, A. S., et al. 1999, ApJ, in press (astro-ph/9902236)

9. Fryer, C., Woosley, S. E., & Hartmann, D. H. 1999, ApJ, in press (astro/ph-9904122)

10. Galama, T. J., et al. 1998, Nature 395, 670

11. Goodman, J. 1997, New Astronomy 2, 449

12. Harrison, F. A., et al. 1999, ApJ, submitted (astro/ph-9905306)

13. Hartmann, D. H., & Woosley, S. E. 1995, Advances in Space Research 15(5), 143

14. Hartmann, D. H. 1999, Proc. Natl. Acad. Sci. USA, 96, 4752

15. Höflich, P., Wheeler, J. C., & Wang, L. 1998, ApJ, in press (astro-ph/9808086)

16. Iwamoto, K., *et al.* 1998, Nature 395, 672

17. Janka, H.-T., & Ruffert, M. 1996, A&A 307, L33

18. Klebesadel, R. W., Strong, I. B., & Olson, R. A. 1973, ApJ 182, L85

19. Kulkarni, S. R., et al. 1998, Nature 393, 35

20. Lee, H. K., Wijers, R. A. M. J., & Brown, G. E. 1999, astro-ph/9906213

21. Livio, M., Ogilvie, G. I., & Pringle, J. E. 1999, ApJ 512, 100

22. MacFadyen, A. I., & Woosley, S. E. 1999, ApJ, in press (astro-ph/9810274)

23. Meszaros, P., & Rees, M. J. 1997, ApJ 476, 232

24. Metzger, M. R., et al. 1997, Nature 387, 878

25. Paciesas, W. S., et al. 1999, ApJ Suppl., in press (astro-ph/9903205)

26. Paczynski, B. 1998, ApJ 494, L45

27. Piran, T. 1999, Physics Reports, in press

28. Piro, L., et al. 1998, A&A 329, 906

29. Popham, R., Woosley, S. E., & Fryer, C. 1998, ApJ, submitted (astro/ph-9807028)

30. Reichart, D. E. 1999, ApJ Letters, in press (astro-ph/9906079)

31. van Paradijs, J., et al. 1997, Nature 386, 686

32. Woosley, S. E. 1993, ApJ 405, 273

33. Woosley, S. E., Eastman, R., & Schmidt, B. 1998, ApJ 516, 788

ELSEVIER

Nuclear Physics B (Proc. Suppl.) 80 (2000) 143–151

NUCLEAR PHYSICS B
PROCEEDINGS
SUPPLEMENTS

www.elsevier.nl/locate/npe

Microquasars

I. F. Mirabel[a] & L. F. Rodríguez[b]

[a]DAPNIA/SAP, CEA Saclay, 91191 Gif/Yvette, France

[b]Instituto de Astronomía, UNAM, Morelia, Michoacán 58090, México

Microquasars are stellar-mass black holes in our own Galaxy that mimic, on a smaller scale, the remarkable relativistic phenomena observed in remote quasars. Their discovery opens new perspectives for understanding the connection between the flow of matter into black holes and the genesis of relativistic jets.

Discovered more than 30 years ago, quasars remain as one of the most mysterious objects in the Universe. It is widely believed that they are powered by black holes of several million solar masses or more that lie at the centers of remote galaxies. Their luminosities are much larger than ordinary galaxies like the Milky Way, yet originate from regions smaller than the size of the solar system. Occasionally, quasars spout jets of gas that appear to move on the plane of the sky with velocities exceeding that of light. The extreme distance of quasars introduces many uncertainties into the interpretation of the source of energy and nature of the ejecta that appear to be moving with superluminal speeds.

The recent finding in our own galaxy of *microquasars*[1,2,3,4], a class of objects that mimic -on scales millons of times smaller- the properties of quasars, has opened new perspectives for the astrophysics of black holes (see Figure 1). These scaled-down versions of quasars are believed to be powered by spinning black holes[5] but with masses of up to a few tens that of the Sun. The word *microquasar* was chosen to suggest that the analogy with quasars is more than morphological, and that there is an underlying unity in the physics of accreting black holes over an enormous range of scales, from stellar-mass black holes in binary stellar systems, to supermassive black holes at the centre of distant galaxies. Since the characteristic times in the flow of matter onto a black hole are proportional to its mass, variations with intervals of minutes in a microquasar correspond to analogous phenomena with durations of thousands

of years in a quasar of 10^9 $M_\odot$, which is much longer than a human life-time. Therefore, variations with minutes of duration in microquasars could be sampling phenomena that we have not been able to study in quasars.

The repeated observation of two-sided moving jets in a microquasar[2,6] has led to a much greater acceptance of the idea that the emission from quasar jets is associated with moving material at speeds close to that of light[6]. Furthermore, simultaneous multiwavelength observations of this microquasar[7,8] are revealing the connection between the sudden disappearance of matter through the horizon of the black hole, with the ejection of expanding clouds of relativistic plasma.

1. Superluminal sources

Superluminal motions have been observed in quasars for more than 20 years[9]. In the past, these motions had been used to argue that quasars can not be as remote as believed, and that the use of redshifts and the Hubble expansion law to determine their distances was not fully justified. In the extragalactic case the jets are only observed moving on one side of the source of ejection, and it is not possible to know if superluminal motions represent the propagation of waves through a slowly moving jet, or if they reflect the actual motion of the sources of radiation.

In the context of the microquasar analogy, one then may ask if superluminal motions could be observed from black hole binary systems known to lie in our own Galaxy. Among the handful of black holes of stellar mass known so far, the X-ray sources GRS 1915+105[10] and GRO J1655-40[11] have indeed been identified at radio waves as transient sources of superluminal jets[2,3,4]. The jets in the two sources move at 0.92c and still hold the speed record in the Galaxy. GRO J1655-40 is known to be at a distance of 10,000 light-years and the apparent transverse motions of the jecta in this source are the largest observed until now from an object beyond the solar system.

2. Special relativity

The observation of superluminal double-sided jets in GRS 1915+105 proved beyond doubt the existence of highly relativistic ejections of matter in the Universe and gave support to the microquasar analogy. Figure 2 shows a major ejection event observed from GRS 1915+105 in 1994 March-April. The brighter cloud (to the left) appears to move faster than the fainter cloud (to the right). This asymmetry in apparent motion and brightness can be explained in terms of relativistic aberration (see Figure 3). In all five major ejections observed so far from GRS 1915+105 the clouds move with the same proper motions and approximately along the same direction on the sky[4]. The observation of counterjets in this microquasar allows a comparison of the parameters of the approaching and receding ejecta, that shows that the emission arises in moving material and rules out the possibility of other explanations[4].

As shown in inset 1, if the proper motions of the twin ejecta and the Doppler shift of spectral lines could be measured in the approaching and/or receding ejecta, the parameters of the system, in particular its distance could be obtained. The main observational difficulty resides in the detection of lines that are strongly Doppler broadened since they arise from plasma clouds that not only move but also expand at relativistic speeds.

3. The central engine

Multiwavelength monitorings of the galactic superluminal sources have shown that the hard

X-ray emission is a necessary but not sufficient condition for the formation of collimated jets of synchrotron radio emission. In GRS 1915+105 the relativistic ejection of pairs of plasma clouds have always been preceded by unusual activity in the hard X-rays[12]. More specifically, the onset of major ejection events seems to be related to the sudden drop from a luminous state in the hard X-rays[13,14]. However, not all unusual activity and sudden drops in the hard X-ray flux appear to be associated with radio emission from relativistic jets. In fact, in GRO J1655-40 there have been several hard X-ray outbursts observed without following radio flare/ejection events[15].

What is the current belief as to how these relativistic jets are generated? Black holes, both in quasars and microquasars are surrounded by an accretion disk. In the case of microquasars the disk is fed by magnetized gas drawn from a star that accompanies the black hole, forming a gravitationally bound binary system (Figure 1). Creating a collimated flow that approaches the speed of light is a theoretical challenge faced more than two decades ago when extragalactic radio jets were first discovered. Such models are now being revived and tested in the context of the microquasar phenomenon. One possibility is magnetohydrodynamic acceleration in the accretion disk itself[16]. An alternative model is the one that taps the enormous rotational energy of a spinning black hole[17], where the energy drawn from the black hole accelerates the magnetized plasma coming from the disk, and ultimately propels it into jets.

Why do some black hole systems produce powerful jets while others apparently do not? The answer may reside in the spin of the black hole. In this unification scheme[18] black hole sources of powerful collimated jets would be rotating black holes with threading magnetic fields maintained by co-rotating accretion disks. Ejection of relativistic jets will take place only in those sources where the spin of the black hole is close to its maximum value.

4. Accretion into the black hole

The X-ray power of the superluminal source GRS 1915+105 exhibits a large variety of quasi-periodic oscillations (QPOs). Of particular interest is a clas of oscillations with a maximum stable frequency of 67 Hz that has been observed many times, irrespective of the X-ray luminosity of the source[19]. It is believed that this fix maximum frequency is a function of the fundamental properties of the black hole, namely, its mass and spin. However, this particular frequency could be related either to the last stable circular orbit around the black hole[19], to a g-mode in general relativity disk seismology[20], or to the relativistic dragging of the inertial frame around the fast rotating collapsed object. Theoretical work to distinguish between these alternatives will be important to estimate the spin of the black holes with masses that have been independently determined.

The episodes of large amplitude X-ray flux variations in time-scales of seconds and minutes, and in particular, the abrupt dips observed in GRS 1915+105 (see Figure 4) are believed to be a strong evidence for the presence of a black hole[21,22,23]. These variations could be explained if the inner part of the accretion disk goes temporarily into an advection-dominated mode[24]. In this mode, the time for the energy transfer from ions (that get most of the energy from viscosity) to electrons (that are responsible for the radiation) is larger than the time of infall to the compact object. Then, the bulk of the energy produced by viscous dissipation in the disk is not radiated (as it happens in standard disk models), but instead is stored in the gas as thermal energy. This gas, with large amounts of locked energy, is advected (transported) to the compact object. If the compact object is a black hole, the energy quietly disappears through the horizon, and one can observe sharp decays in the X-ray luminosity. In constrast, if the compact object is a neutron star, the thermal energy in the superheated gas is released as radiation when it collides and heats up the surface. The cooling time of the neutron star photosphere is relatively long, and in this case a slow decay in the X-ray flux is observed. Then, one would expect the luminosity of black hole binaries to excursion over a much larger range of values than that of neutron star binaries for an interval of time of a few seconds.

5. The formation of jets

During such large amplitude variations in the X-ray flux of GRS 1915+105, remarkable flux variations in time-scales of minutes have also been reported at radio[25,26] and near infrared[27] wavelengths (see Figure 4). The rapid flares at radio and infrared wavelength come from expanding magnetized clouds of relativistic particles. Simultaneous observations in the X-ray, infrared and radio wavelengths show that the ejection of relativistic clouds of plasma[7,8] take place when the matter of the inner accretion disk suddenly disappears through the horizon of the black hole. Current interferometric observations at radio waves with the Very Long Baseline Array[28], carried out during episodic ejections may show in time scales of hours and with spatial resolutions of tens of astronomical units, how the material from the disk is re-directed onto collimated jets.

At present, microquasars offer new opportunities to gain a general understanding of the relativistic jets seen elsewhere in the Universe. It is expected that in the future they will also become probes for the physics of strong-field relativistic gravity near the event horizon of black holes.

6. References

[1] Mirabel, I.F., Rodríguez, L.F., Cordier, B. et al. A double-sided radio jet from the compact Galactic Centre annihilator 1E1740.7-2942. Nature **350**, 215-218 (1992).

[2] Mirabel, I.F. & Rodríguez, L.F. A superluminal source in the galaxy. Nature **371**, 46-48 (1994).

[3] Tingay, S.J., Jauncey, D.L., Preston, R.A. et al. A relativistically expanding radio source associated with GRO J1655-40. Nature **374**, 141-143 (1995).

[4] Hjellming, R.M. & Rupen, M.P. Episodic ejection of relativistic jets by the X-ray transient GRO J1655-40. Nature, **375**, 464-467 (1995).

[5] Shapiro, S.L. & Teukolsky, S.A. *Black Holes, White Dwarfs, and Neutron Stars*, Wiley & Sons,

New York, page 441 (1983).

[6] Rodríguez, L.F. & Mirabel, I.F. Repeated relativistic ejections in GRS 1915+105. Submitted to Astrophys. J. (1997).

[7] Mirabel, I.F., Dhawan, V., Chaty, S. et al. Accretion instabilities and jet formation in GRS 1915+105. Submitted to Astron. Astrophys. (1997).

[8] Eikenberry, S.S., Matthews, K., Morgan, E.H. et al. Evidence for disk-jet interaction in the microquasar GRS 1915+105. Submitted to Astrophys. J. (1997).

[9] Pearson, T.J. & Zensus, J.A. in *Superluminal Radio Sources* (eds Zensus, J.A. & Pearson, T.J.) 1-11 (Cambridge Univ. Press, 1987).

[10] Castro-Tirado, A.J., Brandt, S., Lund, N. et al. Discovery and observations by Watch of the X-ray transient GRS 1915+105. Astrophys. J. Suppl. Ser. 92, 469-472 (1994).

[11] Zhang, S.N., Wilson, C.A., Harmon, B.A., Fishman, G.J., & Wilson, R.B. X-ray nova in Scorpius. IAU Circ. 6046, 1 (1994).

[12] Harmon, B.A., Deal, K.J., Paciesas, W.S. et al. Hard X-ray signature of plasma ejection in the galactic jet source GRS 1915+105. Astrophys.J., **477**, L85-L90 (1997)

[13] Foster, R.S., Waltman, E.B., Tavani M., et al. Radio and X-ray variability of the galactic superluminal source GRS 1915+105. Astrophys. J., 467, L81-L84 (1996)

[14] Mirabel, I.F. Rodríguez L.F., Chaty S., et al. Infrared observations of an energetic outburst in GRS 1915+105. Astrophys. J. **472**, L111-114 (1996).

[15] Zhang, S.N., Mirabel, I.F., Harmon, B.A., Kroeger, R.A., Rodríguez, L.F., Hjellming, R.M. & Rupen, M.P. Galactic black hole binaries: multofrequency connections. In *Proceedings of the Fourth Compton Symposium*, in press (1997).

[16] Blandford, R.D. & Payne, D.G. Hydromagnetic flows from accretion discs and the production of radio jets. Mon. Not. R. Astron. Soc. **199**, 883-894 (1982).

[17] Blandford, R.D. & Znajek, R.L. Electromagnetic extraction of energy from Kerr black holes. Mon. Not. R. Astron. Soc. **179**, 433-440 (1977).

[18] Zhang, S.N., Cui, W., & Chen, W. Black hole spin in X-ray binaries: observational con-sequences. Astrophys. J. **482**, L155-158 (1997).

[19] Morgan, E.H, Remillard, R.A. & Greiner, J. RXTE observations of QPOs in the black hole candidate GRS 1915+105. Astrophys, J. **482**, 993-1010 (1997).

[20] Nowak, M.A., Wagoner, R.V., Begelman, M.C. & Lehr, D.E. The 67 Hz feature in the black hole candidate GRS 1915+105 as a possible "diskoseismic" mode. Astrophys. J 477, L91-94 (1997).

[21] Greiner, J, Morgan, E.H. & Remillard, R.A. Rossi X-ray timing explorer observations of GRS 1915+105. Astrophys. J. **473**, L107-110 (1996).

[22] Belloni, T., Méndez, M., King, A.R., van der Klis, M. & van Paradijs, J. An unstable central disk in the superluminal black hole X-ray binary GRS 1915+105. Astrophys. J. **479**, L145-148 (1997).

[23] Chen, X., Swank, J.H. & Taam, R.E. Rapid bursts from GRS 1915+105 with RXTE. Astrophys. J. **477**, L41-44 (1997).

[24] Narayan, R., Garcia, M.R. & McClintock, J.E. Advection-dominated accretion and black hole event horizons. Astrophys. J. **478**, L79-82 (1997).

[25] Pooley, G. & Fender, R.P. Quasi-periodic variations in the radio emission from GRS 1915+105. Mon. Not. R. Astron. Soc., in press (1997).

[26] Rodríguez, L.F. & Mirabel, I.F. Sinusoidal oscillations in the radio flux of GRS 1915+105. Astrophys. J. **474**, L123-125 (1997).

[27] Fender, R.P., Pooley, G.G., Brocksopp, C. & Newell, S.J. Rapid infrared flares in GRS 1915+105: evidence for infrared synchrotron emission. Mon. Not. R. Astron. Soc. **179**, L65-L69 (1997).

[27] Dhawan, V., Mirabel, I.F. & Rodríguez, L.F. In progress (1997).

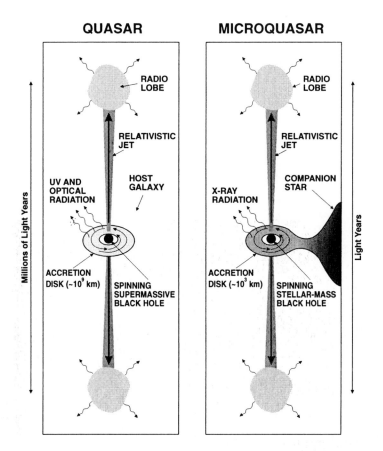

Figure 1. Diagram illustrating current ideas concerning quasars and microquasars (not to scale). As in quasars, in microquasars are found the following three basic ingredients: 1) a spinning black hole, 2) an accretion disk heated by viscous dissipation, and 3) collimated jets of relativistic particles. However, in microquasars the black hole is only a few solar masses instead of several millon solar masses; the accretion disk has mean thermal temperatures of several millon degrees instead of several thousand degrees; and the particles ejected at relativistic speeds can travel up to distances of a few light-years only, instead of the several millon light-years as in some giant radio galaxies. In quasars matter can be drawn into the accretion disk from disrupted stars or from the interstellar medium of the host galaxy, whereas in microquasars the material is being drawn from the companion star in the binary system. In quasars the accretion disk has sizes of $\sim 10^9$ km and radiates mostly in the ultraviolet and optical wavelenghts, whereas in microquasars the accretion disk has sizes of $\sim 10^3$ km and the bulk of the radiation comes out in the X-rays. It is believed that part of the spin energy of the black hole can be tapped to power the collimated ejection of magnetized plasma at relativistic speeds. This analogy between quasars and microquasars resides in the fact that in black holes the physics is essentially the same irrespective of the mass, except that the linear and time scales of phenomena are proportional to the black hole mass. Because of the relative proximity and shorter time scales, in microquasars it is possible to firmly establish the relativistic motion of the sources of radiation, and to better study the physics of accretion flows and jet formation near the horizon of black holes.

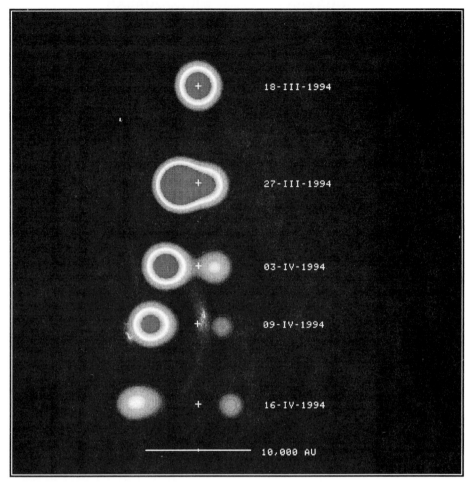

AIPS User 1934

Figure 2. Pair of plasma clouds expelled from the microquasar GRS 1915+105, the first superluminal source detected in the Galaxy[2]. The cloud to the left appears to move away from the centre of ejection (white cross) at 125% the speed of light. An observer in the frame of one of these clouds would see the other receding at 0.997c. It has been demonstrated[2] that the asymmetries in velocity and brightness between the cloud to the left and the cloud to the right can be explained in terms of an antiparallel ejection of twin pairs of plasma clouds moving with bulk speeds of 92% that of light, at an angle of 70 degrees with respect to the line of sight. These maps were obtained about once a week with the Very Large Array of NRAO at $\lambda 3.5$ cm. At a distance of 40,000 light-years from Earth, in one month the clouds moved apart in the plane of the sky a distance equivalent to 10,000 astronomical units (1 AU equals the mean radius of the Earth's orbit around the Sun.)

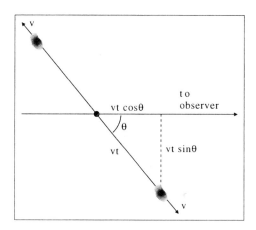

Figure 3. The *superluminal motions* can be fully understood in terms of a relativistic illusion in the observation of ejecta moving close to the speed of light. The ejecta is moving so fast that it nearly catches up with its own radiation. After a time t from the moment of ejection, the condensations, with a true velocity v, have moved a distance vt. As seen in projection by the observer, the displacement seems to be $vt \sin\theta$, where θ is the angle between the line-of-sight and the axis of ejection. However, since the approaching condensation is now closer to the observer by a distance $vt \cos\theta$, the time t' in which the observer sees the condensation move from the origin to its present position is smaller than t and is given by $t' = t - (vt \cos\theta/c)$. The apparent velocity of the approaching condensation is then $v_a = v \sin\theta/(1 - (v \cos\theta/c))$, that can exceed c. By a similar reasoning, the apparent velocity of the receding condensation is then $v_r = v \sin\theta/(1 + (v \cos\theta/c))$. Additionally, for an object moving at relativistic speeds, the emitted radiation "focuses" in the direction of motion (the so-called relativistic beaming), an effect that makes the approaching condensation look brighter than the receding condensation. In remote objects, such as quasars, the approaching ejecta can be observed only when their apparent brightness is very highly enhanced due to beaming to the observer. This Doppler-favoritism for the approaching jet implies the opposite effect for the receding jet, rendering it undetectable in practice.

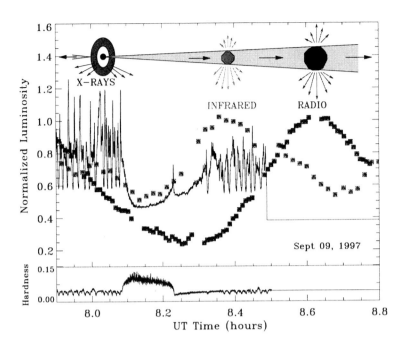

Figure 4. Oscillations in the luminosity at X-rays, infrared, and radio wavelengths of the galactic source of superluminal jets GRS 1915+105 (ref. 7). The disappearance of the inner accretion disk (marked by the X-ray dip) and the beginning of the ejection of relativistic plasma clouds (marked by the start of the infrared-radio flares) occur simultaneously at about 7 minutes after the start of the set of observations shown in this Figure (ref. 7)

DETERMINING DISTANCES WITH SPECIAL RELATIVITY

If one can measure the proper motions in the sky of the approaching and receding ejecta. μ_a and μ_r. two independent equations are obtained:

$$\left[\frac{\mu_{a,r}}{radians\ sec^{-1}}\right] = \frac{v\ sin\theta}{(1 \mp (v/c)\ cos\theta)\ D} \ .$$

where v is the true velocity of the ejecta. θ is the angle between the line of sight and the ejection axis. and D is the distance to the source.

Measuring the wavelength $\lambda_{a,r}$ of spectral lines (with rest wavelength λ_{rest}) arising in either the approaching or receding jets. a third equation can be taken from:

$$\frac{\lambda_{a,r}}{\lambda_{rest}} = \frac{(1 \mp (v/c)\ cos\theta)}{[1 - (v/c)^2]^{1/2}} \ .$$

One can then resolve the system of 3 equations and find the three unknowns: v. θ. and D. the distance to the source.

ELSEVIER

Nuclear Physics B (Proc. Suppl.) 80 (2000) 153–161

PROCEEDINGS
SUPPLEMENTS

www.elsevier.nl/locate/npe

Type Ia supernovae at high z

P. Ruiz–Lapuente[a]

[a]Department of Astronomy, U. Barcelona, Martí i Franques 1, E-08028 Barcelona, Spain
and
Max–Planck Institut für Astrophysik, Karl–Schwarzschild–Str. 1, D-85740 Garching, Germany

Type Ia supernovae are very powerful probes for cosmology and clear tracers of the past history of the universe. Two independent high-redshift supernova collaborations (the High-Z Team and the Supernova Cosmology Project) have presented this year evidence that we live in a low–matter density universe whose expansion is being accelerated by the presence of a dominant vacuum energy density. The Supernova Cosmology Project by using the searches performed at various z for cosmological purposes has measured high-z supernova rates as well. Those measurements provide unvaluable information on the cosmic star formation history, on the efficiency in producing supernovae out of stars, and on the involved timescale to explosion. The cosmic background due to supernova emission can also be calculated in agreement with the measured rates.

1. INTRODUCTION

Among the many open lines of research on Type Ia supernovae (SNe Ia), I have selected to address in this overview some new developments in the field of their cosmological applications. Other than the well known results on Ω_M and Ω_Λ provided by the use of SNe Ia light curves, which tell us about the expansion history of the universe, supernovae have started to provide information on the history of star formation along cosmic time. The results that come from supernovae can be compared to those obtained with other methods and with those predicted by different galaxy formation schemes.

At this point even questions such as the background produced by the supernovae themselves can start to be addressed with empirical basis as we have measurements of rates of supernovae at high z.

Moreover, searches of supernovae at high-z are teaching us about the supernova phenomenon itself by indicating what is the efficiency in forming SNe Ia progenitors and what can be the typical timescale to explosion.

Big unknowns still remain on the evolutionary

and explosive path giving rise to this kind of explosions. But, the bulk of information obtained on these phenomena is growing and the efforts placed both from theory and observations might fructify soon.

2. COSMOLOGY

2.1. The magnitude–redshift diagram

Supernovae of Type Ia are not "standard candles" since they all do not have the same intrinsic luminosity. However, they show a strong correlation between their peak brightness and the rate of decline of the brightness (i.e. light curve), which can be calibrated with high accuracy and enable distance measurements with 5% accuracy (Phillips 1993; Hamuy et al. 1996; Riess et al. 1996). There are different ways to parameterize this correlation and use all the known SNe Ia sample in the magnitude–redshift diagram (Riess et al. 1998; Perlmutter et al. 1998).

The luminosity distance as a function of z for an astrophysical object of known absolute magnitude inform us on the matter density Ω_M and on the cosmological constant density Ω_Λ ($\Omega_\Lambda = \Lambda / 3H_0^2$).

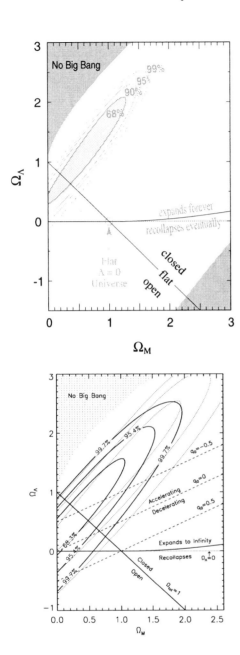

The magnitude–redshift relationship m(z) as a function of Ω_M and Ω_Λ is given in the form:

$$m(z) = M + 5\ log\ d_L(z, \Omega_M, \Omega_\Lambda) - 5\ log\ H_0 + \quad (1)$$
$$K_c + 25 \quad\quad\quad\quad\quad\quad\quad\quad\quad\quad\quad\quad (2)$$

where K_c is the K–correction and $d_L = (1+z)\,d_M$ is the luminosity distance

$$d_L = \frac{(1+z)}{H_0} \int_0^{z_1} \left[(1+z)^2(1+\Omega_M z) - \quad (3\right.$$
$$\left. z(2+z)\Omega_\Lambda\right]^{-1/2} dz \quad\quad\quad\quad\quad (4$$

As noted by Goobar & Perlmutter (1995) by obtaining SNe Ia at various z, it is possible to draw a simultaneous measurement of Ω_Λ and Ω_M. The results of those measurements are plotted in the Ω_M–Ω_Λ plane (see Figure 1). In the diagram obtained during the last years by the two independent collaborations (the High-Z Team and the Supernova Cosmology Project), a universe with $\Omega_M = 1$, $\Omega_\Lambda = 0$ is strongly excluded. The two collaborations find $\Omega_\Lambda > 0$ at a 3σ confidence level. For a flat universe ($\Omega_T = 1$), the results by the Supernova Cosmology Project imply $\Omega_M = 0.28^{+0.09}_{-0.08}(\text{stat})^{+0.05}_{-0.04}$ (syst), and the High-Z Team obtains for a flat universe $\Omega_M = 0.24 \pm 0.1$ (see Perlmutter et al. 1999 and Schmidt et al. 1999, in this volume). The outcoming picture of our universe is that it contains about 20–30% density content in matter and 70–80% in cosmological constant. According to the allowed Ω_M and Ω_Λ values, the universe will expand forever, accelerating its rate of expansion.

2.2. Number counts and volume test

The number density of Type Ia supernovae at high z should also reflect the model of universe, and could provide a complementary test of the Λ-dominancy (Ruiz–Lapuente & Canal 1998). Moreover, it could be a way to estimate the role of dust at high-z obscuring Type Ia supernovae (unidentified dust might lead to an overestimate of Ω_Λ through the magnitude-z test, while

Figure 1. Upper panel: Confidence regions in the Ω_M–Ω_Λ plane from the high–z supernovae observations by the Supernova Cosmology Project (see Perlmutter et al. 1999). Lower panel: Confidence regions in Ω_M–Ω_Λ plane from observations of high-z supernovae by the High-Z Team (Riess et al. 1999).

unidentified dust would play in the opposite way in the number counts test, lowering the Ω_Λ estimate).

The number counts of SNe Ia at $z \sim 1$ are almost two times higher for $\Omega_\Lambda \sim 0.7$ than for $\Omega_\Lambda \sim 0$ due to the comoving volume dependence with z:

$$\frac{dV}{dz d\Omega} = \frac{d_M^2}{(1 + \Omega_k H_0^2 d_M^2)^{1/2}} d(d_M) \qquad (5)$$

The results on this Ω_Λ test, however, will only become clear when the evolution of the star formation rate up to $z\sim 1$ is well determined and when we start to know the efficiency and timing of SNe Ia explosions. The study of both the cosmic star formation history and frequencies of SNe Ia at high z is starting to disentangle these unknowns (see section 3).

3. EFFICIENCY in TYPE IA SUPER-NOVAE AT HIGH Z

We start to have information about the efficiency in producing SNe Ia in our universe. As it has been pointed out (Ruiz–Lapuente, Burkert & Canal 1995; Ruiz–Lapuente, Canal & Burkert 1997; Madau, Della Valle & Panagia 1998; Sadat et al. 1998; Yungelson & Livio 1998), the study of Type Ia supernovae at high z provides clues on the progenitors. Several programmes are providing measurements of the rate of SNeIa both in the local and in the high–z universes. Cappellaro et al.(1998) present some estimates of the rates from various searches and Hamuy & Pinto (1999) have provided values for the rates from the Calan/Tololo search. At high z, the values so far obtained come from the Supernova Cosmology Project (Pain et al. 1996; 1999). The better knowledge of SNeIa rates at different redshifts, combined with that of the average star formation rate, allows us to investigate whether the efficiency in producing SNe Ia has changed along cosmic time. This can shed light on the binary path to the explosion.

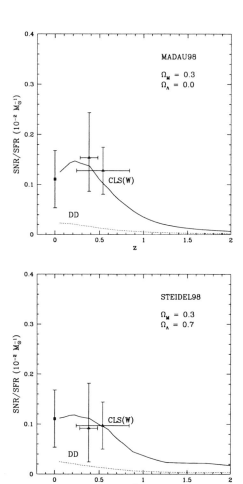

Figure 2. Predictions for the "instantaneous efficiency" in producing SNe Ia (SNe Ia at a given z per unit mass in forming stars) for two SNe Ia candidates systems with different timescales to explosion and range of stars ending as SNeIa. The dotted line is a modeling of merging of double degenerate pairs (ploted DD), the solid line is a modeling of the single degenerate candidate (Algol-type binary pairs) with winds effects (plotted CLSW). The data points have been derived using the SNIa measurements (Pain et al. 1996, 1999) and the star formation rates (Madau et al. 1998; Steidel et al. 1998). The key measurements would come at $z \sim 1$. The quantity plotted is independent of the cosmology chosen, and only sensitive to the timescale and efficiency of SNe Ia explosions.

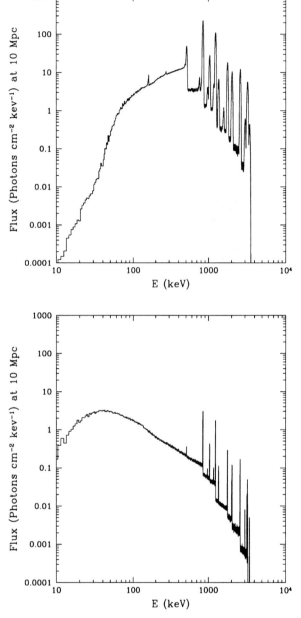

Figure 3. Upper panel: Calculated γ-ray spectrum of a Type Ia supernova integrated over two years (from Ruiz-Lapuente, Cassé & Vangioni-Flam 1998). Lower panel: The same as above but for a Type II supernova.

Little direct evidence is so far available on the nature of the binary system which is responsible for the Type Ia SN explosions. We know that the exploding star is a carbon–oxygen white dwarf (C+O WD) and that there is no compact object left in the explosion. Stellar evolution arguments tell us as well that those explosions take place in binary systems. But so far, it has not been possible to point clearly to a unique type of binary system as responsible for the Type Ia phenomenon (Ruiz–Lapuente, Canal & Isern 1997 and articles therein). Two candidate systems seem favored. One of them is a pair of C+O WDs in a binary system that merge as their orbit shrinks due to the emission of gravitational wave radiation. This system is refered as to *double degenerate system*. The other one is a WD which accretes material from a Roche–lobe overfilling non-WD companion (WD plus Roche–lobe filling subgiant or giant). The system looks as an Algol-type binary. It contains only a WD and therefore is a *single-degenerate system* (Branch et al. 1995; Ruiz–Lapuente, Canal & Burkert 1997).

The peak or characteristic timing of those explosions is different: merging of WDs happen typically a few times 10^8 yrs after star formation whereas the accretion of material from a non-WD companion involves less massive companions which take longer to leave the main sequence to become subgiant and giants. This second alternative evolutionary path to SNe Ia takes about a few times 10^9 yrs.

Let us define an "instantaneous efficiency" in producing SNe Ia as the rate of SNe Ia at a given z (in SN yr^{-1} Mpc^{-3}) divided by the star formation rate at the same z (in $M_\odot^{-1}$ Mpc^{-3}). This value is independent of the cosmology. Any change in this "instantaneous efficiency" (given in number of SNe Ia per unit mass going into star formation) for producing SNe Ia can be a clue as to the progenitor. The effect of the timescale to explosion by different SNe Ia progenitors can be traced in that ratio. SNe Ia coming from the merging of WDs follow more closely the star formation history and would have an approximately constant

Table 1
Type Ia supernova rates along z

Redshift $<z>$	τ_{SNu} SNu h^2	ρ_{Ia} Ia Mpc^{-3}yr^{-1}h^3	counts Ia yr$^{-1}\Delta z^{-1}$sqdeg^{-1}	Search Search/Author
0.	$0.50^{+0.71}_{-0.31}$	$8.2^{+12.3}_{-5.}\,10^{-5}$	-	Calan/Tololo[1]
	0.35 ± 0.12	-	-	5 combined searches[2]
0.15	$0.49^{+0.39}_{-0.32}$	$9.4^{+4.1}_{-1.1}\,10^{-5}$	-	EROS2[3]
0.32	$< 0.75\,(1\sigma)$			INT search [4]
0.4	$0.82^{+0.65}_{-0.45}$	$30.70^{+21.3}_{-14.5}\,10^{-5}$	$160.7^{+111.7}_{-75.7}$	SCP [5]
0.55	0.88	$32.14 \pm 3.5\,10^{-5}$ *	213.6 ± 40.6	SCP [6]

* For the cosmology $\Omega_M = 1$ $\Omega_\Lambda = 0$. [1]Hamuy & Pinto (1999); [2]Cappellaro et al. (1997); [3]Hardin et al. (1999); [4]Hamilton (1999); [5]Pain et al.(1996); [6]Pain et al.(1999)

"instantaneous efficiency". On the contrary, if SNe Ia come from the alternative candidate system, taking longer time to explode, then they will accumulate with cosmic time and one would see a sizeable increase in the number of SNe Ia per unit mass in forming stars at a given z, towards lower z.

In addition, if metallicity effects do not play any important role in SNe Ia, the number of SNe Ia per unit mass in forming stars should remain insensitive to the progressive enhancement in metals of the interstellar medium along cosmic history. However, in the Algol–type scenario, it has been suggested that metallicity plays an important role as the material transfered onto the WD forms a strong wind close to the WD surface which prevents the accretion rates to become very large (Hachisu, Kato & Nomoto 1996). The role of this wind is claimed to be very sensitive to metallicity, and calculations predict that the number of SNeIa at low metallicity (z > 1) should drastically decrease (Hachisu, Kato & Nomoto 1996). A drop at z > 1 would thus confirm the metallicity effects suggested by Hachisu et al. (1998) and Kobayashi et al. (1998).

Has the efficiency in producing SNe Ia changed along cosmic time? What is the local efficiency in producing SNe Ia? Gallego et al. (1995) measured the local star formation rate from Hα emission galaxies and derived a value of $\rho_\star = 3.7 \times 10^{-2}\ M_\odot\ h^2\ Mpc^{-3}$. A more recent estimate by Treyer et al. (1998), from a UV–selected galaxy redshift survey, suggests a slightly higher star formation rate. Their estimate of the dust–corrected star formation rate is $\rho_\star = 4.3 \times 10^{-2}\ M_\odot\ h^2\ Mpc^{-3}$. If we take the SNeIa rates obtained by Hamuy & Pinto (1998) from the Calan/Tololo survey ($\sim 2.2 \times 10^{-5}\ SNeIa\ yr^{-1}\ Mpc^{-3}$), and divide it by the local star formation rate, we obtain a local efficiency in SNeIa production of 2×10^{-3}. This means that, locally, every 500 $M_\odot$ going into star formation give 1 SNIa.

Interestingly enough, the efficiency in producing SNIa appears to be very similar towards high z (up to z$\sim$ 0.5) as can be derived from Table 1 and seen in Figure 2. A measurement of SNIa rate around z$\sim$ 1 would be crucial to check whether there is a progenitor taking about 10^8 yr to explode instead of a progenitor taking a few times 10^9 yr to explode. If this constancy of the SNeIa production stays up to z $\sim$ 1, it will favor the merging of WDs against the Algol–type accreting WDs as the main path to SNeIa explosions.

4. COUNTS AND OTHER SNIA FREQUENCIES

The measured rates at high redshift are quoted in various ways suitable for stressing one or another aspect of the cosmological information contained. The most direct number is the *sur-*

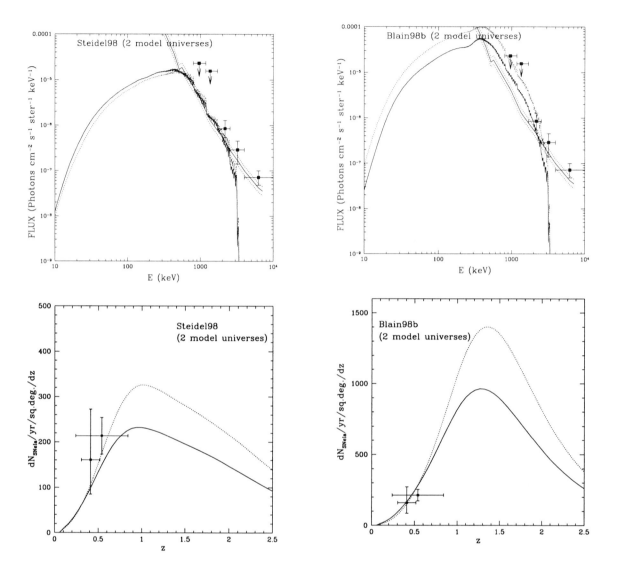

Figure 4. Type Ia rates and their cosmic background. SN rates at high-z (lower panel) are calculated using Steidel et al. (1998) star formation. The corresponding γ-ray background (upper panel) is calculated according to those rates (see Ruiz–Lapuente, Cassé & Vangioni–Flam (1998). The two models of universe shown are an open model with $\Omega_M=0.3$, $\Omega_\Lambda=0$ (solid line) and a flat model with $\Omega_M=0.3$, $\Omega_\Lambda=0.7$ (dotted line).

Figure 5. Type Ia rates and their cosmic background. SN rates at high-z (lower panel) are calculated using one of the star formation histories suggested by Blain et al. (1998) (model C in that paper). The corresponding γ-ray background (upper panel) is calculated according to those rates (see Ruiz–Lapuente, Cassé & Vangioni–Flam (1998)). The two models of universe shown are an open model with $\Omega_M=0.3$, $\Omega_\Lambda=0$ (solid line) and a flat model with $\Omega_M=0.3$, $\Omega_\Lambda=0.7$ (dotted line).

vey rate or *number counts per unit redshift*, which is calculated from the number of supernovae detected during the search by knowing the area of the sky covered and the depth in redshift reached, as well as the detection efficiency and the time spent observing each field.

$$survey\ rate(z_1 < z < z_2) = \frac{N_{SN}}{\sum_i area_i \times \Delta T_i} \quad (6)$$

where ΔT_i is the control time, which estimates the weighted sum of days during which the SN can be detected, the weighting factor being the detection efficiency ϵ (see Pain et al. 1996).

That number has the raw information, and in fact, it has a high potential to make the cosmological comoving volume test (Ruiz-Lapuente & Canal 1998). It can provide a different test of the Λ dominancy as we mentioned in section 2.

In the surveys one can also give the numbers as density of SNe Ia explosions at a given z per comoving volume and year, ρ_{SNIa}. In that case we need to calculate the effective volume covered by the survey, a quantity requiring the choice of a cosmology ($\Omega_m=1$, $\Omega_\Lambda=0$, for instance).

Finally, one can also calculate supernova rates per blue luminosity of their host galaxies. Such unit for measuring supernova rates is called SNu and expresses the number of SNe Ia per century per 10^{10} $L_\odot^B$ (van den Bergh and Tammann 1991). It refers to the number of SNe Ia per young population of stars. This unit was normally used to measure rates in our Galaxy or other galaxies in our Local Group (Tammann et al. 1996), but it can be extended up to high z by estimating the luminosity density at each z. Table 1 shows estimates of the supernova rates along z.

5. OVERALL ABSOLUTE RATES AND THE BACKGROUND CONTRIBUTION

Background light produced by supernovae basically arise from supernovae up to z~ 1. With the information we have, of similar rates of SNe Ia

per unit mass of star formation at low and high z, we can compute the background due to those supernovae. The rates predicted by different cosmic star formation histories can be examined in the optical and compared to observed rates at various z. Those rates can be tested in the γ–ray domain through the prediction of their supernova cosmic γ–ray background.

5.1. Star formation history

Much has been learnt recently on the evolution of star formation. The approach by Madau et al. (1996) of deriving star formation rates from the UV emission density at high z in the Hubble Deep Field revealed an increasing star formation towards higher z. The star formation history reconstructed by Madau, Pozzetti & Dickinson (1998) peaks at z~ 1.5 with a range of 0.12–0.17 M$\odot$ yr^{-1} Mpc^{-3} to fall again at higher z. Soon that decrease at z > 1.5 was questioned from other different approach: the derivation of the star formation history from Lyman break galaxies (Steidel et al. 1998). Steidel et al. (1998) found that the spectroscopic properties of the galaxy samples at z $\sim$ 4 and z $\sim$ 3 are indistinguishable, as are the luminosity function shapes and the total UV luminosity density between z $\sim$ 3 and $\sim$ 4. From that work it is suggested that the star formation rate does not decrease at z > 2 but that it levels off. That conclusion of a star formation rate at z > 2 higher than the one obtained by Madau et al. (1996) is also obtained from results at long wavelengths. The study of star formation with the submillimeter SCUBA array by Hughes et al (1998) reveals significant dust enshrouded star formation at high z. According to those authors, the star formation rate density over the range 2< z < 4 would be at least five times higher than the inferred from the UV emission of the HDF galaxies.

Blain et al.(1998) reconstruct the star formation histories compatible with the observations at various wavelengths incorporating the SCUBA results and the derivations from chemical evolution at high z. Some of their compatible star formation histories peak at a z closer to 2 (rather than 1 or 1.5). We have found through our predictions of rates and γ-ray background that some of the proposed star formation histories by Blain et al. (1998) are disfavored against the observations, whereas the cosmic star formation history with a peak at $z \sim 1$–1.5 of the level found by Madau et al. (1998) and Steidel et al. (1998) seems to fall well within the observational measurements. This question will be briefly addressed in the next section.

5.2. The extragalactic diffuse gamma–ray background and SNe

Supernovae are very important contributors to the gamma–ray background in the MeV. Among supernovae, SNIa superseed by far the contribution SN II and SN Ibc. In Type Ia supernovae the fraction of γ-rays which escapes increases with time, and the peak of the γ-ray flux is found about 3 months after explosion. Type II SNe synthesize a lower mass in radioactive nuclei than Type Ia SNe and their larger overall mass and smaller expansion velocities provide a larger optical depth for Componization of the emission. In Type II supernovae the peak of their emission shifts to the hard X–ray domain. (see Figure 3).

Clayton & Silk first proposed in 1969 that γ-ray emission from supernovae would give rise to a diffuse extragalactic background. Indeed, the observations performed by various satellites in the MeV range (APOLLO, SMM, COMPTEL) detect a particular slope from hard–X rays up to a few MeVs (Kappadath et al. 1996). The discussion of the origin of the background at those wavelengths is tied to the density of sources beyond the Galaxy able to provide photons at those wavelengths. Emitters in the MeV range could be blazars, instead of supernovae. But the shape of those are not well suited to reproduce the spectra. Supernovae (The et al. 1993, Zdziarski 1996, Watanabe et al. 1998) adjust well the background spectral shape at those wavelengths. But do we have enough supernovae at high-z to account for such intense background? The answers given (Watanabe et al. 1998; Ruiz-Lapuente, Cassé &

Vangioni–Flam 1999) are affirmative. The background level can be tested by computing the supernova rates at high-z and integrating the emission along redshift (Figure 4). The agreement of what we see in the optical and what we detect in the γ-ray domain places constraints on the amount of dust obscuring Type Ia supernovae up to $z \sim 1$. It is unlikely that dust would obscure a major fraction of SNe Ia events up to $z \sim 1$. If it were so, we would have a higher level of SNe Ia rates than the one derived by the optical searches and this would lead to a γ-ray background in excess of that observed.

6. CONCLUSIONS

The use of Type Ia supernovae at high-z as "calibrated candles" has provide a measurement of the cosmological parameters in the Ω_M-Ω_Λ plane. For a flat universe ($\Omega_T = 1$), the results by the Supernova Cosmology Project imply $\Omega_M = 0.28^{+0.09}_{-0.08}(\text{stat})^{+0.05}_{-0.04}$ (syst). Similar results are obtained by the High-Z Team.

The evidence for a positive Ω_Λ coming from the magnitude-z diagram of supernovae can also be cross-checked in the number counts of SNe Ia along z, once we learn about the cosmic star formation history and about the efficiency and timing in producing Type Ia supernovae.

Type Ia supernovae searches at high-z have already provided results which suggest the rate of Type Ia supernovae at a given z does not change from z=0 till $z \sim 0.55$ when expressed in SNu: supernovae per century per 10^{10} $L_\odot^B$. That value being about 0.5 h^2 SNu (see references in Table 1). This result is of importance because points to a similar efficiency in the binary processes giving rise to Type Ia explosions at high and low z. A major test is still awaiting through the estimate being done by the Supernova Cosmology Project at $z \sim 1$. Several mechanisms suggest a drop in the production of Type Ia supernovae at z> 1 or even an absence of supernovae (Kobayashi et al. 1998; Hachisu et al. 1998). The discovery of a SNIa at z∼ 1.2 (Aldering et al. 1999) argues against those predicted effects.

To conclude we just want to mention that searches of supernovae with the HST, VLT and in the future with the NGST will enable to enlarge our measurements up to z well beyond 1. Measurements of rates will then allow us to clarify the questions raised here and the supernovae discovered in those searches will become even deeper cosmological probes.

I would like to thank my partners of the Supernova Cosmology Project for all the shared excitement in the project. Thanks go as well to Brian Schmidt, former partner in supernova research, for communicating the High-Z results.

REFERENCES

1. Aldering et al. 1999 (Supernova Cosmology Project), in preparation
2. Blain,A.W., Smail, I., Ivison, R.J. & Kneib, J-P. 1999, MNRAS, 302, 632
3. Branch, D., Livio, M., Yungelson, L.R., Boffi, T.R. & Baron, E. 1995, PASP 107, 1019
4. Cappellaro, E., Turatto, M., Tsvetkov, D. Y., Bartunov, O.S., Pollas, C., Evans, R., & Hamuy, M. 1997, A & A 322, 431
5. Clayton, D.D. & Silk, J. 1969, ApJ, 158, L43
6. Gallego, J., Zamorano, J., Aragon-Salamanca, A., & Rego, M. 1995, ApJ, 361, L1.
7. Goobar, A. & Perlmutter, S. 1995, ApJ, 450, 14
8. Hachisu, I., Kato, M., & Nomoto, K. 1996, ApJ, 470, L97
9. Hamilton, J.C. 1999, (Ph.D thesis. U. Paris XI)
10. Hamuy, M & Pinto, P. A. 1999, ApJ (submitted)
11. Hamuy,M. et al. 1996, AJ, 112, 2391
12. Hardin, D. et al. 1999, A & A (submitted)
13. Hughes, D. et al. 1998, Nature, 394, 241
14. Kappadath et al. 1996, A & AS, 120, 619
15. Kobayashi, C., Tsujimoto, T., Nomoto, K., Hachisu, I., & Kato, M. 1998, ApJ, in press
16. Madau, P., et al. 1996, MNRAS, 283, 1388
17. Madau, P. 1997, PASP Conf. Ser., in press, and preprint
18. Madau, P., Della Valle, M. & Panagia, N. 1998. MNRAS 297, L17
19. Madau, P., Pozzetti, L. & Dickinson, M. 1998, ApJ 498, 106
20. Perlmutter, S. et al. 1998 (Supernova Cosmology Project), ApJ, in press, astro-ph/9812133
21. Perlmutter, S. et al. (Supernova Cosmology Project), in this volume
22. Pain, R., et al. 1996, ApJ, 473, 356
23. Pain, R., et al. 1999 (preprint)
24. Phillips, M.M. 1993, ApJ, 413, L75
25. Riess, A., Press, W. & Kirshner, R. 1996, ApJ 473, 88
26. Riess, A. et al. 1998 (High-Z Team), AJ, in press, astro-ph/9805201
27. Ruiz–Lapuente, P., Burkert, A., & Canal, R. 1995, ApJ, 447, L69
28. Ruiz–Lapuente, P., Canal, R., & Burkert, A. 1997, in Thermonuclear Supernovae, ed. P. Ruiz-Lapuente, R. Canal, & J. Isern (Dordrecht: Kluwer), 205
29. Ruiz-Lapuente, P. & Canal, R. ApJ 497, L57
30. Ruiz-Lapuente, P., Cassé, M. & Vangioni-Flam, E. 1998 (preprint)
31. Sadat, R., Blanchard, A., Guiderdoni, B. & Silk, J. 1998, A & A, 331, L69
32. Schmidt, B. et al. (High-Z Team) 1999, (in this volume)
33. Steidel, C.C., Giavalisco, M., Pettini, M., Dickinson, M.E., & Adelberger, K. 1996, ApJ 467, L17
34. Tammann, G.A., Löffler, W., Schröder, A. 1994, ApJS, 92, 487
35. The, L.-S., Leising, M.D. & Clayton, D.D. 1993, ApJ, 403, 32
36. Treyer, M. A., Ellis, R.S., Millard, B., Donas, J. & Bridges, T.J. 1999. MNRAS (in press)
37. Umeda, H., Nomoto, K., Yamaoka, H. & Wanajo, S. 1998, astro-ph/9806336
38. van den Bergh, S. & Tammann, G.A. 1991 ARA& A 29, 363
39. Watanabe, K., Hartmann, D.H., Leising, M.D., & The, L.-S, 1998, astro-ph/9809197
40. Yungelson, L., Livio, M., Truran, J.W., Tutukov, A., & Fedorova, A. 1996, ApJ, 466, 890
41. Yungelson, L. & Livio, M. 1998, ApJ 497, 168
42. Zdziarski, A.A. 1996, MNRAS, 281, L9

ELSEVIER

Nuclear Physics B (Proc. Suppl.) 80 (2000) 163–172

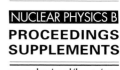

PROCEEDINGS
SUPPLEMENTS

www.elsevier.nl/locate/npe

First Detection of Gamma Rays from the Crab Nebula with the CELESTE "Solar Farm" Cherenkov Detector

D.A. Smith[a]*, R. Bazer-Bachi [b], H. Bergeret [c], P. Bruel [d], A. Cordier[c], G. Debiais [e], M. de Naurois[d], J-P. Dezalay[b], D. Dumora[a], P. Eschstruth[c], P. Espigat [f], B. Fabre[e], P. Fleury[d], B. Giebels[a,f],N. Hérault[c],J. Holder[c], M. Hrabovsky [g], R. Legallou[a], I. Malet[b], B. Merkel[c], F. Münz[f], A. Musquère[b], J-F. Olive[b], E. Paré[d†], J. Québert[a], T. Reposeur[a], L. Rob [h], T. Sako[d], P. Schovanek[g], A. Volte[f]

[a]CEN de Bordeaux-Gradignan, Le Haut Vigneau, F-33175

[b]CESR, Toulouse F-31028

[c]LAL, Université Paris Sud and IN2P3/CNRS, Orsay, F-91405

[d]LPNHE, Ecole Polytechnique, Palaiseau, F-91128

[e]GPF, Université de Perpignan, F-66860

[f]PCC, Collège de France, Paris, F-75231

[g]J.L. Optics, Academy of Sciences and Palacky Univ., Olomouc, Czech Republic

[h]Nuclear Center, Charles University, Prague, Czech Republic

We have converted the THEMIS solar array (French Pyrenees) into an atmospheric Cherenkov telescope, called CELESTE, sensitive to astrophysical gamma rays above 30 GeV (7×10^{24} Hz). In early 1998 the Crab nebula was detected at 80 GeV with a preliminary 18 heliostat setup. The full 40 heliostat array has since been commissioned. The STACEE experiment using the same technique in New Mexico is also analysing their first data. Thus, the window between the EGRET instrument and the Cherenkov imagers has been opened. We describe the CELESTE detector and the data analysis, and discuss the prospects for studying AGN (specifically, blazars) and galactic sources in this energy range.

DEDICATION : CELESTE *is the brainchild of Eric Paré, who died at the age of 39 in an automobile accident, two weeks after finding our first gamma ray signal. We dedicate this work to his memory.*

1. INTRODUCTION

Around 1990 two breakthroughs revolutionized very high energy astrophysics. First, ground-based atmospheric Cherenkov detectors became sensitive, reliable gamma ray detectors above a few hundred GeV, led by the Whipple imager [1] and followed by the Themistocle and ASGAT wavefront samplers [2,3]. Second, the EGRET instrument on the Compton satellite measured the spectra of over 150 point sources in the energy range $0.1 < E_\gamma < 10$ GeV [4]. Since then, a number of Whipple-class imagers have begun operation [5–7], and imager arrays with increased sensitivity are being planned [8,9]. GLAST will be a high performance successor to EGRET when launched in 2005 [10].

An energy gap existed between the Cherenkov telescopes and EGRET. The former "turn on" above 250 GeV; the latter have a significant counting rate only below 10 GeV. The ground devices have good flux sensitivity: the sources

---
*e-mail: *smith@cenbg.in2p3.fr*
†deceased.

detected on the ground are amongst the weakest seen by EGRET (*e.g.* Mrk 421), or were undetectable by EGRET (*e.g.* Mrk 501). And so it is quite striking that the large majority of EGRET sources have *not* been seen by the Cherenkov imagers. Two possible explanations, both of great physical interest, are : the acceleration mechanisms distort the energy spectra between the EGRET and Cherenkov ranges; and/or gamma ray absorption by the diffuse infrared background in the intergalactic medium comes into play.

The motivation to explore the intermediate energy region, $30 < E_\gamma < 300$ GeV, is thus strong. For the CELESTE experiment, we transformed the Electricité de France central receiver solar power plant at Thémis (N. 42.50°, E. 1.97°, 1650 m. a.s.l.) into a gamma ray telescope [16–18]. We adapted the atmospheric cherenkov wavefront sampling technique to the geometry imposed by the power plant, in order to proceed rapidly and economically. The presence of the CAT imager on the same site, as well as the Themistocle and ASGAT experiments, also speeded this project.

After a discussion of the scientific goals of CELESTE, we will describe the experimental apparatus and we present the first detection of the Crab nebula in this new energy range. References [14,15] are comprehensive reviews of very high-energy gamma-ray astronomy.

2. SCIENCE

2.1. Extragalactic gamma rays: Blazars

Figure 1 shows a remarkable result obtained in 1997. A flare of the blazar Mrk 501 was detected simultaneously by the BeppoSAX X-ray satellite and the Cherenkov imagers Whipple, HEGRA, and CAT [19]. During the flare, the X-ray and TeV intensities increased by two orders of magnitude. According to the models this coincides with shifts of the peak positions shifted to higher energies. The dark curves are from a Synchrotron Self-Compton model (SSC) for gamma ray production. Models with an external source of photons can better reproduce the inverse compton peak. The rapid variability is relatively difficult to explain with models where energetic protons dominate gamma ray production.

The answer to the question raised above, why have so few EGRET blazars been seen by the Cherenkov imagers, lies partially in the figure: EGRET detects preferentially those blazars, such as Mrk 421, for which the inverse compton peak is near 10^{23} Hz (400 MeV). The sensitivity range of the Cherenkov imagers is then beyond the peak. Inversely, Mrk 501 (and 1ES2344+514, the third Whipple blazar [20]) has its peak in the range accessible by the Cherenkov imagers, and EGRET finds itself "in the hole".

To constrain the blazar models further, and thus AGN models in general, requires further detections beyond the EGRET energy range. This is the goal of the CELESTE experiment: peak sensitivity is near 40 GeV (10^{25} Hz) allowing detection of many EGRET blazars (mainly radio-selected BL Lac objects) as well as X-ray selected BL Lac objects as have been seen by the Cherenkov imagers. Once the spectra have been measured for a larger variety of blazars, a major step towards understanding AGN in general will have been made.

2.2. Diffuse infrared gamma absorption

Mrk 421 is the dimmest blazar seen by EGRET, while Mrk 501 and 1ES2344+514 are even fainter. Naive extrapolation of the other EGRET sources to imager energies suggests that many should have been seen. Gamma-ray production arguments could suffice to explain the shortfall. However, one notes that Mrk 421, Mrk 501, and 1ES2344+514 are the closest known blazars (redshifts from $z = 0.031$ to 0.044). An alternative explanation is then that gamma rays are absorbed via $\gamma\gamma \rightarrow e^+e^-$ in the intergalactic medium, where the target photons are from the diffuse infrared background [12]. The threshold energy of the target photons is m^2/E_γ (m is the electron mass). The density of diffuse extragalactic light is a decreasing power law from 3 K to UV frequencies and beyond. For TeV photons (and thus infrared target photons), one optical depth corresponds to a redshift of about $z = 0.1$, as shown in figure 2. At EGRET energies the effect is negligible. The intermediate energy region covered by CELESTE should permit observation of the spectral cut-off due to infrared absorption.

Distinguishing intrinsic spectral cut-offs from

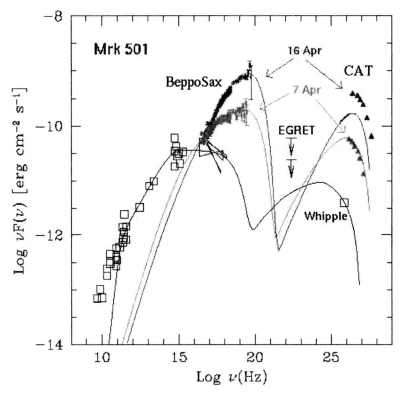

Figure 1. Spectrum of the blazar Mrk 501. The CAT Cherenkov imager (inverse compton peak, to the right) and the X-ray satellite SAX (synchrotron peak, to the left) both saw large changes in the spectral shape and intensity during a flare in April 1997. CELESTE, under construction at the time, covers the decade in energy just below CAT. The model is from Pian *et al.*

intergalactic absorption will be tricky, at best. And CELESTE, with coverage of a single decade in energy, will need information from the adjacent energy ranges as well as a larger sample of sources than has presently been observed from the ground. Nevertheless, the scientific gain could be considerable: galaxy formation theory predicts that the extragalactic *IR* density is a good estimator of the epoch of galaxy formation. Galactic *IR* foregrounds render direct measurements of the extragalactic *IR* density difficult. The CELESTE energy range is thus critical.

2.3. Supernova remnants

Supernova explosions are the most plausible sites for the acceleration of charged particles to

the energies observed in the cosmic radiation: the observed spectrum is a power law up to the "knee" near 10^{15} eV. (The steeper power law beyond the knee could emanate from extragalactic acceleration sites. The blazar studies discussed above should cast light on this problem.)

Nevertheless, almost a century after the discovery of cosmic rays, only scant direct evidence pinpoints supernova remants as cosmic ray accelerators. It's a tough business - for the high-energy charged particles to produce a bright secondary gamma ray source requires a target such as a molecular cloud. The number of candidate SNR's is thus small. The intensity is near the limits of EGRET's sensitivity; the Cherenkov imagers

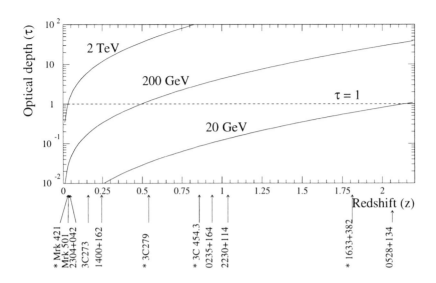

Figure 2. *The optical depth τ for gamma ray absorption by $\gamma\gamma \to e^+e^-$ pair production, as a function of the redshift of the source, for three gamma ray energies. The curves are calculated using a power law approximation of the diffuse extragalactic infrared photon density wavelength dependence. Absorption become appreciable in the energy range in which CELESTE is sensitive.*

energy range may be higher than the spectral cut-off. Furthermore, SNR's tend to be extended spatially and all Cherenkov telescopes have highest performance for point sources. CELESTE operates in an energy region where non-observations may sufficiently constrain models. Reference [13] discusses the implications of Whipple's upper limits on the gamma ray flux from supernova for theories of the orgin of cosmic rays.

3. THE ATMOSPHERIC CHERENKOV TECHNIQUE

An energetic particle incident on the atmosphere initiates a particle cascade. Below a few tens of TeV the only part of the shower reaching the ground is the Cherenkov light emitted by the ultrarelativistic charged shower particles. Photomultiplier tubes mounted in the focal plane of one or several mirrors sample a fraction of this Cherenkov light, from which the energy and direction of the primary particle are deduced. Primary particle identification is sufficient to enhance the gamma ray signal above the hadron background on a statistical basis.

Two major classes of Cherenkov telescopes exist. "Wavefront samplers" use several mirrors distributed over a field of the size illuminated by a shower (about 200 meters). Each mirror has one to several phototubes, to sample the spatial distribution and arrival time of the Cherenkov light on the ground. The only wavefront samplers to have seen a gamma ray source (the Crab nebula) are ASGAT and Themistocle, both at the Themis site.

"Imagers" use many phototubes mounted on a single mirror, and thereby sample the angular distribution of the Cherenkov light at one point in the field. (A variant is to use additional imagers for stereoscopic vision). Whipple remains the premier imager, but the CAT telescope at Themis has comparable flux sensitivity and energy threshold, compensating for a smaller mirror (16 m^2 vs. 75 m^2) with a smaller pixel size (2.2 mr vs. 4 mr) and faster optics and electronics [5].

In general, Cherenkov telescopes trigger when the signals from n phototubes exceed m photoelectrons, in a coincidence time gate τ. The phototubes are illuminated by the constant diffuse night sky light ϕ (photons per unit time, area, and solid angle), and the threshold m has to be high enough above the average light level so that fluctuations above the mean yield a rate of accidental coincidences much lower than the Cherenkov trigger rate. For a given phototube type (speed, dispersion, and dark current are the key parameters) the fluctuations scale as the square root of the mean light level per phototube, which is $\phi\tau\epsilon\Omega A$. Ω is the solid angle per pixel, ϵ is the photon collection efficiency, and A is the mirror area. The number of Cherenkov photons scales as $\epsilon E_\gamma A$, where E_γ is the primary gamma ray energy, and hence the energy threshold for a Cherenkov gamma ray telescope is

$$E_\gamma^{threshold} \propto \sqrt{\frac{\Omega\tau\phi}{A\epsilon}}. \qquad (1)$$

The current generation of imagers have pushed τ and Ω to their practical limits. The CAT imager is an excellent example: an isochronous mirror and very fast phototubes and electronics match the detector response to the intrinsic time spread of the Cherenkov signal (a few nanoseconds). While waiting for technological progress to improve quantum efficiencies and hence ϵ, the race to lower energies amounts essentially to a search for larger mirror areas, A.

The MAGIC project is the most ambitious effort to extend the imaging technique to 30 GeV [21]. MAGIC should be operational in 2003. CELESTE and STACEE (see below) have already reached the new energy regime, exploiting solar plant infrastructures to obtain very large mirror areas. Two Cherenkov projects dominate the horizon : HESS is a franco-german imager array that will be built in Namibia [8]; VERITAS is

the American version [9]. Operation will begin in 2002.

4. IMAGERS AND WAVEFRONT SAMPLERS

The key parameters defining the performance of a gamma ray telescope are: flux sensitivity; angular resolution; and energy resolution. Flux sensitivity depends on the collection area, background rejection, and angular resolution of the instrument. (The collection area here is not the mirror area sampling the shower but the area over which air showers trigger the telescope.) The angular resolution further determines the precision with which a source can be located. About half of the EGRET sources are unidentified, in part due to EGRET's $\sim 1°$ source localisation. The typical angular resolution of a Cherenkov telescope is about a tenth of that, and so Cherenkov telescopes should be able to contribute to the identification problem.

Imagers have worked best above 200 GeV, since the fine sampling of their cameras allows excellent hadron rejection. But below 100 GeV, hadron backgrounds are naturally suppressed because the Cherenkov yield of the hadron showers decreases faster than that of the gamma showers. Hence, the advantage of the imagers is less.

Below 100 GeV single cosmic ray muons replace hadron showers as the most important background for high precision Cherenkov imagers. A single muon passing near the mirror appears as a circle or an arc in the imager camera and can be easily rejected. But muons farther from the mirror produce arcs so short that some are indistinguishable from a gamma ray signature. Modern imagers thus incorporate subsystems to help measure this background, and the trend is towards multiple (e.g. stereo) imagers in part to decrease muon sensitivity. Wavefront imagers, and especially CELESTE, are inherently insensitive to muons. This is because the small field-of-view (10 mr) combined with the separation between heliostats makes it nearly impossible for a single Cherenkov cone to illuminate more than a few phototubes.

At low energies, the background due to cosmic

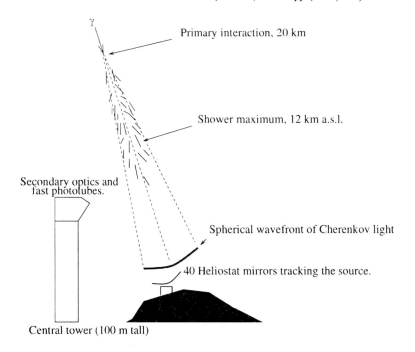

Figure 3. *Principle of a "solar farm" Čerenkov gamma ray telescope. Heliostats and secondary mirrors are not to scale. The tower is 100 meters tall.*

ray electrons is also more important than above 200 GeV. Their energy spectrum is steeper than the hadron spectrum, and as stated, the efficiency for primary hadrons falls off rapidly below 100 GeV. Since electron induced showers are practically indistinguishable from gamma ray showers, angular resolution is the key to suppressing the electron background.

Angular resolution, especially at lower energies, tends to be limited by the shower development rather than by instrument response. Scattering in the first few generations of the shower, and deflection of low energy secondaries in the geomagnetic field, are the factors limiting angular precision. Hence, a high-granularity imager does not necessarily outperform a wavefront sampler. Both should achieve resolutions around 0.1°.

5. SOLAR FARMS

The central receiver solar power plants, built in the early 1980's in several countries and then abandoned, consist of thousands of square meters of mirrors designed to focus light from a celestial body onto the top of a central tower. The heliostats collect the Cherenkov light just as in the wavefront sampling detectors. Figure 3 sketches the basic idea. The major difference with the "classic" wavefront samplers is the focal length of the mirrors: the heliostats reflect the light to the top of the tower, rather than onto phototubes mounted directly on the mirrors. Correcting for path length differences which vary as a source is tracked, to permit fast coincidence electronics, requires secondary optics that image each heliostat onto a single photomultiplier tube.

Modern imagers have several hundred photomultipliers. The most ambitious heliostat array projects foresee only one or two hundred channels, and we believe that good science can be done with as few as 40 channels. Hence, the cost is reduced.

The heliostat array approach has weaknesses. For example, since the heliostat mirrors are oriented so as to reflect the Cherenkov light onto the tower, a source is in general off the heliostat axis, increasing light losses due to aberrations. The angles vary as the source crosses the sky. Thus, the dependence of the energy threshold and flux sensitivity on source position is complex. The relative low cost and speed with which a solar plant can be used to explore 30 GeV gamma rays outweigh the disadvantages.

6. THE CELESTE EXPERIMENT

At Themis we removed the 30-ton heat receiver from the top of the tower, thus liberating a five-by-five meter opening for the secondary optics. We installed a counting house in the tower, just below the focal area, with electronics and acquisition computers.

Forty heliostats are in use (2000 m^2). Measurements of their optical properties using sun and moon images projected on the tower confirm that the mirror quality is adequate for our purposes. The heliostats have been aligned by recording the phototube currents while scanning bright stars.

The secondary mirrors are spherical, and are bigger than the ~meter spot-size in order for the phototube camera to view the entire heliostat field with small aberrations. The mirror is divided into six subsections to decrease shadowing by the phototube camera: the lowest subsection views the farthest heliostats while the top three subsections view the heliostats at the foot of the tower. Each phototube sees a single heliostat. The field-of-view of 10 mr (total) is sharply defined using solid Winston cones on each phototube. The CAT and CELESTE mirrors as well as the CELESTE Winston cones were made by our collaborators in the Czech Republic. The small field-of-view requires us to aim the telescope not at the gamma ray source itself, but at the region in the atmosphere where the Cherenkov light is generated (about 10 km above the site when tracking a source near the zenith).

The phototube signals are recorded with 1 Ghz VME flash ADC's: in this way variable delay, charge measurement, and timing measurement are made by the same circuit. The Flash ADC's were developed specifically for CELESTE and provide an excellent compromise between cost and performance [22]. The trigger is conceived to reach the lowest possible energy threshold: programmable analog delays allow the Cherenkov signals to be summed in five groups of eight heliostats each. The resulting signal is discriminated, path length differences between the groups are compensated for using programmable logic delays, and the trigger is a threefold coincidence amongst the five groups. We trigger at 3 photoelectrons per heliostat, which according to detailed Monte Carlo simulations corresponds to a gamma ray primary energy near 30 GeV. The trigger rate is 16 Hz.

The data acquisition system is rather elaborate. An HP workstation orchestrates tasks amongst secondary computers: heliostat tracking; phototube anode currents; phototube high voltage; and weather monitoring are each on a separate PC. The VME crates are controlled by Motorola processors running the Lynx variant of real-time unix. All exchanges amongst the computers use standard TCP/IP protocol and the data files are stored on the central HP. The night's data is transferred each morning to major computer facilities in Lyon for subsequent analysis.

7. THE STACEE EXPERIMENT IN NEW MEXICO

The modern incarnation of the idea to re-cycle a heliostat array into a Cherenkov gamma-ray telescope came from Tumay Tümer at U.C. Riverside, and initial work focussed on the Solar-1 facility near Barstow. Solar-1 work continues. But besides CELESTE, the project that has progressed furthest is STACEE, at Sandia Laboratories [23].

Form follows function, and the real differences between the two experiments are few. As examples, STACEE's secondary optics are off-axis parabolas; the trigger scheme is purely digital ; the mirror area with 32 heliostats is 1200 m^2. The current telescope is still in many respects a prototype and performance will improve. The gamma ray trigger threshold is near 80 GeV. The STACEE collaboration is currently analysing data acquired during this past winter and results should be forthcoming.

Crabe

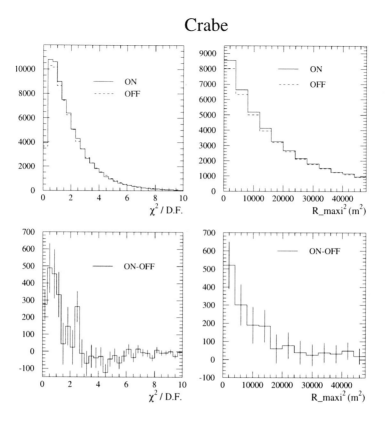

Figure 4. $\chi^2/D.F.$ *(left) and R^2 (right) distributions for Crab data. R is the distance from the center of the field-of-view to the reconstructed shower position. Top: superposition of ON and OFF source histograms. Bottom: Difference between the ON and OFF source data. With the cuts described in the text the signal is 5.6σ after 3.5 hours of observation.*

8. FIRST HELIOSTAT ARRAY DETECTION OF GAMMA RAYS

8.1. Crab data sample

CELESTE tracked the Crab for the first time in February 1998 with 18 of the current 40 heliostats. The optics and photomultipliers were definitive. The trigger electronics was as well, configured in three groups of 6 heliostats each. The thresholds were 5.1 photoelectrons per heliostat, or 55 GeV. The trigger rate was 8 Hz.

Readout electronics and the overall acquisition architecture was temporary. In particular the Flash ADC's then in use contributed electronic noise to the readout. More importantly, only three FADC channels were available, forcing us to record three phototubes per ADC thereby increasing the fluctuations of the night sky noise underlying the Cherenkov signal by a factor of $\sqrt{3}$, for only 9 heliostats. By the time the telescope had been aligned and calibrated, the Crab was visible for only a short time after sunset, and the data set is small.

Cherenkov telescopes demonstrate a gamma ray signal by comparing data with other data taken while tracking the same path in the sky, just after (or before) the gamma ray source. In CELESTE each 30 minute "ON" run is followed by a 30 minute "OFF" run, and the difference between the ON and OFF rates, after applying cuts intended to reject hadronic showers while preserving electromagnetic showers, gives the gamma ray signal. The atmospheric conditions must thus be stable during an hour. After data quality cuts, 211 minutes of Crab data remained.

8.2. Analysis method

The first step in analysis is to find peaks in the FADC data, and thereby to estimate the Cherenkov amplitude and wavefront arrival time for each heliostat, on a shower-by-shower basis. Next, we use the pulse times to reconstruct a spherical wavefront of radius $(11km)/\cos\theta$ where θ is the zenith angle. The amplitude-dependent timing resolution was estimated by injecting software pulses into real data, and then reconstructing them. We imposed the following cuts:

- The sum of the peaks for all heliostats exceeds 90 photoelectrons (total energy cut).

- At least 8 heliostats have a peak of at least 4 photoelectrons.

- $\chi^2/D.F. < 2$ after a second iteration of the wavefront timing fit.

- $R < 125$ meters, where R is the distance of the reconstructed sphere's center from the line-of-sight.

With only 9 phototubes digitized, and with the increased night sky light per channel, the spatial distribution of the pulseheights does not allow an estimation of the shower impact point. For this reason no explicit shower direction reconstruction was attempted.

The raw data rates varied by a few percent between the ON and the OFF data, with sign depending on the source, due to small differences in the night sky background. The first two cuts reduce this effect, as will be discussed further, below. The remaining cuts are intended to reject hadron showers while preserving the gamma signal. For the Crab data, the raw OFF rate exceeded the ON rate by 5%.

Figure 4 shows the χ^2 and R distributions after the first cuts. The excess of events at small values in the ON data are as expected for gamma rays. The significance of the excess is 5.6 sigma in 3.5 hours and the rate is 6 gammas per minute. (Naive expectations are for 11 gammas per minute, assuming 100% detection efficiency.) The minimum energy after analysis cuts is 80 GeV. We emphasize that these results were obtained with an incomplete detector and that the work necessary to properly quantify the results will not be done, since we are now analysing data from the 40 heliostat detector, obtained in much improved experimental conditions. This result is therefore merely a "proof-of-principle".

8.3. Cross-checks

Two other data samples from the 18 heliostat array were analysed. 167 minutes of "OFF-OFF" data were obtained by tracking a direction near, but not on, Mrk 421 i.e. away from any known gamma source, as well as a corresponding OFF region. A 1% excess in the raw ON rate was reduced to 0.1% by the first two cuts. After the complete analysis, a -0.5σ non-signal was obtained.

Similarly, we analysed 234 minutes of Mrk 421 data. This analysis is trickier: a 6th magnitude star coincident with Mrk 421 increases the raw ON source trigger rates to several percent more than the OFF. After analysis, an artifact of this initial difference remains and our interpretation is that there is no signal. The CAT imager reported that the average intensity of Mrk 421 was one-third of the Crab during this time (March 1998), below the sensitivity of the 18 heliostat telescope.

9. CONCLUSIONS AND PROSPECTS

We have argued that the solar farm variant of the wavefront sampling technique is a good way to measure gamma ray spectra from astrophysical sources in the unexplored energy range between satellites and current ground-based devices. The advantages of the imaging technique as compared to wavefront sampling at 1 TeV are not as strong when one descends below 100 GeV. The cost is relatively modest because we exploit the substantial infrastructure available at solar central receiver power facilities.

We have thus built the first gamma ray telescope sensitive between 30 and 300 GeV. The full 40 heliostat array is taking data and analysis is underway. This energy range is critical to test AGN models in the extreme case of blazars, and to constrain the extragalactic diffuse infrared background via gamma-photon absorption.

Data taken when only half of the heliostats and one tenth of the digitization channels were available has been analysed, and we have presented the first signal from the Crab nebula at 80 GeV. The result in of itself is modest - systematic uncertainties remain large and the energy threshold and flux sensitivity are short of those expected with the full detector - nevertheless it is a satisfying validation of the experimental approach.

The 40 heliostat array has a 30 GeV gamma ray energy threshold, and the Crab nebula should be detectable in a matter of minutes. Sources with $1/20^{th}$ of the Crab's intensity should be seen in 20 hours. Combining extrapolations of the power law source spectra measured by the EGRET instrument on board the Compton Gamma Ray Observatory satellite with the predictions of spectra for known X-ray selected BL Lac objects not seen by EGRET, CELESTE should detect 20 blazars before the next generation of telescopes comes on line beginning in 2003.

L. Rob gratefully acknowledges the support of the Czech Grant Agency.

REFERENCES

1. J. Quinn *et al*, Ap. J. Lett. **456** (1996) 83.
2. P. Baillon *et al*, Astropart. Phy. **1** (1993) 341.
3. P. Goret *et al*, A&A **270** (1993) 401.
4. D.J. Thompson *et al*, Ap.J.S. **101** (1995) 259.
5. A. Barrau *et al*, Nucl. Instr. Meth. **416A** (1998) 278.
6. N. Bulian *et al*, Astropart. Phy. **8** (1998) 223.
7. M.D. Roberts *et al*, Astron. Astrophys. **337** (1998) 25.
8. http://www-hfm.mpi-hd.mpg.de/HESS
9. http://venus.physics.purdue.edu/veritas
10. http://glast.gsfc.nasa.gov
11. C. von Montigny *et al*, Ap. J. **440** (1995) 525.
12. S.D. Biller *et al*, Ap. J. **445** (1995) 227.
13. J.H. Buckley *et al*, Astron. Astrophys. **329** (1998) 639.
14. R.A. Ong, Phys. Rep. **305** (1998) 93-202.
15. C.M. Hoffman, C. Sinnis, P. Fleury, M. Punch, Rev. Mod. Phys. (in press, March 1999).
16. CELESTE proposal, available at http://wwwcenbg.in2p3.fr/Astroparticule, or from the author.
17. B. Giebels *et al*, Nucl. Instr. Meth. **412A** (1998).
18. D.A. Smith *et al*, Nucl. Phys. **54B** (Proc. Suppl.) (1997) 362-367.
19. E. Pian *et al*, Ap. J. Lett. **492** (1998) p. 17.
20. M. Catanese *et al*, Ap. J. Lett. **501** (1998) p. 616.
21. R. Mirzoyan, Nucl. Phys. **54B** (Proc. Suppl.) (1997) 350-361.
22. Etep 301c 2-channel 1 GHz Flash ADC, see http://www.etep.com.
23. M.C. Chantell *et al*, Nucl. Instr. Meth. **408A** (1998) 468.

ELSEVIER

Nuclear Physics B (Proc. Suppl.) 80 (2000) 173–181

NUCLEAR PHYSICS B
**PROCEEDINGS
SUPPLEMENTS**

www.elsevier.nl/locate/npe

A Black Hole Spectral Signature

Lev Titarchuk [a] and Philippe Laurent [b]

[a]NASA/GSFC/GMU, code 661, Greenbelt, MD 20771, USA

[b]CEA/DSM/DAPNIA/SAp, CEA Saclay 91191, Gif sur Yvette, France

An accreting black hole is, by definition, characterized by the drain. Namely, the matter falls into a black hole much the same way as water disappears down a drain - matter goes in and nothing comes out. As this can only happen in a black hole, it provides a way to see "a black hole", an unique observational signature. The accretion proceeds almost in a free-fall manner close to the black hole horizon, where the strong gravitational field dominates the pressure forces. In this paper we present analytical calculations and Monte-Carlo simulations of the specific features of X-ray spectra formed as a result of upscattering of the soft (disk) photons in the converging inflow (CI) into the black hole. The full relativistic treatment has been implemented to reproduce these spectra. We show that spectra in the soft state of black hole systems (BHS) can be described as the sum of a thermal (disk) component and the convolution of some fraction of this component with the CI upscattering spread (Greens) function. The latter boosted photon component is seen as an extended power-law at energies much higher than the characteristic energy of the soft photons. We demonstrate the stability of the power spectral index over a wide range of the plasma temperature $0 - 10$ keV and mass accretion rates (higher than 2 in Eddington units). We also demonstrate that the sharp high energy cutoff occurs at energies of 200-400 keV which are related to the average energy of electrons $m_e c^2$ impinging upon the event horizon. The spectrum is practically identical to the standard thermal Comptonization spectrum when the CI plasma temperature is getting of order of 50 keV (the typical ones for the hard state of BHS). In this case one can see the effect of the bulk motion only at high energies where there is an excess in the CI spectrum with respect to the pure thermal one. Furthermore we demonstrate that the change of spectral shapes from the soft X-ray state to the hard X-ray state is clearly to be related with the temperature of the bulk flow. In other words the effect of the bulk Comptonization compared to the thermal one is getting stronger when the plasma temperature drops below 10 keV. We clearly demonstrate that these spectra emerging from the converging inflow are a inevitable stamp of the BHS where the strong gravitational field dominates the pressure forces.

1. INTRODUCTION

Do black holes interact with an accretion flow in such a way so a distinct observational signature entirely different from those associated with any other compact object exists? In other words can the existence of a black hole be solely inferred from the radiation observed at infinity? These are the crucial questions where theoreticians and observers are confronting nowadays. Even though there have been by now accumulated enormous observational evidences in favor of black hole existence, still it is fair to say that their existence has not been firmly established. The proof of their existence would have been a much easier task if, for instance, an argument would have been advanced, which would: (i) single out the generic component (or components) of a black hole which is responsible for shaping up a unique observed feature associated with black holes and (ii) prove that indeed this generic component always results in the same observable feature independent of the environmental conditions that the black hole finds itself.

The lack of such an argument may be traced in the plethora of various accretion flows advanced in the literature: accretion in a state of free fall, optically thin or optically thick, accretion disks with or without relativistic corrections, shocked flows, advection flows *etc.* Of course, this diversity of accretion models is highly justified. On physical grounds one expects accretion flows describing a solar-mass black hole accreting interstellar medium to be distinct from those flows describing accretion onto a black hole in a close binary system or from a supermassive black hole at the center of an AGN. Viewed from this angle, the detectability of a black hole appears to be a rather frustrating issue since it is not clear a priori what type of the existing accretion models (if any) would describe a realistic accreting black hole. *In the present talk we will show that may be not the case. The distinct feature of a black hole spacetime, as opposed to the spacetimes due to other compact objects is the presence of the event horizon.* Near the horizon the strong gravitational field is expected to dominate the pressure forces and thus to drive the accreting material into a free fall. In contrast, for other compact objects the pressure forces are becoming dominant as their surface is approached, and thus free fall state is absent. *We argue that this difference is rather crucial, resulting in an observational signature of a black hole.* Roughly, the origin of this signature is due to the inverse Comptonization of low energy photons from fast moving electrons. The presence of low-energy photon component is expected to be generic due, for instance, to the disk structure near a black hole or to bremstrahlung of the electron component from the corresponding proton one. The boosted photon component is characterized by a power law spectrum, and is entirely independent of the initial spectrum of the low-energy photons. The spectral index of the boosted photons is determined by the mass accretion rate and the bulk motion plasma temperature only. A key ingredient in proving our claim is the employment of the exact relativistic transfer describing the Compton scattering of the low-energy radiation field of the Maxwellian distribution of fast moving electrons. We will prove that the power law is always present as a part of the black hole spectrum in a wide energy range. We investigate the particular case of a non-rotating accreting Schwartzchild black hole, leaving the case of a rotating one for future analysis. The presence of the power law part in the upcomptonized spectra was rigorously proven by Titarchuk & Lyubarskij 1995 (hereafter TL95) [1]. It has been demonstrated there that for the wide class of the electron distributions the broken power law is the Greens function

of the full kinetic equation. The importance of Compton upscattering of low-frequency photons in an optically thick, converging flow has been understood for a long time. Blandford and Payne were the first to address this problem in a series of papers (Blandford & Payne 1981 [2] and Payne & Blandford 1981 [3]). They derived the Fokker-Planck radiative transfer equation which took into account photon diffusion in space and energy, while in the second paper they solved the Fokker-Planck radiative transfer equation in the case of the steady state, spherically symmetric, super-critical accretion into a central black hole with the assumption of a power-law flow velocity $v \propto r^{-\beta}$ and neglecting thermal Comptonization. For the inner boundary condition they assumed adiabatic compression of photons as $r \to 0$. Thus, their flow extended from $r = 0$ to infinity. They showed that all emergent spectra have a high-energy power-law tail with index $\alpha = 3/(2 - \beta)$ [for free fall $\beta = 1/2$] Titarchuk, Mastichiadis & Kylafis (1996) paper [4] (hereafter TMK96 and see also the extended version in TMK97 [5]) presents the exact numerical and approximate analytical solutions of the problem of spectral formation in a converging flow, taking into account the inner boundary condition, the dynamical effects of the free fall, and the thermal motion of the electrons. The inner boundary has been taken at *finite* radius with the spherical surface being considered as a fully absorptive. Titarchuk, Mastichiadis & Kylafis [4] have used a variant of the Fokker-Plank formalism where the inner boundary mimics a black-hole horizon; no relativistic effects (special or general) are taken into account. Thus their results are instructively useful but they are not directly comparable with the observations. *By using numerical and analytical techniques they demonstrated that the extended power laws are present in the resulting spectra in addition to the blackbody like emission at lower energies.* In our talk we present the results of the full relativistic treatment in terms of the relativistic Boltzmann kinetic equation without recourse to the Fokker-Planck approximation in either configuration and energy space. The relativistic radiative equation in the curve spacetime was derived by Lindquist (1966) [6]. For the completeness of our talk we delineate some important points of that theory related with the application to the radiative transfer in the electron atmosphere which was developed by Titarchuk & Zannias 1998 (hereafter TZ98 [7]). In particular, we demonstrate that the power-law spectra are produced when low-frequency photons are scattered in the Thomson regime (i.e. when the dimensionless photon energy $z' = E/m_e c^2$ measured in the electron rest frame satisfies $z' \ll 1$). Also we will present some of our results of the extensive Monte Carlo simulations of the X-ray spectral formation in the converging inflow (see Laurent & Titarchuk 1999, hereafter LT99 [8], for more details). We will compare these results with ones obtained by different (analytical and numerical) methods. We will show that the observed spectral transitions in black hole systems (BHS) is clearly related to the bulk inflow temperature. We will demonstrate the relevance of the converging inflow model to the recent high-energy observations.

2. The Full Relativistic Treatment

2.1. The main equation

We begin with considering background geometry, described by the following line element:

$$ds^2 = -f dt^2 + \frac{dr^2}{f} + r^2 d\Omega^2 \qquad (1)$$

where, for the Schwarzschild black hole, $f = 1 - r_s/r$, $r_s = 2GM/c^2$, and t, r, θ, φ are the event coordinates with $d\Omega^2 = d\theta^2 + \sin^2\theta d\varphi^2$. G is the gravitational constant and M the mass of the black hole. In order to describe the photon radiation field we shall employ the concept of the distribution function N. The distribution function $N(x,\mathbf{p})$ (the occupation number) describes the number dN of photons (photon world lines) which cross a given spacelike volume element dV and whose 4− momenta $\mathbf{p}$ lie within a specified 3-surface element dP in momentum space. It is desirable to choose dV and dP to be coordinate invariants. Thus dN would be invariant as well and the same would be true of $N(x,\mathbf{p})$. In [7], the authors presented a derivation of the relativistic radiative transfer equation relating the distribution function $N(x,\mathbf{p})$ and the interaction den-

sity function $S(N)$ (see [6] for more details of the equation derivation). The full relativistic equation for the occupation number $N(x,\mathbf{p})$ can be written in the following form:

$$\mu\sqrt{f}\frac{\partial N}{\partial r} - \nu\mu\frac{\partial\sqrt{f}}{\partial r}\frac{\partial N}{\partial \nu} - (1-\mu^2)\left(\frac{\partial\sqrt{f}}{\partial r} - \frac{\sqrt{f}}{r}\right)\cdot\frac{\partial N}{\partial \mu}$$

$$= \int_0^\infty d\nu_1 \int_{4\pi} d\Omega_1$$

$$\left[\left(\frac{\nu_1}{\nu}\right)^2 \sigma_s(\nu_1,\nu,\xi)N(\nu_1,\mu_1) - \sigma_s(\nu,\nu_1,\xi)N(\nu,\mu)\right] \qquad (2)$$

$$\sigma_s(\nu \to \nu_1, \xi, \Theta) = \frac{3}{16\pi}\frac{n_e\sigma_T}{\nu z}\int_0^\pi \sin\theta d\theta \int d^3\mathbf{v}$$

$$\frac{F(r,P_e)}{\gamma}\left\{1 + \left[1 - \frac{1-\xi}{\gamma^2 DD'}\right]^2 + \frac{zz'(1-\xi)^2}{\gamma^2 DD'}\right\}$$

$$\times \delta(\xi - 1 + \gamma D'/z - \gamma D/z'), \qquad (3)$$

where $h\nu/m_e c^2$ is a dimensionless photon energy, $D = 1 - \mu V$, $D_1 = 1 - \mu' V$, $\gamma = (1-V^2)^{-1/2}$, $\xi = \mathbf{\Omega}'\cdot\mathbf{\Omega}$ is the cosine of scattering angle and $F(r, P_e)$ stands for the local Maxwellian distribution (see TZ98, Eqs. 2, 5, 14).

2.2. Analytical solution

As long as the ejected low energy photons satisfy $z'h\nu'/m_e c^2\gamma \ll 1$, the integration over incoming frequencies ν' is trivially implemented provided that the explicit function of $N(r,\nu',\nu,\mathbf{\Omega})$ is known. Thus, we need to describe the main properties of the Green function $N(r,\nu',\nu,\mathbf{\Omega})$ in a situation when the low-energy photons are injected into an atmosphere with bulk motion. The power-law part of the spectrum ([9], [1]) occurs at frequencies lower than that of Wien cut-off ($E < E_e$, where E_e is the average electron energy). In this regime the energy change due to the recoil effect of the electron can be neglected in comparison with the Doppler shift of the photon. Hence we can drop the third term in parenthesis and the term $\xi - 1$ of the delta-function argument in the scattering kernel (3) transforming that into the classical Thomson scattering kernel [1] (see also Giesler & Kirk 1997 [10]). Now we seek the solution of the Boltzmann equation (2) with the aforementioned simplifications, in the form

$$N(r, \nu, \mathbf{\Omega}) = \nu^{-3+\alpha} J(r, \mu). \qquad (4)$$

Then we can formally get from (2) that

$$\ell J = \mu \sqrt{f} \frac{\partial J}{\partial r} + (\alpha + 3)\mu \frac{\partial \sqrt{f}}{\partial r} J$$

$$-(1 - \mu^2)\left(\frac{\partial \sqrt{f}}{\partial r} - \frac{\sqrt{f}}{r}\right)\frac{\partial J}{\partial \mu}$$

$$= n_e \sigma_T \left[-J + \frac{1}{4\pi}\int_{-1}^{1} d\mu_1 \int_0^{2\pi} d\varphi R(\xi) J(\mu_1, \tau)\right] \qquad (5)$$

Where the phase function $R(\xi)$ is:

$$R(\xi) = \frac{3}{4}\int_0^\pi \sin\theta d\theta \int d^3\mathbf{v} \frac{F(r, P_e)}{\gamma^2}$$

$$\times \left(\frac{D_1}{D}\right)^{\alpha+2}\frac{1}{D_1}[1 + (\xi')^2], \qquad (6)$$

where ξ' is the cosine of scattering angle between photon incoming and outgoing directions in the electron rest frame. The reduced integro-differential equation is two dimensional and it can be treated and solved much easier than the original equation (2). The whole problem is reduced to the eigenvalue problem for equation (5). For the converging inflow in the low temperature limit ($kT_e \ll m_e c^2$) the electron distribution is the delta-function $F(r, P_e) = \delta(\mathbf{v} - \mathbf{v}_b)$ defined in the velocity phase space in the way that

$$\int_0^\pi \sin\theta d\theta \int d^3\mathbf{v} F(r, P_e) = 1$$

In this case the phase function is

$$R_b(\xi) = \frac{3}{4}\frac{1}{\gamma_b^2}\left(\frac{D_{1b}}{D_b}\right)^{\alpha+2}\frac{1}{D_{1b}}[1 + (\xi_b')^2], \quad (7)$$

where the subscript "b" is related with the bulk velocity direction (the case of arbitrary temperature was calculated using Monte Carlo simulations). Our goal is to find the nontrivial solution $J(r, \mu)$ of this homogeneous problem and the appropriate spectral index α for which this solution exists. This problem can be solved using

the iteration method which involves the integration of the integro-differential equation (5) with the given boundary conditions along the characteristics of the differential part of equation (5) by using Runge-Kuttas method (see TZ98 for more details of the method). There are two boundary conditions that our solution satisfy. The first one is that there is no radiation being scattered outside the atmosphere. And the second boundary condition is that we have an absorptive boundary at the black hole horizon radius, r_s. Namely, there is no radiation which emerges from the horizon.

2.3. Photon Trajectories in the Curve Space as the Characteristics of the Space Operator, ℓ

The characteristics of the differential operator ℓ are determined by the following differential equation

$$\left[-\frac{1}{2x^2(1 - x^{-1})} + x^{-1}\right] dx = d[\ln(1 - \mu^2)^{-1/2}], \qquad (8)$$

where $x = r/r_s$ is a dimensionless radius. The integral curves of this equation (the characteristic curves) are given by

$$\frac{x(1 - \mu^2)^{1/2}}{(1 - x^{-1})^{1/2}} = \frac{x_0(1 - \mu_0^2)^{1/2}}{(1 - x_0^{-1})^{1/2}} = p, \qquad (9)$$

where p is an impact parameter at infinity. p is determined at a given point in a characteristic by the cosine of an angle between the tangent to the photon trajectory and the radius vector to the point and by the given point position x_0.

In the the flat geometry, the characteristics are just straight lines

$$x(1 - \mu^2)^{1/2} = p, \qquad (10)$$

where the impact parameter p is the distance of a given straight line to the center. We can resolve equation (9) with respect of μ to get

$$\mu = \pm(1 - p^2/y^2)^{1/2} \qquad (11)$$

where $y = x^{3/2}/(x - 1)^{1/2}$. The graph of y as a function of x is presented in TZ98 (Fig. 1 there) which allows to comprehend the possible range of radii for the given impact parameter p through the inequality $p \leq y$. For example, if $p \leq \sqrt{6.75}$, then the photon can escape from the inner boundary (the black hole horizon) toward the observer or vice versa all photons going toward the horizon having these impact parameters are gravitationally attracted by the black hole. However, if $p > \sqrt{6.75}$, then finite trajectories are possible with the radius range between $1 \leq x \leq 1.5$, or infinite trajectories with $p \leq y(x)$ (x is always more than 1.5).

2.4. Spectral indices and source distribution of the upscattered photons in the converging inflow atmosphere

We are assuming a free fall for the background flow where the bulk velocity of the infalling plasma is given by $v = c(r_s/r)^{1/2}$. In the kinetic equations (2, 5) the density n is measured in the local rest frame of the flow and it is $n = \dot{m}(r_s/r)^{1/2}/(2r\sigma_T)$. Here $\dot{m} == \dot{M}/\dot{M}_E$, $\dot{M}$ is mass accretion rate and $\dot{M}_E \equiv L_E/c^2 = 4\pi G M m_p/\sigma_T c$ is the Eddington accretion rate. TZ98 (Fig. 2 there) presented the results of the calculations of the spectral indices as a function of mass accretion rates. It is clearly seen that the spectral index is a weak function of mass accretion rate in a wide range of $\dot{m} = 3 - 10$. The asymptotic value of the spectral index for high mass accretion rate is 1.75 which is between $\alpha = 2$ that was found by Blandford & Payne 1981 [2] for the infinite medium and $\alpha \approx 1.4$ which was found by TMK96 [4] for the finite bulk motion atmosphere. It is worth noting that the Monte Carlo simulated spectral indices obtained in [8] for the nonrelativistic case are very close to TMK97 results (see Figs. 6 there) made in the nonrelativistic Fokker-Planck approximation with an assumption of a purely Newtonian geometry. The efficiency of the hard photon production in the cold bulk motion atmosphere is quite low because of two effects of General Relativity: photon bending and gravitational redshift, thus the emergent spectra are much steeper than those which can be expected with the same mass accretion (or optical depths) in the absence of strong gravity. We compare the results of the Monte Carlo simulations [8] with the above solution of the eigen-

value problem for the relativistic equation and we found the Monte Carlo simulated spectral indices are in agreement with the TZ98 results. Particularly important, that the spectral index value is around 1.7-1.8 for the super Eddington mass accretion rate. *It is worth noting here that the observed power law index of Narrow Line Seyfert 1 (NLS1) galaxy population - which may represent the extragalactic analogue of the BHS in the high-soft state - is around 1.8.* This is close to what we have obtained in the low kT_e regime, as it is expected from Comptonization on electrons whose motion is dominated by the bulk free-fall. This is not the case if the plasma temperature is of order of 10 keV or higher. The coupling effect between the bulk and local Maxwellian motion occurs when the bulk motion velocity is very close to the speed of light, i.e. when the matter is very close to the horizon. The upscattering effect increases significantly in the latter case [8]. As an example, TZ98 (Fig. 3 there) showed the zeroth moment of the eigenfunction source distribution (the hard photon production) over radius. It can be seen there that the distribution has a strong peak around 2 r_s. This means that the vicinity of the black hole is a place where the hard photons are produced by upscattering of the soft photons off the converging electrons. Two effects are responsible for the formation of this particular distribution: The Doppler effect which is getting stronger closer to the horizon (with the increase of the free fall velocity) and the photon bending which prevents to see most of the hard photons producing in the proximity of the horizon. This statement regarding the shape of eigenfunction source distribution is generic and it is independent of the initial soft photon source distribution. Thus because the extended power law of the bulk motion spectrum is formed very close to the horizon $(1 - 3)r_s$ and the source soft photons come from the the innermost part of the disk one should expect two characteristic timescale for variability of this feature: one is associated with the crossing timescale $t_{cross} \sim 10^{-5} ms$ and another one is related with the oscillation frequency of the inner edge of the disk ν_d which is of order of $(800/m)$ Hz [11] where $m = M/M_\odot$. Borozdin et al. 1999 [12] found that the BH mass in GRS 1915+105 is

$18.6 \pm 2.2 M_\odot$ and consequently $t_{cross} \sim 2 \times 10^{-4}$s and $\nu_d \sim 45$ Hz in this source. In fact, the detection of 67 Hz QPOs from GRS 1915+105 by RXTE was recently reported by Morgan et al. 1997 [13]. It was clearly demonstrated that this feature is associated with the high-energy component (extended power law component) visible in the PCA detector of RXTE. This can be readily explained in terms of a QPO of the soft photon component originated in the innermost part of the disk. The variations of the soft component should lead to the variations in the hard component, since significant changes in the illumination geometry of converging inflow site can occur. The assumption of Thomson scattering accepted in our solution restricts the relevant energy range to $E < m_e c^2$. Our analytical approach cannot determine accurately the exact position of the high energy cutoff which is formed due the downscattering of the very energetic photons in the bulk motion electron atmosphere. Additional efforts are required to confirm the qualitative estimates of the high energy cutoff position as of order $m_e c^2$ [5].

3. Bulk Motion Spectra and Their Relevance to the Recent High Energy Observations of High-Soft State in BHS

Laurent & Titarchuk 1999 [8] using Monte Carlo simulations calculated the resulting spectra from the bulk motion atmosphere, checked and confirmed the results of the analytical theory [7,4,5]. In Figure 1 we present the typical example of the photon spectrum for a cloud with a temperature of 5 keV and $\dot{m}$ equal to 2, spectrum formed in the converging atmosphere due to the upscattering of the soft photons of the disk. Arrows point out the particular places in the system where the appropriate photons are produced. In comparison we show in Figure 2 what would happen if the accretion with high mass accretion rate (close to Eddington) occurs in the neutron star system. The high radiation pressure finally stops matter falling in and the blackbody-like radiation with effective temperature of order 1 keV would be formed (Fig. 2). *Thus our results of Monte Carlo simulations strongly support the idea*

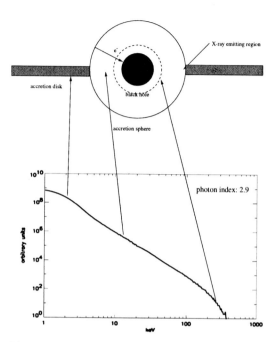

HIGH STATE OF BLACK HOLE SYSTEM

Figure 1. Soft state of a black hole system.

that the bulk motion Comptonization might be responsible for the extended power-law spectra seen in the black-hole X-ray sources in their soft state. On the other hand, during the soft states of neutron star systems, when the mass accretion rate is close to the Eddington limit, the bulk-motion infall is not present. The radiation pressure caused by such a mass accretion rate prevents the flow from free-falling towards the neutron star surface.

The effect of thermal motion for electron temperatures higher than 10 keV is clearly seen in spectra which are getting harder, i.e the index decreases, and the high-energy cutoff increases. Furthermore, the spectra obtained at high kT_e and $\dot{m}$ greater than 2 have a spectral index around 0.8 which is what is observed from BHS in the hard state, where the thermal Comptonization is thought to dominate (Ebisawa, Titarchuk, and Chakrabarti, 1996 [14]). This is also shown in Figure 3 where we present the results of the computation for $kT_e = 50$ keV, and $\dot{m} = 4$, com-

HIGH STATE OF NEUTRON STAR SYSTEM

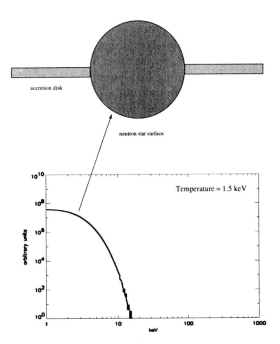

Figure 2. Soft state of a neutron star system.

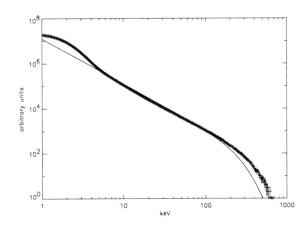

Figure 3. Emergent Comptonized spectrum in the general relativistic case with $kT_e = 50$ keV, and $\dot{m} = 2$. The high energy part of the spectrum is compared to the thermal Comptonization spectrum [15].

pared to the analytical Greens function of Hua & Titarchuk 1995 [15] (Eq. 6) derived for the pure thermal Comptonization case. The spectrum is practically identical to the standard thermal Comptonization spectrum. In this case the effect of the bulk motion can be seen only at high energies where there is an excess in the converging inflow spectrum due to coupling of the thermal and bulk motion velocities. It is worth noting here that this excess is really detected in the observation of the hard state of Cyg X-1 source (e.g. Grove et al. 1998 [16]). So, the change of spectral shapes from the soft Xray state to the hard Xray state is clearly to be related to the temperature of the bulk flow. In other words the effect of the bulk Comptonization compared to the thermal one is getting stronger when the plasma temperature drops below 10 keV. *The hard-soft states transition is regulated by the plasma temperature of the converging inflow into a black hole.* LT99 (Eq. 8 there) derived a generic formula for the temperature of the emitting region (CI) that de-

pends on the energy release in this very region and in the disk. Using this formula, they demonstrated that the temperature of the emission region in the hard state of Black Hole Systems is approximately 2 times higher than the ones of neutron star systems in the hard state which is confirmed by recent *RXTE* and *Beppo-SAX* observations of the hard state.

3.1. X-ray Observations of BHS in the High-Soft State

Application of the bulk motion Comptonization (BMC) model demonstrates good agreement with the data, even though the spectrum of accretion disk is approximated by a single-temperature blackbody. For spectra extending below several keV, replacement of the single black-body by a multicolor disk component within the framework of BMC model is expected to be necessary, but given the limited soft sensitivity of RXTE, the single black body approximation is adequate for most spectra we have analyzed [12].

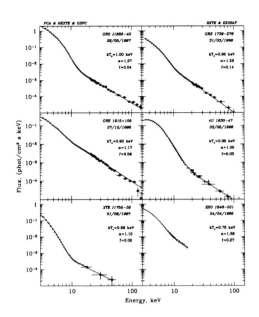

Figure 4. Application of the BMC model to the spectra of six soft X-ray transients observed by RXTE and Exosat [12].

are very well represented by thermal Comptonization models as is the case for the low-hard state black holes. Non-bursting, high-luminosity low-mass X-ray binaries containing weakly magnetized neutron stars, on the other hand show no indication of sustaining high-energy emission. *This seems to be a clear indication that the high-soft state, and in particular the high-energy power-law associated with that state, is directly tied to the physics of the black hole environment.* This, further, strongly suggestive that the phenomena that underlie it are associated with the innermost probable regions of the black hole system. The recent timing and spectral analysis of the data from GRS 1915+105 [13], GRO 1655+105 [20,22] and XTE J1550-564 [21,23] clearly support this claim. It has been shown that the frequencies of the fast QPOs in these sources, (67, 300, 200 Hz respectively) are closely related to the hard power-law component of the energy spectrum. Thus one can conclude that the timing and spectral features of the high-soft states are tied with the close vicinity of the central object and *these spectra are the inevitable stamp of black hole systems where strong gravitational forces dominate.*

4. Conclusions

Our theoretical analysis and comparison with the broad-band X-ray spectral data for several soft X-ray transients, representing a large fraction of the publicly available RXTE archive for this type of sources, demonstrate that the bulk-motion theory, even in its simplified analytical form are very relevant with all of the data (Fig.4). The bulk motion theory allows one to explain, in a self-consistent manner, the origin of the two-component spectra in high state on the basis of simple geometrical assumptions and basic physical principals. Quite distinct from the high-soft state, the low-hard spectral state does not appear to uniquely associated with black hole binaries. Neutron star binaries, when observed in low-luminosity states, in fact exhibit spectra which are strikingly similar to the black-hole low-hard state [17–19]. The best demonstrations of this come from observations of X-ray bursters during periods of low-intensity. The high-energy spectra

REFERENCES

1. Titarchuk L.G., & Lyubarskij Yu.E. 1995, ApJ, 450, 876 (TL95)
2. Blandford, R. D., & Payne, D. G. 1981, MNRAS, 194, 1033
3. Payne, D. G., & Blandford, R. D. 1981, MNRAS, 196, 781
4. Titarchuk, L. G., Mastichiadis, A., Kylafis, N. D. 1996, A&AS, 120, C171 (TMK96)
5. Titarchuk, L. G., Mastichiadis, A., Kylafis, N. D. 1997, ApJ, 487, 834 (TMK97)
6. Lindquist, R., 1966, Annals of Physics 37, 487
7. Titarchuk L. & Zannias T., 1998, ApJ, 493, 863 (TZ98)
8. Laurent, P., & Titarchuk, L. G., 1999, ApJ, 511, 289 (LT99)
9. Sunyaev, R.A., Titarchuk, L.G. 1980, A&A, 86, 121 (ST80)
10. Giesler, U. & Kirk, J. 1997, A&A, 323, 259
11. Titarchuk, L., Lapidus, I. & Muslimov, A. 1998, ApJ, 499, 315

12. Borozdin K., Revnivtzev, M, Trudulyubov, S., Shrader, C., & Titarchuk, L. 1999, ApJ, in press

13. Morgan, E.H., Remillard, R.A., & K.I. Griener, J. 1997, ApJ, 482, 993

14. Ebisawa K., Titarchuk L., and Chakrabarti S.K., 1996, PASJ, 48, 59

15. Hua X.M. & Titarchuk L., 1995, ApJ, 449, 188

16. Grove et al., 1998, ApJ 500, 899

17. Barret D., & Vedrenne G., 1994, ApJS, 92, 505

18. Barret D., & Grindlay J., 1995, ApJ, 440, 841

19. Barret D., et al., 1999, A&A, 341, 789

20. Remillard.,R.A., Morgan, E.H., McClintock, J.E., Bailin, C.D., & Orosz, J.A. 1999a, ApJ, in press; astro-ph/9806049

21. Remillard.,R.A., et al., 1999b, ApJLetters, in press; astro-ph/9903403

22. Sobczak, G., McClintock, J.E., Remillard.,R.A., Bailin, C.D., & Orosz, J.A. 1999a, ApJ, in press; astro-ph/9809195

23. Sobczak, G., et al., 1999b, ApJLetters, in press; astro-ph/9903395

ELSEVIER

Nuclear Physics B (Proc. Suppl.) 80 (2000) 183–194

NUCLEAR PHYSICS B
**PROCEEDINGS
SUPPLEMENTS**

www.elsevier.nl/locate/npe

The Solar Neutrino Puzzle

S. Turck-Chièze[a]

[a]DAPNIA/Service d'Astrophysique, CEA Saclay, 91191 Gif sur Yvette cedex, FRANCE

Neutrino fluxes coming from the Sun are now measured on earth with improved accuracy. The predicted neutrino emissions in the solar core depend on fundamental nuclear and astrophysical properties which are nowadays better constrained by experiments. This paper is focused on the highlights in these fields and on their perspectives.

1. Introduction

Since the first detection of solar neutrinos in 1968, different communities have contributed to solving the discrepancy between predictions and detections of solar neutrinos, showing the complexity of the interpretation of such measurements. The problem is harder than thought at the beginning, due to the difficulty of detecting neutrinos on earth and to the fact that we have not yet been able to measure separately neutrinos of different sources as listed in figure 1.

One could say that during these 30 years, the aim of this research has evolved, shifting from a direct check of the central temperature and proton proton nuclear energy generation, to a double objective: a detailed verification of the solar nuclear plasma and a possible manifestation of neutrino oscillation, leading to an estimate of the electron neutrino mass. The Sun appears today as a natural laboratory of nuclear, plasma and atomic physics and gives access to a range of neutrino oscillation parameters ($10^{-6} < m_{\nu_e} < 10^{-2} eV$) which is not accessible to ground laboratories. Proving a very small mass oscillation of the neutrinos is difficult as we do not measure the neutrino emitted fluxes, so, all the different aspects of the problem must be examined with great care before reaching definitive conclusions. In these new perspectives, helioseismology is a useful tool to check the theoretical solar structure estimate, and consequently, to partly consolidate the estimated neutrino emission.

In this review, we will recall the present status and the weak points (see also Bahcall 1989, Turck-Chièze et al. 1993).

2. The main actors

The solar neutrino puzzle is at the interface of three disciplines: the nuclear physics, the astrophysics and the particle physics.

- The nuclear part

The Sun is the only hydrogen-burning star which produces neutrinos detectable on earth, and the detection of these neutrinos is the signature that weak interaction exists in our star, which stabilizes its structure, contributing to our life on earth by the extended hydrogen burning phase duration. The context is radically different from the neutrino emission of a supernova where neutrinos leave the star with practically the entire generated energy. In the case of the Sun, or of any solar-like stars in hydrogen burning phase, most of the neutrinos escape with a small amount of energy. The largest part is carried by photons which interact with ions, atoms inside the star and slowly escape through the photosphere, downgraded in the visible range. Unfortunately, the most productive source of neutrinos ($1.68\,10^{38}\nu_e/s$), the $p + p \rightarrow D + e + \nu_e$ reaction, creates neutrinos with an energy smaller than 0.4 MeV, whereas it is easier to detect the high part of the neutrino spectrum corresponding to ^8B (typically $1.35\,10^{34}\nu_e$ emitted/s) and hep neutrinos; 25 years of experimental efforts have been necessary to achieve the required sensitivity, on ground to detect all the different sources of neutrinos (see figure 1). It is clear from this fact that as far as we cannot check separatively the different sources of neutrinos, the nuclear balance between the different channels (including CNO cycle) is crucial in the puzzle, this implies good nuclear laboratory experiments (Aldelberger et al. 1998) a precise understanding of the screening effect in the laboratory (Engstler et al., 1992) at low energy and a correct determination of the screening effect in the stellar plasma for the different species (Dzitko, Turck-Chièze, Delbourgo and Lagrange, 1995; Gruzinov and Bahcall 1998).

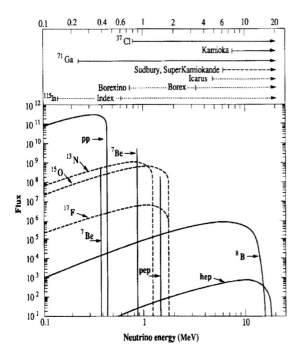

$p + p \rightarrow D + e^+ + \nu_e$ (pp neutrinos)

$p + p + e^- \rightarrow D + \nu_e$ (pep neutrinos)

ppII chain: $^7Be + e^- \rightarrow^7 Li + \nu_e$ (^7Be neutrinos)

ppIII chain: $^7Be + p \rightarrow^8 B$ $^8B \rightarrow^8 Be^* + e^+ + \nu_e$

$^8Be^* \rightarrow 2^4He$ (^8B neutrinos)

$p +^3 He \rightarrow \nu_e + e^+ +^4 He$ (hep neutrinos)

CNO   I cycle: $^{13}N \rightarrow^{13} C + e^+ + \nu_e$ (^{13}N neutrinos)

$^{15}O \rightarrow^{15} N + e^+ + \nu_e$ (^{15}O neutrinos)

CNO   II cycle: $^{17}F \rightarrow^{17} O + e^+ + \nu_e$ (^{17}F neutrinos)

Figure 1. *Solar neutrino energy spectrum and the sensitivity of the different detectors above the threshold. The neutrino fluxes from continuum sources are in $cm^{-2}s^{-1}MeV^{-1}$. The line fluxes are in $cm^{-2}s^{-1}$. Below are all the nuclear processes leading to neutrino emission.*

A second difficulty comes from the precise knowledge of the low absorption cross sections of the neutrino on detectors, amplified by the fact that 2 experiments over 3 produce radiochemical elements with rather short lifetime and extractions every 3 or 4 weeks with no more than 5 or 6 events per cycle. Nowadays this problem is partly solved by the calibration of the gallium radiochemical experiments and the capability of the real time experiments to get high statistics.

But the chlorine experiment is no longer calibrated. Moreover, the absorption cross section $\sigma^8 B$: $\nu_e +^{37} Cl \rightarrow e +^{37} Ar$ must be known for the excited states to extract the ^8B high energy neutrinos from this experiment and is more than 1000 times greater than the other components $\sigma(pep)$, $\sigma^7 Be$ ... If the uncertainty on this key cross section, recently reestimated at $11.4\,10^{-43}cm^2$ (Bahcall et al. 1996) was larger than thought, it would contribute to the understanding of the chlorine and SuperKamiokande results as far as the ^8B neutrinos are concerned.

Finally, the different fluxes have different behaviours. The pp neutrino flux is directly dependent on the solar luminosity; this luminosity is equal to the nuclear energy produced by p-p chain and CNO cycle (less than 2%) to which one needs to subtract the energy of the neutrinos. Therefore the neutrino pp flux is practically independent on the central temperature ($\Phi_{pp} \approx T^{-1.2}$) and consequently independent on the solar model details. The two other important neutrino contributions have different sensitivities on the temperature, typically $\mathbf{T}^8$-$\mathbf{T}^{10}$ for Be^7 neutrinos and $\mathbf{T}^{18}$-$\mathbf{T}^{24}$ for B^8 neutrinos. These neutrinos are less and less numerous due to the respective branching ratio, so one can deduce that the corresponding flux is more and more dependent on the detailed solar calculations. It is also the case for ^{13}N and ^{15}O neutrinos. Today, except for the case of (Super)Kamiokande, which might be sensitive mainly to ^8B neutrinos, one cannot experimentally separate the main neutrino sources. Consequently the discrepancy between predictions and detections does not have a direct interpretation. This is *the third difficulty* which requires complete control of the different parts of the calculations and experiments (see Turck-Chièze et al., 1993).

- **The astrophysical part**

The astrophysical community has the responsability to produce theoretical predictions for the different sources of neutrinos described above. Without any condition on the kinematics of the reactions (the energy dependence of the neutrino fluxes described by figure 1 is deduced from the nuclear processes), these predictions are extracted from solar models which describe the temporal and radial evolution of the Sun in the classical framework of stellar evolution. By solving the four structure equations, one performs an ab initio complete calculation which requires a good description of the nuclear and atomic processes and has no free parameter.

The first equation assumes hydrostatic equilibrium (each gas shell is balanced by the competition between the downward gravitational force and the outward pressure force):

$$\frac{dP}{dr} = -\frac{M(r)G}{r^2}\rho \qquad (1)$$

where P and ρ are the pressure and the density. $M(r)$ represents the mass enclosed within a sphere of radius r:

$$\frac{dM}{dr} = 4\pi r^2 \rho \qquad (2)$$

One further assumes thermal equilibrium. The energy $(4\pi r^2 \rho \epsilon)$ produced by nuclear reactions, balances the energy flux $L(r)$ emerging from the sphere of radius r. To this energy flux must be added the energy loss by neutrinos. Taking into account quasistatic gravitational readjustment and composition variation, one must incorporate a heat transfer term TdS where S is the total entropy per gram of the gas and the energy bookkeeping yields:

$$\frac{dL}{dr} = 4\pi r^2 \rho \left(\epsilon_{nucl} - \frac{TdS}{dt} \right) \qquad (3)$$

At the end of hydrogen burning, the nuclear energy vanishes and contraction again contributes to the luminosity.

Finally, the temperature gradient depends on the luminosity and the physical process of the energy transport. In a radiative region of a star, the diffusion approximation is appropriate, and the relation between temperature gradient and luminosity is:

$$\frac{dT}{dr} = \frac{-3}{4ac} \frac{\kappa \rho}{T^3} \frac{L(r)}{4\pi r^2} \qquad (4a)$$

When the opacity coefficient κ increases too much, like in the external part of smaller stars with $M \leq 1.5 M\odot$ or when the luminosity is very high (in the internal part of stars with mass $> 1.5 M\odot$), the radiative gradient increases so much that matter becomes convectively unstable. The resulting temperature gradient is then nearly adiabatic:

$$\frac{dT}{dr} = \left(\frac{dT}{dr}\right)_{ad} = \frac{\Gamma_2 - 1}{\Gamma_2} \frac{T}{P} \frac{dP}{dr} \qquad (4b)$$

The adiabatic exponent Γ_2 is defined by

$$P^{1-\Gamma_2} T^{\Gamma_2} = const.$$

From these structural equations, we extract temperature, density, pressure and composition for the different shells of the present Sun and consequently the neutrino emitted fluxes. One may notice that the energy sumrule which allows to extract without detailed calculations the pp neutrino flux is explicitly contained in the third equation of stellar evolution.

So, if one supposes no magnetic interaction or change of flavour, $5.98 10^{10} \nu_e$ reach the earth /s coming from this fundamental pp interaction which produces the $1.68 10^{38} \nu_e$/s discussed above. This is, of course, the main source of neutrinos coming from the Sun but unfortunately, the corresponding neutrino energy is small and this flux has never been measured totally and alone.

The other sources are smaller: typically $1.32 10^{37}$/s for ^7Be neutrinos $1.35 10^{34}$/s for ^8B, $1.31 10^{36}$/s for ^{13}N, $1.12 10^{36}$/s for ^{15}O. These sources are directly proportional to the density in the range of emission and are largely temperature dependent as noticed above, so they significantly depend on the solar structure.

This is the reason why the solar model has received a lot of attention these last 10 years. The recent efforts may be summarized as follows:

- The reanalysis, including remeasurement of some of them, of all the nuclear cross sections participating to the hydrogen burning (Adelberger et al. 1998 and references therein). The estimated $^{7}Be(p,\gamma)$ cross section has been reduced by 15 % but stays the most badly determined (20 %).

- The improvement of the equation of state and opacity calculations in quality (checked by experimental laser measurements) and completness (Rogers, Swenson and Iglesias, 1996; Iglesias and Rogers, 1996).

The effect of these improvements, together with the effect of the age, is discussed in Turck-Chièze et al. (1998a) together with a large comparison between different models.

- The development of helioseismology which allows us to deduce from the acoustic mode frequency spectrum, a radial dependence of the solar sound speed, density and differential rotation down to 0.08 $R\odot$. Presently the measurement of more than 3000 modes leads to an accuracy of some 10^{-4} on the radial sound speed (Turck-Chièze et al. 1997, 1998) which is what we need (see below), to validate the existing stellar processes and introduce macroscopic motions.

The results of these experimental and theoretical efforts are summarized in figures 2a) and 2b) (Brun, Turck-Chièze and Morel (1998); Brun, Turck-Chièze and Zahn (1999)). In these papers, we show that the solar structure has been substantially improved by the introduction of some extra processes not included in classical stellar evolution 10 years ago and prescribed by the helioseismic constraints.

These progresses concern the evolution of the elemental composition with time. Three terms are added in the equation describing the time evolution of the nuclear species X_i :

$$\frac{\partial X_i}{\partial t} = \frac{\partial X_i}{\partial t}_{nucl} - \frac{\partial \left[(4\pi\rho r^2 (D_i + D_T)\frac{\partial X_i}{\partial m} - v_i X_i) \right]}{\partial m}$$

$$(5)$$

In the Sun, the very slow microscopic diffusion of the elements, dominated by the gravitational settling, modifies the composition along the radial profile and the progress done by the introduction of the first and third term is clearly visible on figure 2a (Proffitt and Michaud 1991). It has some impact on the hydrogen burning duration (reduction by about 0.6-1 Gyr). The presence of this process is the only way to interpret

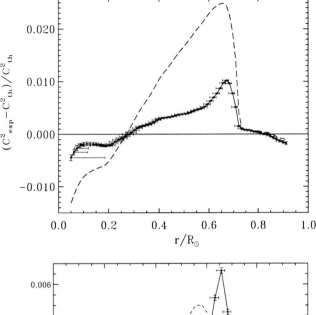

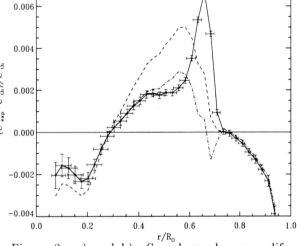

Figure 2. a) and b): Sound speed square difference between the Sun seen by GOLF (Gabriel 1995, 1997) + MDI (Scherrer 1995, Kosovishev et al., 1997) instruments aboard SOHO (Turck-Chièze et al. 1997,1998) and solar models (Brun, Turck-Chièze, Morel 1998; Brun, Turck-Chièze, Zahn 1999). The full line corresponds to a reference model where the microscopic diffusion is included. In the first figure, the model without this process is represented in dashed line. In the second figure, models including turbulent terms, the dashed line corresponds to a model where the photospheric Z/X=0.0245 constraint of the reference model is kept, in the model with dot-dashed line, this constraint is relaxed, the photospheric Z/X stays in the error bars of the observations.

a rather cosmological photospheric helium determined by helioseismology (0.249 ± 0.003). Contrary to what was assumed in the classical stellar evolution, the initial abundance is not strictly equal to the present photospheric one. The photospheric composition (relative to hydrogen) is reduced by about 10 % during the solar lifetime.

The model including the microscopic diffusion is considered today as our standard model, it is obtained with a constraint on the luminosity, radius and photospheric heavy element composition. The introduction of such a process increases the sum of the neutrino predictions for the chlorine and water detectors by about 20%. But one knows for a long time that such a process cannot act freely in stars as predicted anomalies on more evolved star than the Sun are not observed (Vauclair, Vauclair, Michaud 1978). It is the reason for the introduction of the D_T term (T for turbulent) in eq. 5 which must partly inhibit the previous process. Up to recently, astrophysicists had treated this term in an ad hoc manner.

The internal rotation measured by helioseismology has demonstrated a sharp transition between differential rotation in the external convective zone and a rather rigid rotation in the radiative zone (Kosovishev et al. 1997, and figure 3). The thickness of such a transition (Corbard et al., 1998) corresponds to the observed peak in the sound speed profile, clearly visible in figure 2b. This suggests that the inhibition of the microscopic diffusion at the radiative-convective transition could be due to some instability induced by the differential rotation of the Sun (Spiegel and Zahn 1992). Taking into account this dynamical process leads to a better understanding of the lithium deficit problem in solar-like stars, and improves the photospheric helium determination (Brun, Turck-Chièze and Zahn 1999). In limiting the microscopic diffusion, the net effect is a slight relative reduction of the neutrino fluxes but we have noticed that if we relax the imposed constraint of Z/X=0.0245 (Z is here the heavy element mass fraction, X, hydrogen mass fraction)at the surface and accept a slight increase of 5%, always compatible with the present observation of the heavy element composition, one gets a very good agreement with the solar sound speed pro-file (dot-dashed line in figure 2b) with practically the same neutrino predictions than the standard model as the solar core is not modified.

With all these progresses, the detected ^8B flux by SuperKamiokande is 50 % of the predicted flux obtained by the best solar model including the processes checked by helioseismology, the chlorine result is 0.37 of the predicted value, and the gallium one about 60 % (see table 1). One needs to recall that there is no free parameter in these detailed neutrino estimates. **Following the initial objective of these measurements, we can guarantee that the central temperature of the Sun is under control within 2%.** This is a real success which confirms the basis of stellar evolution and the major hydrogen burning stage. But this is certainly not sufficient to solve the neutrino puzzle.

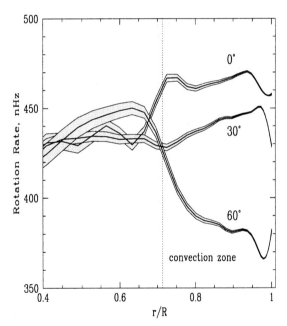

Figure 3. : Radial rotation frequency obtained by MDI instrument aboard SOHO for 3 latitudinal angles, 0 corresponds to the equator (Kosovishev et al. 1997).

- The particle physics part

The three types of neutrino experiments do not give a coherent picture compatible with standard physics, since we cannot easily understand why the chlorine experiment leads to a greater reduction than the SuperKamiokande experiment (see table 1). Is there a problem with this experiment or with the knowledge of the absorption cross section ? With the recent gallium results, a large reduction of ^7Be neutrino flux could be a way to reconcile these two sets of data.

So, may be the solar neutrino puzzle is a manifestation of neutrino properties and not of stellar interior ? This is a suggestion of the data but we have no proof that it is reality. Nevertheless, this new challenge is extremely interesting, and the improved technology is demonstrating its capability to answer the present questions and motivates all the future coherent efforts in the different fields described below.

3. The facts

- The neutrino detection

The pioneering radiochemical experiment of Homestake has shown, in the seventies, a detection deficit of practically a factor of 4. This situation triggered complementary tasks listed below:

- The construction of calibrated radiochemical experiments at low energy sensitive to pp energy neutrinos: GALLEX and SAGE detectors (Hampel et al. 1996, Abdurashitov et al. 1996),

- The construction of real time measurements with high statistics, energy neutrino spectrum and directivity of the events: Kamiokande and SuperKamiokande detectors, about 5000 ev/yr (Fukuda et al. 1998).

- Some complementary checks which will be offered by the SNO experiment which is sensitive to ν_μ detection (next year) and by Borexino which must detect ^7Be neutrinos.

- The sound speed in the nuclear region

We can extract more and more properly the solar sound speed profile from the helioseismic data, consequently, if we may demonstrate that there is a significant deviation between the Sun and the theoretical model, this could be the opportunity to go beyond this theoretical model and deduces neutrino fluxes directly from our observation of the Sun. Before that, it is interesting to compare the sensitivity of the two present probes: neutrinos and sound speed face to the neutrino puzzle. We have recalled the ^8B neutrino flux dependence on the central temperature of the star. One can remark that if the Sun was exactly at the beginning of hydrogen burning the chlorine prediction would be 0.57 SNU (instead of the present 7 SNU for the present prediction) and the gallium one 67 SNU (instead of 127 SNU) that means a factor 15 and 2 smaller. On the other side, as it is shown on figure 4, one may noticed that the central sound speed has only varied by 9 % during the same period of time. Effectively, as

$$\Delta c^2/c^2 = \Delta T/T - \Delta\mu/\mu \qquad (6)$$

and $\Delta T/T = 13.5\%$ ($T_{init}= 13.5 \ 10^6 K$), $\Delta\mu/\mu = 32 \%$ (the mean molecular weight passes from 0.31 to 0.41), so $\Delta c/c= -9\%$, which is exactly what we see on the figure. So one concludes that we need to know the sound speed profile with an extremely good accuracy (better than 10^{-3}) if we would like to observe some deviation of this quantity at the level of 1%, which is exactly what is required for questionning the present solar structure with a doubt of less than 2 % on the central temperature. This is the demanding challenge of the SuperKamiokande result and approximatively what we can wait from the helioseismic results.

It is also important to note that there is no direct determination of the temperature through acoustic modes and that $\Delta c^2/c^2$ could be different from $\Delta T/T$ as it is shown on the above example.

If we believe the figures 2, we may consider that we have reached a quantitative check of the theoretical solar structure which gives confidence in the present solar neutrino estimates and that we can exclude any astrophysical problem. Effectively, the residual present discrepancy in the core, seems small to put in evidence some astrophysical new phenomenon. But extensive comparisons show that the present accuracy of the sound speed determination in the solar core is probably underestimated. We get different results with different experiments at the level of some 10^{-3}. This is due to the natural nature

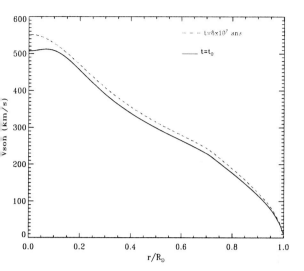

Figure 4. Evolution of the sound speed along the hydrogen burning

of the acoustic modes and the method used to extract the sound speed. These modes are extremely adapted to check the thermodynamical quantities of the low density layers. On the 3000 detected acoustic modes, only 100 penetrate the nuclear region and less than 30 reach the center. As a radial mode penetrates down to the core, one may calculate its acoustic radius which gives the time this mode passes through the different parts of the Sun (Turck-Chièze and Brun 1997):

$$\tau(r) = \int_r^R dr/c(r) \qquad (7)$$

This mode passes about 70 % of its time in the region above $0.6\ R_\odot$ (this is a priori even better for high degree modes) this region is now very well described and has largely questionned the simple representation of the standard model which does not consider effect of rotation and complex convective flows, ignoring also the role of the magnetic field which produces the well observed solar cycle. But the same mode passes only 3 % of its time between 0.2 and 0.1 $R_\odot$ and less than that below. If we suspect a phenomenon which may modify the central temperature by 1 or 2 %, we are looking for effect ¡ 3 10^{-4} of the frequency observed (less than 1 μ Hz at 3000 μ Hz). This is the

kind of accuracy we normally reach today with the networks on ground (BiSon, IRIS, GONG) and SOHO satellite (GOLF, MDI) but we are still looking for inacceptable effect as technique of data analysis or bias due to the bad knowledge of the surface. It is why extensive studies (Basu et al. 1999) are connected to the way we extract the sound speed which may summarize as follows:

$$\frac{\delta\omega_i}{\omega_i} = \int K^i_{c^2,\rho}(r)\frac{\delta c^2}{c^2}(r)dr + \int K^i_{\rho,c^2}(r)\frac{\delta\rho}{\rho}(r)dr$$

$$+\frac{F_{\text{surf}}(\omega_i)}{E_i} \qquad (8)$$

In this equation $\delta\omega_i$ is the difference between the observed and theoretical frequency of the mode i and the kernels $K^i_{c^2,\rho}$ and K^i_{ρ,c^2} are known functions of the reference model which relate the changes in frequency to the change in c^2 and ρ. The last term comes from the difficulty to describe the solar near surface layers. Effectively, one notes a systematic biais between absolute predicted values and observed ones illustrated by figure 5 for different solar models (Turck-Chièze et al. 1997). This problem is reasonably well taken into account by eq (8) but may perturb the extraction of the information of the solar core as a very small amount of the total frequency is concerned by the solar core. Impressive progresses have been realised these last years to understand the excitation of the modes and some spurious effect as the asymmetry of the line (Toutain et al., 1998, Thiery et al. 1999), to develop more and more techniques of inversion... but the conclusion on the real difference between the Sun and the model in the core supposes a proper understanding of this surface effect and of the solar cycle effect which is also of the same order than any modification of the core (about 0.5 μ Hz). One may notice that going to the low part of the frequency spectrum will help because the modes of low order reflect lower at the surface and are less sensitive to all these perturbations (typically below 2000 μ Hz, figure 5). It is why we are confident that we shall progress on the accuracy of the sound speed in the solar core with longer observations and good instruments.

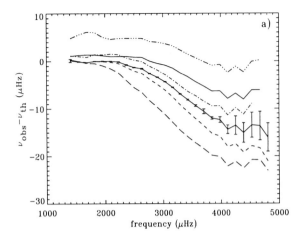

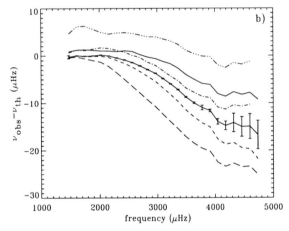

Figure 5. Frequency differences between GOLF observations and different solar models for radial modes (a) and modes with $l = 1$ (b), in the sense (sun) - (theory): $-\cdot\cdot\cdot-$: Turck-Chièze and Lopes (1993); full line: Gabriel and Carlier (1997); $-\cdot-$: Guzik et al. (1996); full and experimental error bars: Christensen-Dalsgaard et al. (1996); $- - -$: Morel, Provost and Berthomieu (1997); large $- - -$: Brun et al. (1997).

- **The nuclear uncertainties**

We note that the nuclear uncertainties are still large and seem to dominate the theoretical neutrino predictions. Figure 6, dashed line, corresponds to a modification of the pp reaction rate by $+2\%$ and the $(^3He, ^4He)$ reaction rate by -8%, as a demonstration of the present sensitivity of the sound speed, and to stress that these values may be favoured by seismic constraints.

We note also in Brun, Turck-Chièze and Morel 1998, that an uncertainty on $(^7Be, p)$ of 20% exists, leading to an uncertainty of about 23% on the predicted ^8B neutrino flux. This persistent large error justifies all the present laboratory nuclear efforts in this field (Hammache et al. 1998). One may consider two contributions to the present error bar of this cross section: the experimental one coming from laboratory experiments and the uncertainty on the extrapolated value at 17 keV which is the energy where this interaction acts in the Sun. The difficulty in the theoretical extrapolation comes from the knowledge of the wave functions of the ^8B ion which has an asymmetric number of protons (5) and neutrons (3).

Table 1 shows a minimal neutrino flux predictions taking into account the nuclear uncertainties (Brun, Turck-Chièze and Morel 1998). This model is compatible with seismic results and cannot be rejected by statistical considerations as the uncertainties on the nuclear part are not gaussian; it reduces the discrepancy by a factor 2.

The analysis of this section shows that a continuous effort to improve the quality of the astrophysical neutrino predictions is useful for the understanding of the present solar neutrino puzzle.

4. The different interpretations of the neutrino puzzle

4.1. the particle physics solution

The most popular idea for explaining the solar neutrino discrepancy is that part of the electronic neutrinos is transformed in muonic neutrinos in the Sun (MSW effect) or in vacuum. From the different experiments one can determine directly two parameters: the mass difference of the mass eigenstates ν_1 ν_2 and their mixing an-

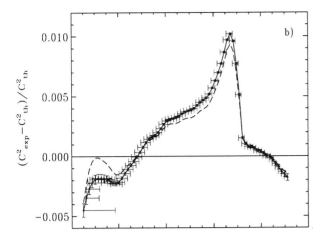

Figure 6. : Sound speed square difference between the Sun seen by GOLF + MDI (Turck-Chièze et al. 1997,1998) and solar models. The full line corresponds to a model where the microscopic diffusion is included, the dashed line to a model where the nuclear reaction rates have been adjusted inside their error bars to get a neutrino minimal. nuclear model

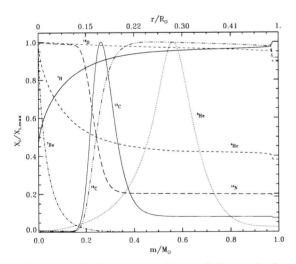

Figure 7. Radial composition of the main important elements normalised to their maximum value: H=0.7386, ^{3}He=0.00309, ^{4}He=0.6350, ^{7}Be=1.718 10^{-11}, ^{12}C=0.0030, ^{13}C=0.00044, ^{14}N=0.0045, ^{16}O=0.0085.

gle. This would be a beautiful solution, probably the most interesting one, awaiting for 20 years (Wolfenstein, 1978). The best estimates were $\Delta m^2 = 5\ 10^{-6} eV^2$, $sin^2 2\theta = 5.5\ 10^{-3}$ for active neutrinos (Bahcall, Krastev and Smirnov, 1998). This solution emerges if the Sun is standard, the absorption cross sections correct and all the experiments well understood. On this basis, one can extract from the three type of experiments the three major components ν_{pp}, ν_{Be}, ν_B. In this simplified approach, one supposes that CNO neutrinos are well estimated. Doing so, it appears easily that the beryllium neutrino flux must be largely suppressed. This idea can be partly checked by the energy neutrino spectrum of SuperKamiokande. But this spectrum (Suzuki 1998, Totsuka, these proceedings) is in debate today: there is no clear deviation at relatively low energy (7-10 MeV) as was suggested by the MSW solution, and an important increase above 13 MeV. It is not easy to reconcile such behaviours retaining only neutrino oscillations (Bahcall, Krastev 1998). An other way to solve the puzzle is to evoke the existence of a sterile neutrino or oscillation in vacuum (Vallé 1998, Berezinsky, these proceedings) .

4.2. Some astrophysical and nuclear aspects of the puzzle

The seismic probe currently plays a crucial role in this game, as was illustrated in the previous section. We know from density, sound speed, and rotation profile, that the solar structure is reasonably well understood down to 0.6 R⊙. We are presently solving the major discrepancies noticed in stellar evolution for this kind of stars (photospheric helium, lithium, beryllium abundances). Below this region of the Sun, that is for 0.94 M⊙, it is still premature to conclude firmly as we have not a complete understanding of the history of the solar angular momentum and still some problem to believe and interpret the effects observed in the core. It is why we cannot exclude a smooth mixing process due to internal waves and some very central effects connected to residual higher rotation (which is not yet properly extracted, Corbard et al. 1999) or to the presence of a magnetic field but we have already reject large mixing evoked previously. One cannot exclude that one way to interpret the SuperKamiokande result is to imagine a reduction by a factor 2 of the ^7Be abundance in the very central region of the Sun (figure 7), this possibility will be difficult to demonstrate with acoustic modes today. Such a model is proposed in table 1 (^7Be model).

An estimate of the present uncertainties on the neutrino fluxes leads to an error of about 7% on gallium, 25 % on chlorine experiment and water detectors. In this estimate, it appears that uncertainties on gallium predictions due to ^{13}N and ^{15}O neutrino fluxes (which represent 17% of the total flux) are at a level of 20-30%. We cannot exclude that the role of these neutrinos may result in a misleading interpretation of the gallium and chlorine experiments. Part of such uncertainties comes from plasma effects which do not influence the solar structure and cannot be checked by present seismology. In this estimated uncertainties, astrophysical uncertainty is not definitively established for the reasons described above.

In parallel to the points already mentionned, several activities are under investigations with the objective to check the following ideas:

- Is ^7Be abundance (see figure 7) sufficiently well described by the nucleosynthesis generally used in stellar evolution ? (Turck-Chièze, Chièze and Dzitko 2000). Of course present seismology is unable to check this point.

- Is the hep cross section correctly estimated? (Bahcall and Krastev 1998). In fact, it seems difficult to imagine an error of a factor 20 on this cross section. It would be useful to see if helioseismology is able to check at least the abundance of ^3He at its maximum (figure 7) near 0.3 $R_\odot$. It is a way to progress on the idea of small mixing in the radiative interior (Turck-Chièze, 1998).

Some of the different points discussed here may contribute to decrease the present discrepancies.

5. The way ahead

The recent experiments have shown a reduced discrepancy between predicted and detected neutrinos. These discrepancies have not yet a clear understanding. Near future results in seismology aboard the satellite SOHO and in neutrino detection (SNO and BOREXINO) will give some crucial answers to the questions previously asked:

- on the ability to determine some key abundances: heavy elements and ^3He,
  - on the very internal rotation profile,
  - on some gravity modes may be,
- on the ratio neutral/charge currents and some evidence of oscillation,
- of the order of magnitude of the ^7Be neutrino flux suppression.

But it is clear today that no presently programmed experiments or calculations will definitively solve the present solar neutrino puzzle and its consequences, because the near future neutrino experiments are at too high energies, and because the very central Sun is not accessible to acoustic modes. The only way to conclude on the neutrino puzzle is to measure the neutrino low energy spectrum and to detect gravity modes, except if we can demonstrate soon that it is a pure nuclear problem.

5.1. the gravity modes

The present sensitivity of the sound speed to fundamental ingredients is very high: 2% variation on pp reaction is visible: effect of $2\ 10^{-3}$ on the sound speed, 2% variation of the opacity coefficients locally is also measurable (same effect). Paradoxically, acoustic modes are indirectly sensitive to the solar core and cannot check the absolute value of the opacities, so the measurement of the gravity modes would be very useful to be sure of what we extract and to discriminate different physical effects.

Effectively, the frequency of these modes is proportional to:

$$\int_0^r N(r)/r dr \qquad (9)$$

where N is the Brunt Väisälä frequency:

$$N^2 = g\left(\frac{1}{\Gamma_1}\frac{d\ln P}{dr} - \frac{d\ln\rho}{dr}\right)$$

and enhances the information of the core (60 % from the inner 0.2 $R_\odot$) contrary to the acoustic mode frequency (5%). This is an important objective of the helioseismic experiments aboard SOHO. Some possible candidates are under investigation, which could help to progress on this field and disentangle the different previously suggested ideas (Gabriel et al. 1998, Turck-Chièze et al. 1998b). Larger effort is in preparation to reduce the inherent granulation noise seating in the range of frequencies where they are waiting.

5.2. the solar energy neutrino spectrum: the ultimate answer?

The measurement of the neutrino energy spectrum of figure 1 is the key for a non-ambiguous interpretation of the present detections.

The four different ideas evoked in this review, have different energy spectrum signatures:

- the oscillation of neutrinos produces a deformed spectrum, amplified for ^7Be neutrinos.

- an astrophysical solution implies the reduction of the different sources of neutrinos according to the temperature dependence (larger for ^8B and CNO neutrinos) without any deformation of the corresponding spectrum.

- a nuclear solution, as an incorrect reaction rate or a different screening enhancement, induces a reduction of the corresponding neutrino flux without any deformation of the spectrum.

- a nuclear solution, inducing an incorrect ^7Be abundance, implies the same reduction for ^7Be and ^8B spectrum, without affecting the other sources.

The technical challenge for this new step is difficult and justifies the present different ways engaged to reach a definitive solution (HELLAZ, SUPERMUNU, LENS solar neutrino projects). The difficulty in the next generation of experiments measuring the energy spectrum below 1 MeV will be to disentangle all these solutions by an appropriate energy resolution. This step must be prepared very seriously, integrating all the new results along years.

6. Acknowledgements

I would like to address a great thank to the whole helioseismic community. Many works, intercomparisons and interpretations come from GOLF and MDI helioseismic experiments aboard the satellite SOHO, which is an ESA/NASA project and have also take benefit of the ground networks BiSON, IRIS, GONG and LOWL results.

7. Bibliography

Abdurashitov et al. (SAGE collaboration), 1996, Phys. Rev. Lett., 77, 4708.

Adelberger, E. et al. 1998, Rev. Mod. Phys., 70 (4), 1265.

Bahcall, J. N., Neutrino Astrophysics (Cambridge University Press), 1989.

Bahcall, J. N., et al., 1996, Phys. Rev. C, 54, 411.

Bahcall, J. N., Krastev, P., 1998, Phys. Lett. B, 436, 243.

Bahcall, J. N., Krastev, P. I., Smirnov, A. Y., 1999, Phys. Rev. D, 58, 096016-1.

Basu, S., Turck-Chièze, S., Berthomieu, G. et al., 1999, ApJ, in preparation

Brun, S. et al., 1997, in Sounding solar and stellar interiors, Symposium IAU 181, ed J. Provost and F.X. Schmider, poster volume, p 69.

Brun, S., Turck-Chièze, S., Morel, P., 1998, ApJ., 506, 913.

Brun, S., Turck-Chièze, S., Zahn, J.P., 1999, ApJ., to appear.

Christensen-Dalsgaard, J. et al., 1996, Science, 272, 1286.

Corbard, T., Berthomieu, J., Provost, J., & Morel, P. 1998, A&A, 330, 1149.

Corbard, T. et al., 1999, Proceedings of the SOHO6/GONG 98 Worshop, 'Structure and dynamics of the Interior of the Sun and Sun-like Stars', ed. S. Korzennik & A. Wilson, ESA SP-418, 741.

Dzitko, H., Turck-Chièze, S., Delbourgo-Salvador, P., & Lagrange, G. 1995, ApJ., 447, 428.

Fukuda et al., (Super-Kamiokande collaboration), 1998, Phys. Rev. Lett., 81, 1158.

Gabriel, M. and Carlier, F.: 1997, Astron. Astrophys., 317, 580.

Gabriel, A. H., and the GOLF team, 1995, Sol. Phys., 162, 39.

Gabriel, A. H., and the GOLF team, 1997, Sol. Phys., 175 207.

Gabriel, A., Turck-Chièze, S., Garcia, R., et al., 1998, in Structure and Dynamics of the Interior of the Sun and sun-like stars (eds. S.G. Korzennik & A. Wilson), ESA SP-418, 61.

Gruzinov, A. V. & Bahcall J. N. 1998, ApJ, 504, 996.

Guzik, J., Cox, A. N. and Swenson, F. J.: 1996, Bull. Astron. Soc. India, 24, 161.

Hammache, F. et al., 1998, Phys. Rev. Lett., 80, 928.

Hampel, W. et al. (GALLEX collaboration), 1996, Phys. Lett. B388, 384.

Iglesias, C. & Rogers, F. J. 1996, ApJ, 464, 943.

Table 1: Confrontation of observed neutrino detections and respective neutrino predictions for the chlorine, gallium and water detectors in the case of the reference model and two other models which suppose variation of the cross sections in the present error bars or a reduction of the ^7Be abundance in the Sun to respect the SuperKamiokande result.

Neutrino Observations				
Experiments	Chlorine exp.	SuperK exp. $(10^6/cm^2/s)$	Gallium GALLEX	experiments SAGE
	2.55 ± 0.25 SNU	2.44 ± 0.26	76 ± 8 SNU (cal= 0.91 ± 0.08)	70 ± 8 SNU (cal= 0.95 ± 0.12)
Neutrino predictions				
Reference model	7 ± 1.7 SNU	5 ± 1.25	127 ± 8 SNU	127 ± 8 SNU
Nuclear model	5.2 SNU	3.21	119 SNU	119 SNU
^7Be model	3.87 SNU	2.4	105 SNU	105 SNU

1 SNU is the solar neutrino unit which is equal to 10^{-36} capture/atom s.

Kosovichev, A. G., Schou, J., Scherrer, P. H., et al., 1997, Sol. Phys., 170, 43.

Morel, P., Provost, J., and Berthomieu, G., 1997, A&A, 327, 349.

Proffitt, C. R. & Michaud, G., ApJ., 380, (1991), 238.

Rogers, F. J., Swenson, J. & Iglesias, C. 1996, ApJ, 456, 902.

Scherrer, P. H. and the MDI team, 1995, Sol. Phys., 162, 129.

Spiegel, E.A., & Zahn, J.P. 1992, A&A, 265, 106.

Suzuki, Y., 1998, in Solar composition and its evolution, from core to corona, ed. S. Steiger, Space Science Review, 91.

Thiery et al., 1999, submitted to Astron. Astroph.

Toutain, T., Appourchaux, T., Fröhlich, C., Kosovichev, A. G., Nigam, R., & Scherrer, P. H. 1998, ApJ, 506, L147.

Turck-Chièze, S. & Lopes, I. 1993, ApJ, 408, 347.

Turck-Chièze and Brun S., 1997, in Fourth international solar neutrino conference, ed W. Hampel, Max-Planck-Institut für Kernphysik Heidelberg, p 41.

Turck-Chièze, S., 1998, in Solar composition and its evolution, from core to corona, ed. S. Steiger, Space Science Review, 125.

Turck-Chièze, S., Däppen, W., Fossat, E., Provost,

J., Schatzman, E. & Vignaud, D., 1993, Physics Rep., 230 (2-4), 57-235.

Turck-Chièze, S., Basu, S., Brun, A. S., et al., 1997, Sol. Phys., 175, 247-265.

Turck-Chièze, S., Basu, S., Berthomieu, et al., 1998a, in Structure and Dynamics of the Interior of the Sun and sun-like stars (eds. S.G. Korzennik & A. Wilson), ESA SP-418, 555.

Turck-Chièze, S., Brun, S., Chièze, J.P., Garcia, R., 1998b, in Structure and Dynamics of the Interior of the Sun and sun-like stars (eds. S.G. Korzennik & A. Wilson), ESA SP-418, 549.

Turck-Chièze, S., Chièze, J. P., Dzitko, H., 2000, to be published in ApJ.

Vauclair, G., Vauclair, S., Michaud, G., ApJ, 223 (1978), 920.

Valle, J. W., 1998, in New Trends in Neutrino Physics, Tegernsee, Germany.

Wolfenstein, L., Phys. Rev. D, (1978), 17, 2369; (1979), 20, 2634.

ELSEVIER

Nuclear Physics B (Proc. Suppl.) 80 (2000) 195–199

NUCLEAR PHYSICS B
PROCEEDINGS
SUPPLEMENTS

www.elsevier.nl/locate/npe

Neutrinos after Takayama

F. Vannucci[a]

[a]LPNHE-Université Paris 7

The Super-Kamiokande collaboration has provided evidence of atmospheric neutrino oscillations. If confirmed, this would imply that neutrinos have non-zero masses. However, the data favour masses much smaller than those required by cosmology to influence the missing-mass problem of the Universe.

Takayama is the name of a small Japanese town where the Neutrino 98 conference was held. At the conference it was announced that atmospheric neutrinos oscillate. If confirmed, this would provide the first evidence of physics beyond the very successful Standard Model, but would probably discard neutrinos as being candidates for the dark matter of the Universe.

1. Neutrinos before Takayama

What do we know about neutrinos in particle physics?

- There are three and only three neutrinos that have a light mass: ν_e, ν_μ and ν_τ.

- Theorists prefer a hierarchy of neutrino masses with ν_τ much heavier than ν_μ, and ν_μ much heavier than ν_e. This is formalized in the so-called see-saw mechanism which gives a plausible explanation for small neutrino masses. In this model the neutrino of a given generation has a mass proportionnal to the squared mass of the up-quark of the same generation.

If neutrinos have a non-zero mass, they probably mix, giving rise to the oscillation phenomenon, namely spontaneous transformation between neutrinos of different types. The probability of oscillation is given by:

$$\sin^2(2\theta)\sin^2(\pi R/L) \ ,$$

where $\sin^2(2\theta)$ denotes the mixing, R the distance between production and detection, L the oscillation length which depends on the energy E, and

δm^2 the difference of the squared masses between the two oscillating neutrinos:

$$L = 2.5 \ E(\text{GeV})/\delta m^2 \ (\text{eV}^2) \ \text{km} \ .$$

The small masses correspond to experiments carried out over long flight distances.

Before the recent conference, an educated guess about massive neutrinos gave the following plausible scenario:

- ν_μ of some 10^{-3} eV in order to explain the deficit of solar neutrinos.

The sun sends $60 \times 10^9 \ \nu_e/\text{cm}^2/\text{s}$ on the earth's surface, but only part of this theoretical flux is measured in the various experiments. The interpretation of this deficit is that the solar ν_e oscillate before reaching the earth into a new flavour, not seen by the detectors. In fact, a resonant oscillation in the interior of the sun gives the best explanation of the varying deficits seen in the different experiments. The resulting δm^2 is about 10^{-5} eV^2 which gives the quoted mass of the ν_μ when assuming a negligible mass for the ν_e.

- ν_e of a much smaller mass, 10^{-6} eV or less if the see-saw recipe is correct.

- ν_τ of a few eV with the same argument of the see-saw mechanism.

At this point it was observed that such a ν_τ could explain part of the missing mass of the Universe. The argument goes as follows: the Big Bang scenario predicts that neutrinos are several billion times more abundant than other matter particles in the Universe. A mass of a few eV is sufficient to influence the dynamics of the Universe. Furthermore, some kind of non-shining missing mass has been detected through gravitation. It is tempting to advocate massive neutrinos here. The missing mass cannot be composed solely of neutrinos (hot dark matter), but models for structure formations favour a non-negligible part (20%–30%) of these constituents.

The argument was not absolutely compelling but it was considered elegant enough to justify a large effort and the construction of two sophisticated experiments at CERN: CHORUS and NOMAD which were optimized to search for a ν_τ with such a mass.

2. The Super-Kamiokande experiment

Super-Kamiokande is a massive underground detector set in the Kamioka mine in the heart of Japan. It consists of 50 ktons (22.5 ktons of fiducial mass) of purified water observed by 11 000 large photomultiplier tubes. Figure 1 shows a photograph of the detector whilst it is being filled. Technicians on a canoe are checking the phototubes before they are covered by the water.

In water a neutrino interaction produces charged particles which are energetic enough to give Cherenkov light, and this light is detected at the level of the tubes. A charged particle gives a Cherenkov ring, and the information obtained is sufficient to measure the energy of the track by the number of tubes hit, its direction, and the vertex position by the pattern of the hits and the arrival times.

It is possible to discriminate between electrons produced by ν_e interactions and muons produced by ν_μ interactions, as a muon does not form a shower in the water and results in a better defined Cherenkov ring. The analysis is essentially based on the study of quasi-elastic events. For this category, the information on electron or muon energy and direction translates into information about the energy and the direction of the incident neutrino. Because of the large amount of statistics accumulated, the detector is able to give the distributions in the direction of the incident neutrinos, independently for ν_e and ν_μ.

Figure 1. Photograph of the Super-Kamiokande detector during fill-up (ICRR-Tokyo photograph).

3. Oscillations of atmospheric neutrinos

Atmospheric neutrinos are produced by primary cosmic rays impinging on the upper levels of the atmosphere. They produce mesons which decay rapidly into ν_μ and muons, themselves decaying mostly before they reach the earth. Finally, one expects about two ν_μ's for one ν_e.

The uncertainties in the simulation are multiple: primary cosmic ray flux, density of the atmosphere, magnitude of the geomagnetic field, relief of the earth, solar activity. Nevertheless, there are several calculations which agree within 20%. In fact, experimentalists prefer to consider the ratio ν_μ/ν_e for which some systematics cancel out. The ratio is supposed to be known within 5%.

Figure 2 gives the zenith angle distributions of the candidate ν_e and ν_μ events, namely the incident direction of the neutrinos with respect to the vertical axis. The distribution of ν_e events agrees with expectations, whilst there is a loss of ν_μ events, mainly for neutrinos coming from the direction of the antipodes [1]. These distributions are obtained for the so-called sub-GeV contained events for which all the tracks are absorbed in the water. They correspond to a measured average energy of less than 1.3 GeV. Two other categories have been investigated namely up-going muon events coming from interactions of neutrinos in the rock beneath the detector with muons either crossing the detector or stopping in it [2].

Super-Kamiokande measures a rate of ν_e in agreement with calculations, but the rate of ν_μ shows a deficit mainly for the ν_μ's going upwards in the detector. This is found in the three types of events studied, and is well explained by the oscillation of the ν_μ into another neutrino, presumably the ν_τ. Figure 3 shows the very suggestive distribution of distance/energy which extends over three orders of magnitude.

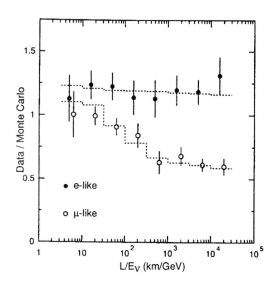

Figure 3. The ratio of the number of data events to Monte Carlo events versus reconstructed L/E_ν. The points show the ratio of observed data to Monte Carlo expectation in the absence of oscillations. The dashed lines show the expected shape for $\nu_\mu \leftrightarrow \nu_\tau$ at $\Delta m^2 = 2.2 \times 10^{-3}$ eV^2 and $\sin^2 2\theta = 1$. The slight L/E_ν-dependence for e-like events is caused by the contamination (2–7%) of ν_μ CC interactions.

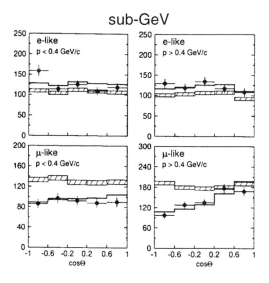

Figure 2. Zenith angle distributions of μ-like and e-like events for sub-GeV data sets. Upward-going particles have $\cos\Theta < 0$, and downward-going particles have $\cos\Theta > 0$. Sub-GeV data are shown separately for $p < 400$ MeV/c and $p > 400$ MeV/c. The hatched region shows the Monte Carlo expectation for no oscillations normalized to the data live-time with statistical errors. The bold line is the best-fit expectation for $\nu_\mu \leftrightarrow \nu_\tau$ oscillations with the overall flux normalization fitted as a free parameter.

The neutrinos produced in the atmosphere directly above the apparatus do not show a deficit, they have only travelled about 10 km. The neutrinos which show the maximum deficit have travelled through the whole planet before being detected. This can be interpreted as evidence for the oscillation of ν_μ's over distances of several 1000 km. The ν_e's do not seem to oscillate over the same distances. The favoured channel could therefore be the conversion of ν_μ's into ν_τ's or a new neutrino having no interactions with matter. With the numbers involved in the process, the oscillation suggests a difference of squared masses between ν_μ and ν_τ

$$\delta m^2 = 3.2 \times 10^{-3} \text{ eV}^2 \text{ [3]} .$$

With a hierarchy of masses, this fixes the mass of the ν_τ at about 50 meV. The two other masses of ν_μ and ν_e are much smaller if one retains the see-saw prescription.

4. Alternative solutions?

Since the announcement of the Super-Kamiokande result, there has been a large number of articles trying to understand the results. Many of them consider the different possible scenarios of neutrino masses and their adequation with theoretical models. Some discuss possible alternatives.

What has been seen by the experiment is a disappearance of ν_μ whilst it traverses the diameter of the earth. Oscillation is the best known process which could be responsible for the deficit, but other processes have been invoked:

- the existence of flavour-changing neutrino matter interactions. This results in a $\nu_\mu \to \nu_\tau$ transformation but does not require neutrino masses [4].

- non-radiative decay of ν_μ's [5], radiative decays being discarded by laboratory experiments.

- a loss of energy in matter for the ν_μ's. A loss of about 100 MeV over 13 000 km would be enough to explain the apparent deficit because of threshold and cross-section effects. An experiment measuring the energy losses in a crystal of germanium carried out in the CERN ν_μ beam excludes a loss of more than 10 keV [6].

All the alternative solutions proposed so far go beyond the Standard Model, and do not seem more economical than the oscillation hypothesis.

5. Conclusion

Although the Super-Kamiokande result is of prime importance, the whole situation with respect to oscillation searches is not completely satisfactory.

Other evidence comes from the solar neutrino deficit and from the LSND result. In the usual mixing vs. δm^2 plot of oscillation searches, there are three sets of regions favoured, as shown in Fig. 4. Three different δm^2 are too many with only three different neutrino masses. This is why a fourth sterile neutrino has been advocated to reconcile all the experimental signals.

Among these various results, Super-Kamiokande provides the strongest evidence, with the signal of atmospheric neutrinos. However, it still needs confirmation. A first test will come from Super-Kamiokande itself, with the study of neutral current interactions measured through π^0 production. These interactions are the same for all flavours of interacting neutrinos and if the oscillation happens between ν_μ and ν_τ, the π^0 signal should not show a deficit. For the time being, this very powerful test is statistically limited and should become constraining with time.

Other tests will come with the oscillation studies of accelerator neutrinos. First there will be the K2K experiment with neutrinos from KEK sent to the Super-Kamiokande detector, 250 km away. Then higher energy long base-line experiments are foreseen in Europe and the United States with beams from CERN and Fermilab pointing to Gran Sasso and Soudan, respectively.

Super-Kamiokande has provided strong evidence of neutrino masses and one can ask if these masses are relevant for cosmology.

With a ν_τ of 50 meV, the sum total of neutrino masses is much smaller than the baryonic component of the Universe, and the cosmological role of neutrinos would be gravitationnally marginal. This conclusion can be evaded if the oscillation takes place in a new sterile neutrino, or if the

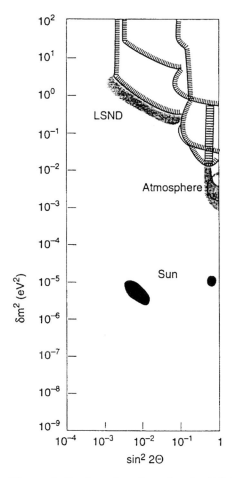

Figure 4. Preferred regions for the LSND, atmospheric and solar results. They do not necessarily apply to the same physical channel.

REFERENCES

1. Y. Fukuda et al., The Super-Kamiokande collaboration, Phys. Lett. **B433** (1998) 9; Phys. Lett. **B436** (1998) 33; Phys. Rev. Lett. **81** (1998) 1562.
2. Y. Fukuda et al., hep–ph/9812014.
3. Y. Totsuka, Texas Symposium, Paris, December 1998, These proceedings.
4. M.C. Gonzales-Garcia et al., hep–ph/9809531.
5. Barger et al., hep–ph/9810121.
6. A. Castera et al., to be published in Phys. Lett.

masses are almost degenerated. Oscillations only fix a difference of masses, and each flavour could have a few eV, even though this is contrary to the see-saw prescription.

With the Super-Kamiokande result, neutrinos are again taking centre stage in particle physics. We may have obtained the first indication of physics beyond the Standard Model. For the experimentalists a lot of difficult measurements lie ahead, and it will take many years to disentangle the whole pattern of neutrino masses and mixings.

Summaries of Mini-symposia

ELSEVIER

Nuclear Physics B (Proc. Suppl.) 80 (2000) 203–206

NUCLEAR PHYSICS B
**PROCEEDINGS
SUPPLEMENTS**

www.elsevier.nl/locate/npe

AGN, QSOs, and Jets

Mitchell C. Begelman[a] and Laura Maraschi[b]

[a]JILA, University of Colorado, Boulder, CO 80309-0440, USA

[b]Osservatorio Astronomico di Brera, via Bianchi 46, I-22055 Merate, Italy

The minisymposium on AGN, QSOs, and jets was highly successful with about 100 participants and more than 60 posters presented. Due to the very limited time available the broad area indicated in the title could only partly be covered by oral presentations. The 9 invited speakers focused on two main topics in AGN research: the formation and structure of disks around massive black holes and particle acceleration and radiation in relativistic jets whose emission extends up to the TeV range.

Below we briefly summarize the oral presentations. Interested readers are urged to consult the CD-ROM for a fuller discussion of these and other topics presented at the minisymposium.

J.M. Moran described the new results obtained from several years of observations of water masers tracing accretion disks in the nuclei of some galaxies (6 disks found from observations of 700 galaxies of which 22 showed water maser emission). In the best studied example, NGC4258, the redshifted and blueshifted maser spots obey a Keplerian rotation curve with high precision, while the maser spots near the systemic velocity appear to be produced at a well-defined distance from the nucleus. The disk is warped in a way that could explain the narrow annular spread of the systemic features. The azimuthal positions of the high velocity features determined from accelerations reveal no evidence for a spiral wave. The magnetic field in the disk and its distance from the sun have been measured (¡300mG and 7.2+/-0.3 Mpc, respectively). Estimates of the accretion rate onto the central black hole seem inconsistent with an ADAF model for the nuclear activity of this source.

F. Mirabel presented mid-infrared and submm maps of the central region of Cen A, the closest radio galaxy, obtained with ISO and SCUBA. The morphology of the dust emission at about 1 kpc (1 arcmin) from the center of this giant elliptical galaxy suggests the presence of a small quasistable barred spiral structure with an infrared luminosity per unit mass of interstellar molecular gas typical of starburst galaxies. This structure could have formed out of the infalling gaseous component of a small tidally disrupted accreted galaxy and could provide through the bar the dynamical instability that serves to feed the activity of the nucleus.

As discussed by **E. Schreier**, the same prototypical galaxy was observed also at higher resolution, corresponding to pc scale length, in the visible and near infrared (1.6 and 2.2 μ), as well as in the lines Paschen α and FeII with HST WFPC2 and NICMOS. While the visible images show more structure in the larger galaxy and the dense dust lane, the continuum IR images show extended emission typical of an elliptical galaxy and the strong unresolved central nucleus of the galaxy. A prominent elongated structure seen in line emission, centered on the nucleus and extended by $\sim 2''$, is interpreted as an inclined, ~ 40 pc diameter, thin nuclear disk of ionized gas. The disk is one of the smallest ever observed at the nucleus of an AGN. It is not perpendicular to the jet, but is consistent with being oriented along the major axis of the bulge. Even on the scale of a few parsecs, the disk is dominated by the galaxy gravitational potential and is not directly related to the symmetry axis of the AGN as traced by the radio jets.

H. Inoue introduced studies of the innermost region of accretion disks around black holes (scale length of $10^{-4} - 10^{-5}$ pc), based on their X-ray emission and in particular on the profile of the fluorescent Fe Kα line. Long observations of the Seyfert galaxy, MCG–6-30-15 with ASCA discovered a broad and skewed iron line structure that is now generally believed to be due to Doppler and gravitational red-shift effects near the last stable orbit of an accretion disk around a massive black hole. The profile can be approximated by two components, a relatively narrow feature at 6.4 keV and a broad feature around 5.5 keV, and it was found that the fluxes of the two components independently changed on a time scale as short as or shorter than 10 ksec. This time scale of the variation strongly supports the innermost-disk-origin of the line photons but the temporal profile-change forces us to consider a complicated configuration of the line emitting region. Those observations also revealed absorption features which are probably due to warm absorbers existing on the line of sight. Constraints on the physical properties of the warm absorbers obtained from the observations were also discussed.

C. Reynolds addressed the prospects for X-ray iron line reverberation mapping of AGN, using time-resolved observations of the relativistically broadened X-ray iron lines seen in many AGN spectra (such as in MCG–6-30-15). Assuming that the line is produced by X-ray illumination and that a rapid flare occurs at the center of the disk, each ring in the disk will reverberate at different times, allowing one to reconstruct the velocities at different distances from the center. Future high throughput X-ray observatories, such as Constellation-X, should be able to track the temporal variability of the line strength and profile in response to X-ray flares above the accretion disk, as demonstrated by an impressive video showing animated simulations. This method is potentially useful for probing black hole masses and spin, as well as the geometry of the accretion disk and corona.

The second theme of the session, concerning relativistic jets, was opened by **G. Ghisellini** with a synthetic overview of the spectral energy distributions of blazars, thought to be a single family of objects whether or not emission lines are detectable in their spectra. In fact, considering complete samples the average SEDs in a ν–νF_ν representation in all cases show two peaks associated with two smooth broad band components which are usually attributed to the synchrotron and inverse Compton mechanisms respectively. The peak frequencies move systematically to higher values for objects of lower bolometric luminosity, suggesting a spectral sequence. This behavior can be understood in terms of a similar powering mechanism for all objects but different local conditions around the jet, leading to higher energies for the accelerated particles when radiation losses are lower. At one extreme of the sequence the synchrotron spectrum peaks in the X–ray band and the inverse Compton spectrum near the TeV range. Ghisellini suggested that even more extreme BL Lacs could exist, whose synchrotron spectrum peaks in the MeV band. These sources should emit a substantial fraction of their power in the TeV band by the inverse Compton process. Theoretical limits to the maximum possible emitted frequencies were discussed.

G. Madejski pointed out the importance of the discovery of GeV and TeV emission from many blazars. The jet-like structures and rapid variability of blazars is best explained by a scenario where the radiating material moves at relativistic velocities close to the line of sight, and thus the entire continuum is Doppler-boosted, with Lorentz factors of the jet Γ_{jet} of $\sim 5 - 20$. This increases the apparent brightness, and shortens the observed variability time scales. The best current models for the radiation from those sources invoke the synchrotron process for the low energy component and Comptonization of lower energy photons by the same electrons that produce the synchrotron radiation as the origin of the high energy component.

In the context of the above scenario, the origin of the "seed" photons appears to depend on the relative density of the ambient radiation field. For blazars that show broad emission lines, these seed photons are most likely from the rescattered broad line light (as in the "External Radiation Compton," or ERC models), while when the emission lines are absent, the internal synchrotron photons (as in the "Synchrotron Self-Compton," or SSC models) appear to dominate. Interestingly, it is the latter class that has spectra extending often to the TeV range, while no TeV emission was so far detected from blazars with emission lines. Modelling of spectra of the former sub-class implies electron Lorentz factors γ_{el} of $\sim 10^3 - 10^4$, while the latter class requires $\gamma_{el} \sim 10^5 - 10^6$. In both classes, the magnetic field B is calculated to be ~ 0.1 to a few Gauss.

The first blazar detected in the TeV band — and so far, perhaps the most extensively studied — is Mkn 421. Many multiwavelength campaigns to observe this object revealed that the variability in the X–ray and the TeV γ–ray bands shows greater amplitude than in other bands. The flux observed in these two bands appears to be correlated, suggesting a common population of particles radiating in both bands. This alone implies γ_{el} of $\sim 10^6$. The detailed studies in the X–ray band reveal spectral variability, where the spectrum hardens as the source brightens, and softens when the source flux drops. This is well explained by the dependence of the lifetime of the radiating particles on the particle energy, where the drop of the soft X–ray flux lags that of the hard X–ray flux. In this context, the data imply the lifetime of 1 keV electrons to be ~ 6000 s, which in turn yields $\gamma_{el} \sim 10^6$ and $B \sim 0.1$ Gauss, nicely consistent with the values derived from the SSC models. A similar general behavior of Mkn 421 was observed in the multiwavelength campaign conducted in April 1998, where a large flare lasting for several days was detected, although the X–ray data suggest that this flare may be a superposition of many smaller, more rapid events.

F. Aharonian presented the temporal and spectral characteristics of TeV radiation from the two brightest sources known, Mkn 421 and Mkn 501, as obtained with the imaging atmospheric Cherenkov telescopes (IACT) HEGRA and CAT. The distinctive feature of the CAT detector is its low energy threshold $E_{\text{th}} \simeq 250 GeV$. The HEGRA stereoscopic system of IACTs operates at higher energies, $E_{\text{th}} \sim 500\,\text{GeV}$, but has superior cosmic ray background rejection power based on the excellent (of about 5 arcmin) angular resolution and high efficiency of γ/hadron separation. Both instruments have reasonable energy resolution $\sim 20 - 25\%$.

The minimum detectable fluence of $\sim 10^{-7}$ erg/cm^2 for burst-like events with duration less than several hours is nicely suited for the search of γ-ray flares from these highly variable sources. The range of timescales and spectral coverage is well matched to the current broad band X-ray detectors like ASCA, BeppoSax, and RXTE. The BL Lac objects Mkn 421 and Mkn 501, discovered by the Whipple collaboration as TeV γ-ray emitters, are ideal laboratories for studying acceleration and radiation processes in AGN jets. Continuous monitoring of Mkn 501 by the HEGRA and CAT telescopes during the extraordinary (both in strength and duration) outburst of the source in 1997 revealed an interesting feature of the source. Namely, the HEGRA data show that the shapes of the daily γ-ray spectra above 1 TeV remained essentially stable throughout the entire state of high activity despite dramatic (more than factor of 10) flux variations. Although at lower energies the CAT measurements do indicate a correlation of the strength of the source and the spectral shape, the effect is not very strong. This makes the determination of the time-averaged spectrum of Mkn 501 astrophysically meaningful. The HEGRA measurements reveal a well-defined, smooth differential spectrum extending deeply into the exponential regime. From 1 to 24 TeV it is well approximated by a power-law with an exponential cutoff: $dN/dE \propto E^{-\alpha} \exp{-(E/E_0)}$, with $\alpha = 1.92 \pm 0.03_{\text{stat}} \pm 0.20_{\text{sys}}$, and $E_0 = (6.2 \pm 0.4_{\text{stat}} (-1.5 + 2.9)_{\text{sys}})$ TeV.

Mkn 421 was monitored by the HEGRA and CAT instruments during a multiwavelength campaign of the source in April 1998 with participation of ASCA, BeppoSax and RXTE. Fortunately the source was in a high state with average TeV flux at the level of half Crab. This allowed us to measure the energy spectrum of the source, which at energies below several TeV appeared to be noticeably steeper than the spectrum of Mkn 501. Again, as in the case of Mkn 501, the preliminary analysis of the HEGRA results does not show any significant correlation between the flux and spectral shape of Mkn 421.

The interpretation of time-resolved spectra of Mkn 501 was discussed by **G. Bicknell** (in collaboration with Stephan Wagner), after presenting the results of near-simultaneous monitoring of Mkn 501 in optical, 2-100 keV X-ray and TeV γ-ray regimes. Time variability implies that the size of the high energy emitting region $< 10^{16}$ cm, close to the limits implied by pair-production opacity constraints. On the basis of the spectra, they convincingly argued that the TeV γ-rays are inverse Compton emission from the electrons responsible for the 10-100 keV emission. A spectral break at about 3 keV in the X-rays and and a synchrotron-self-Compton model of the γ-ray emission imply a sub-equipartition magnetic field of about 0.05 G in the emitting region, a Doppler factor of order 10-20 and a jet energy flux of about 1.5×10^{44} ergs s^{-1}. The slope of the TeV spectrum is consistent with SSC emission in the Klein-Nishina limit and a maximum Lorentz factor of the emitting electrons $\sim 2 \times 10^6$.

ELSEVIEP

·Nuclear Physics B (Proc. Suppl.) 80 (2000) 207–208

NUCLEAR PHYSICS B
PROCEEDINGS SUPPLEMENTS

www.elsevier.nl/locate/npe

Gamma-Ray Bursts

Kevin Hurley[a] and Robert Mochkovitch[b]

[a] Space Sciences Laboratory,
University of California at Berkeley, CA 94720, USA

[b] Institut d'Astrophysique de Paris,
98bis Bd Arago 75014 Paris, France

It has now been established that cosmic gamma-ray bursts are a multi-wavelength, cosmological phenomenon. Indeed, they may be the most intense electromagnetic explosions in the Universe. The interest in studying them, explaining them, and utilizing them as cosmological tools is now at an all-time high. Every month seems to bring a new, exciting, and completely unexpected discovery. Thus it comes as no surprise that the Mini-Symposium on Gamma-Ray Bursts, held in conjunction with the 1998 Texas Symposium in Relativistic Astrophysics, attracted numerous participants. A total of 75 people submitted abstracts for what was initially intended to be a single oral and poster session. Because all of these presentations could not be accomodated into the one-afternoon session allocated by the organizing committee, it was decided to hold an overflow oral session at the Institut d'Astrophysique de Paris. Even with this extra afternoon, the presentations had to be limited to only 10 - 15 minutes.

The abstracts received were organized into the following topics: Global Properties, Temporal Properties, Spectral Properties, Localizations and Host Galaxies, GRB's at Other Wavelengths and Afterglows, Theory - Central Engine, Theory - Emission Processes, and Future Experiments. In the end, 39 oral presentations and about 40 posters were presented, and for 27 of them, manuscripts were received. We review them briefly here, and encourage the interested reader to consult the complete manuscripts in the CD-ROM version of the proceedings.

1. Temporal Properties

Because bursts originate at cosmological distances, it is possible that some fraction of them are gravitationally lensed. Beskin et al. estimate the probability of observing a lensed event and suggest that lensing can be used to study the behavior of GRB sources prior to the burst itself. Cline et al. have analyzed the spectra and durations of a class of bursts and conclude that they may be useful for evaluating the cosmological constant. Yu et al. have reanalyzed the duration distribution for bursts and find that there is evidence for time dilation in the long duration, weak events.

2. Spectral Properties

Gamma-ray bursts have very hard energy spectra, which have been observed up to 20 GeV. Boettcher and Dermer consider radiation mechanisms which can produce time-delayed emission in the GeV-TeV energy range. Brainerd et al. have studied the role of the BATSE instrument response in the observed distribution of the νF_ν peak energy, and conclude that the distribution is not significantly affected by it. Frontera et al. present the results of a study of spectral evolution with BeppoSAX, and show, among other things, that the X-ray afterglows actually begin during the gamma-ray bursts. Mallozzi et al. have analyzed a sample of BATSE bursts which display precursor emission, and find that the characteristics of the initial outburst are essentially identical to those of the main burst which follows.

3. Localizations & Host Galaxies

There is growing evidence that GRB's occur within their host galaxies, and that the sources have therefore not been ejected from them, although this result could be affected by selection effects. Bulik and Belczynski have modeled compact object mergers and find that the majority of them should take place outside the host galaxies; they discuss the selection effects which can reconcile this with the observational data. Motivated by the recent possible association of SN1998bw with GRB980425, Klose and Hudec and Hudcova have searched for a correlation between bursts and cataloged supernovae, but find the results to be consistent with no statistically significant correlation. Marani et al. and Miller et al. have analyzed a subset of bursts for evidence of gravitational lensing, and from the null result, derive an upper limit to the average burst redshift.

4. GRB's at Other Wavelengths & Afterglows

The observations of radio, optical, and X-ray afterglows during and following gamma-ray bursts has led to a breakthrough in understanding the burster distances and emission mechanisms, but the number of such observations is still rather limited. Hudec reviews the instruments, databases, and strategies which can be used to detect optical afterglows. Hudec and Kroll discuss the use of sky patrol data to detect optical transients before, during, and after the bursts.

5. Theory - Central Engine

The nature of the energy release which ultimately gives rise to a gamma-ray burst is unknown. It may be due to a merger of two compact objects (2 neutron stars, or a neutron star and a black hole), or to a hypernova. Beskin et al. have considered the possibility that the energy release comes from accretion onto a single black hole accreting interstellar plasma. Derishev et al. have studied the novel possibility that the energy comes from the collapse of a single neutron star induced by a collision with an orbiting primordial black hole. In another paper, Derishev et al. consider the role of neutrons in the energy release, and suggest that they may be responsible for two classes of bursts. Lee and Kluzniak have studied the final stages of a neutron star – black hole coalescence, leading to beamed energy release. Ruffert et al. have similarly considered neutron star-black hole mergers, and in particular, the role of neutrinos (their presentation was mainly in the form of a video). Daigne and Mochkovitch estimate the mass loss rate and the final Lorentz factor in a magnetically driven wind emitted by a disk orbiting a stellar mass black hole. Ruffini considers the different possible ways to extract energy from a black hole. Salmonson et al. have focussed on the details of the collapse of close binary neutron stars as they approach their last stable orbit.

6. Theory - Emission Processes

There is general agreement that, once the central engine has released its energy, the emission which we observe will come from a relativistic fireball. Brainerd proposes that GRB emission comes from plasma instabilities when a relativistic shell of matter passes through the interstellar medium. Bykov has studied the electron acceleration in the fireball in a model which follows the time evolution of GRB emission.

7. Future Experiments

The breakthrough in our understanding of GRBs has come from the identification of multi-wavelength counterparts, but their detection rate is still quite small. Thus many experiments now being planned are aimed at increasing the detection rate of precisely localized bursts. Beskin et al. have studied a wide field optical telescope which operates in conjunction with a gamma-ray detector. Hudec et al. examine an X-ray lobster-eye telescope and Stacy et al. describe a coded-aperture X-ray/gamma-ray telescope for an ultra long balloon flight.

ELSEVIER

Nuclear Physics B (Proc. Suppl.) 80 (2000) 209

NUCLEAR PHYSICS B
PROCEEDINGS
SUPPLEMENTS

www.elsevier.nl/locate/npe

High Energy Cosmic Rays

G. Schatz[a] and M. Teshima[b]

[a]Forschungszentrum Karlsruhe, Institut für Kernphysik, P. O. Box 3640, D-76021 Karlsruhe, Germany.

[b]Institute for Cosmic Ray Research, University of Tokyo, Midoricho, Tanashi, Tokyo 188, Japan

This Mini Symposium comprised c. 30 contributions spanning a large range of observational and theoretical topics.

On the observational side, Takeda et al. reported about the latest results of the AGASA experiment observing the highest energy cosmic rays. This experiment has now accumulated 7 events above the expected Greisen-Zatsepin-Kuzmin cut-off. This is more than all other experiments combined. The field will soon be joined again by the Fly's Eye group which reported on the start-up of the new HiRes Fly's Eye detector (Loh et al.). Data from the Fly's Eye Stage 1 have been analyzed with respect to signals from neutrinos of extremely high energy (Kieda et al.).

Further contributions presented results from the DICE (Kieda et al.) and KASCADE (Klages et al.) air shower experiments. The CAT group searched (unsuccessfully) for gamma rays from the Super Nova Remnant Cas A (Goret et al.). Such radiation is expected if supernova remnants are the place of acceleration of the bulk of cosmic rays up to c. 10^{16} eV, as is widely accepted. Two contributions described new experimental approaches to TeV gamma ray astronomy. Williams et al. reported about the first observations from the MILAGRO experiment, a novel type of detector registering Cherenkov light produced in a water pool on a mountain top, and also about the STACEE detector, a field of heliostats for solar power applications which will be used for Cherenkov light detection during night time.

New types of detectors have been proposed for the study of extremely high energy cosmic rays (AIRWATCH; Catalano et al. and Giarusso et al.), for neutrino astronomy (ANTARES; Basa et al.) and TeV gamma rays (Miller and Wester-

hoff).

A special highlight was a report by Totsuka et al. on the results of the SUPERKAMIOKANDE experiment which has observed evidence for oscillations of neutrinos from cosmic ray interactions in the atmosphere. This is not only the first evidence for physics beyond the standard model of particle physics but also a come-back of cosmic rays in the study of elementary particles.

The origin of the highest energy cosmic rays remains more enigmatic than before, in view of the non-observations of the Greisen-Zatsepin-Kuzmin cut-off. Correspondingly varied are the theoretical proposals which range from the decay of ultra-heavy relic particles from the early phases of the universe (Berezinsky et al.), gamma ray bursts (Nakamura) to interactions of ultra-high energy neutrinos in the vicinity of the Galaxy (Fargion et al.). Magnetic fields in the local supercluster of galaxies may have a measurable influence on the observed spectrum and angular distribution (Lemoine et al., Olinto and Blasi). No consensus could be reached about limits on the extragalactic neutrino flux (Rachen et al., Waxman and Bahcall), apparently due to different astrophysical assumptions.

Propagation of cosmic rays in the Galaxy is still a subject of investigation. The influence of dust on the injection into the acceleration process and on propagation was studied by Ellison et al. and by Istomin and Barabash. In addition to propagation, the interactions of charged cosmic rays have to be modelled carefully in order to allow reasonable comparison of theoretical models with experiments. In this sense the photoproduction of mesons was considered in two contributions by Muecke et al.

ELSEVIER

Nuclear Physics B (Proc. Suppl.) 80 (2000) 211–213

NUCLEAR PHYSICS B
PROCEEDINGS
SUPPLEMENTS

www.elsevier.nl/locate/npe

Mapping the Universe from Weak Lensing Analysis

Yannick Mellier[a] and Uros Seljak[b]

[a] Institut d'Astrophysique de Paris
98[b], Bd Arago
75014 Paris, France

[b]Center For Astrophysiscs
Harvard University
Cambridge, MA 02138, USA

The first session of the mini-symposium was devoted to weak lensing analysis of clusters of galaxies and its cosmological implications.

M. Lombardi and G. Bertin presented two contributions. In a first part, they discussed how clusters of galaxies can be used to constrain the cosmological parameters. If the total mass of a very massive clusters is known, then by using the redshift distribution of the background sources inferred from observations, the mass distribution and the cosmological parameters (Ω, Λ) can be recovered simultaneously. This technique needs a large sample of very massive clusters in order to obtain a statistically significant result.
In a second part, they discussed the various techniques used to measure the shear and to reconstruct the mass distribution from weak lensing analysis. They summarized the various approaches which were proposed in the past, with particular emphasis on the unbiased estimators. They also discussed the alternative estimator they proposed.

J. Frieman presented their on-going project to survey nearby clusters with weak lensing. Though we expect weak signal from nearby lensing clusters, they permit to use a broad and large sample of background galaxies since they are less sensitive to the redshift distribution. Furthermore, a given angular scale of a low-redshift cluster probes a smaller physical scale than high-redshift clusters. This offers the possibility to observe cluster mass distribution on small-scales

with great details. A dozen a clusters are underway now. This project is a part of the weak lensing analysis of the SLOAN survey. J. Frieman showed the very first slices of the photometric part of the SLOAN survey. From investigation of the image quality, it seems that it is good enough to provide shapes of galaxies with good precision, so that weak lensing analysis should be feasible with the SLOAN data.

The two next talks were devoted to mass reconstructions of high- redshift clusters of galaxies. T. Broadhurst presented very deep multicolor observations done at the Keck telescope of intermediate redshift clusters. The data clearly show the depletions produced by the magnification bias. Because they have multi-color data sets, they can control the redshift distribution of the lensed sources. In particular in the case of the lensing cluster Cl0024+1654, T. Broadhurst showed that the mass profile and the velocity dispersion of the dark matter of the clusters can be recovered with a good accuracy.

D. Clowe presented recent ground-based observations of the very high-redshift clusters MS1054-03 at redshift 0.83 . The deep UH-2.2 m data show that the number of galaxies is extremely large and spread over a very large angular size. With G. Luppino and N. Kaiser produced the shear map and the mass distribution of this cluster. The dark matter distribution they inferred looks very similar to the light distribution of the red galaxies of the clusters. Three clumps of dark matter seem to define the bulk of the system. The substructures have been confirmed by H. Hoestra who shortly presented very recent HST data of this cluster.

Finally, Zakharov presented the lensing properties of non-singular isothermal spheres. He discussed their practical interest to model lenses like clusters of galaxies or those responsible for microlensing events in the various ongoing microlensing experiments.

The second part of the session was devoted to theoretical investigations of weak lensing as well as to the analysis of existing data.

L. Van Waerbeke presented an investigation of the ability of ongoing weak lensing weak lensing surveys to probe the cosmological parameters and the power spectrum: his emphasis was how to define the best observational strategy: what are the optimal depth, size and redshift distribution and how should the data be analyzed?

Next two talks discussed weak lensing effect on the cosmic microwave background. M. Zaldarriaga presented a method to reconstruct the projected density fields from the distortion of the cosmic microwave background. The method allows one to directly reconstruct the projected dark matter density power spectrum over several order of magnitude in scale, as well as search for time dependent gravitational potential using cross-correlation with the cosmic microwave background itself. With upcoming Planck satellite one should be able to measure these signatures, which will significantly constrain the viable cosmological models.

F. Bernardeau addressed the same topic from a different perspective. In the absence of lensing the primary CMB anisotropies are expected to form a 2D Gaussian map. Gravitational lensing generates mode couplings that can be detectable through induced non-Gaussian features. He described the resulting connected four-point corellation function and the effects of lenses on the distribution function of the temperature peak shapes, both of which should be observable with upcoming satellites.

B. Metcalf reviewed some of the properties of the weak lensing point sources. He presented an interpretation of the magnification probability distribution function in terms of the nature of lensing structures. The distribution function differs significantly between lensing produced by large-scale structures and lensing produced by massive compact objects, allowing one to detect the latter if present in the universe with sufficient density. They could be detected using high-redshift type Ia supernovae searches.

L. Kofman presented recent results on supercluster lensing and its interpretation in the context of cosmic web. After a general overview of cosmic web paradigm he presented a result obtained in collaboration with N. kaiser and his group showing a detection of weak lensing in 3 adjacent clusters connected with massive bridges between them. This result provides a quantitative confirmation of the cosmic web idea.

J. Quashnock presented an analysis of the gravitational lensing rate in the Hubble Deep Field. A comparison between the theoretical expectations and the observed number of strongly lensed galaxies constrains the current value of Ω_m, Ω_Λ, where Ω_m is the mean mass density of the universe and Ω_Λ is the normalized cosmological constant. Based on current estimates of the HDF luminosity function and associated uncertainties in individual parameters, his 95% confidence lower limit on Ω_m, Ω_Λ range between -0.44, if there is no strongly lensed galaxies in the HDFn to -0.73, if there are two strongly lensed galaxies in the HDF.

S. Seitz presented preliminary results on the ongoing work to measure cosmic shear on an arcminute scale using the STIS-parallel data. The unprecedented sensitivity and angular resolution of STIS in imaging mode, as well as the demonstrated stability of the PSF, allows the detection of cosmic shear on these small scales, even with a modest number of fields. Such a measurement would yield the amplitude of the density fluctuations and the matter content of the universe and test the structure formation paradigm in the nonlinear regime.

Finally. K. Tomita discussed statistical behavior of cosmological gravitational lensing in various cosmological background models. Evolution of inhomogeneous matter distribution was obtained using N-body simulations, while light ray propagation was derived by solving directly the null-geodesic equation. The behavior of ray bundles was analyzed by calculating their convergence, shear, and magnification. This was then compared analytically and provide appropriate solutions for some special cases.

ELSEVIER

Nuclear Physics B (Proc. Suppl.) 80 (2000) 215–217

NUCLEAR PHYSICS B
PROCEEDINGS
SUPPLEMENTS

www.elsevier.nl/locate/npe

Nucleosynthesis and Gamma Ray-Line Astronomy

Elisabeth Vangioni-Flam[a], Reuven Ramaty[b] and Michel Cassé[c]

[a] Institut d'Astrophysique de Paris,
98bis Bd Arago 75014 Paris, France

[b] GSFC, NASA, Greenbelt, MD 20771, USA

[c] Service d'Astrophysique, CEA,
Orme des Merisiers 91191 Gif/Yvette, France

The most energetic part of the electromagnetic spectrum bears the purest clues to the synthesis of atomic nuclei in the universe. The decay of radioactive species, synthesized in stellar environments and ejected into the interstellar medium, gives rise to specific gamma ray lines. The observations gathered up to now show evidence for radioactivities throughout the galactic disk, in young supernova remnants (Cas A, Vela), and in nearby extragalactic supernovae (SN 1987A, SN 1991T and SN1998bu), in the form of specific gamma ray lines resulting, respectively, from the radioactive decay of ^{26}Al, ^{44}Ti and ^{56}Co. The various astrophysical sites of thermal nucleosynthesis of the radioactive nuclei were discussed: AGB and Wolf-Rayet stars, novae, and type Ia and type II supernovae. Nuclear excitations by fast particles also produce gamma ray lines which have been observed in great detail from solar flares, and more hypothetically from active star forming regions where massive supernovae and WR stars abound. This non thermal process and its nucleosynthetic consequences was reviewed. The 511 keV line arising from $e^+ + e^-$ annihilation also provides important information on explosive nucleosynthesis, as well as on the nature of the interstellar medium where the positrons annihilate. INTEGRAL, the main mission devoted to high resolution nuclear spectroscopy, should lead to important progress in this field.

1. Observational status

The experimental situation in gamma ray line astronomy was summarized by G. Vedrenne. The highlights are: i) the discovery (1) of ^{44}Ti emission from the Vela region (GRO JO852-4642) near a new supernova remnant detected in X rays by the ROSAT satellite (2); ii) the positive detection of a recent SNIa by COMPTEL (SN 1998bu) located at 8.1 Mpc (3); iii) the release of a new ^{26}Al COMPTEL map derived from the observations using a sophisticated technique of data analysis (4); iv) the withdrawal of the Orion gamma ray line data, followed immediately by the announcement of a similar emission from the Vela region (5). In addition, P. von Balmoos presented a review on the origin of galactic positrons, including compact galactic sources and radioactive nuclei (^{26}Al, ^{44}Ti, ^{56}Co).

2. Production of radioactive nuclei in thermal nucleosynthesis

2.1. Non explosive nucleosynthesis: AGB and Wolf-Rayet stars

G. Meynet analyzed the synthesis of ^{26}Al in AGB and WR stars. Production in AGB stars falls short from explaining the required live radioactive aluminum in the galaxy (about $2M_\odot$), but WR stars remain a serious candidate. Indeed, the detailed analysis of the COMPTEL ^{26}Al map and its correlation with the free-free emission of the galactic disk, as observed by COBE, indicates that massive stars are the most likely candidates for ^{26}Al production (4). But at the moment, it is not possible to discriminate between core collapse supernovae and WR stars since neither the Vela SNR nor the γ Velorum WR star coincide with peaks on the 1.8 MeV COMPTEL map. The absence of a clear detection signal implies that the progenitor of WR11 in Vela has a mass less than 40 M$_\odot$.

M. Arnould, broadening the scope, has pointed out the exceptional interest of radionuclide astrophysics at large, since it provides strong links between gamma ray astronomy, chemical evolution of the galaxy, stellar nucleosynthesis, and the physico-chemistry of circumstellar envelopes, the ISM, and the early solar system. After a critical analysis of all nucleosynthetic sites, he concluded that WR modeling is immensely simpler than that of AGB stars, novae and supernovae. He surmised that WR stars might be of interest to cosmochemists since they could provide a wealth of isotopic anomalies, potentially observable in meteorites.

2.2. Explosive nucleosynthesis

A. Core collapse supernovae and their remnants (SNII).

The nucleosynthesis of ^{26}Al and ^{44}Ti by core collapse was critically examined by F-K. Thielemann. Taking for example a 15 M$_\odot$ star, the ^{26}Al yields of (6) and (7) differ significantly (3×10^{-8} against 2.7×10^{-6} M$_\odot$). The origins of differences concern the choice of the still controversial ^{12}C$(\alpha, \gamma)^{16}$O reaction rate, and above all the treatment of convection (Schwarzschild or Ledoux + semiconvection, rotationally induced mixing and so on). Concerning the Fe-group elements, the variations between models are expected to be more acute due to a different simulation of the explosion, affecting the mass cut. Surprisingly, for the 15 M$_\odot$ model, the amounts of ejected ^{44}Ti, ^{56}Ni, ^{57}Ni are similar in both cases (respectively 6×10^{-5}, 0.1 and 4×10^{-3} M$_\odot$). However, the optimism should be tempered since the ^{44}Ti yield varies a lot as a function of mass between the different authors. Anyway, using the new half-life determination (59-62 yr), the amounts of ^{44}Ti ejected by SN 1987A (estimated from the late light curve, roughly 10^{-4} M$_\odot$), Cas A (about

1.3×10^{-4} M$_\odot$ from gamma rays) and JO852-4642 in Vela (5×10^{-5} M$_\odot$) can be explained with a calculation employing spherical symmetry. Concerning the synthesis of Fe, the main question is whether the mass of ^{56}Ni (^{56}Fe) ejected by core collapse supernova decreases or not as a function of the mass of the progenitor above 20 M$_\odot$. Light curve analyses that should help to solve this question are for the moment limited to the low mass range (less than 30 M$_\odot$), unfortunately. So the question remains unsettled.

B. Thermonuclear supernovae (SNIa).

Type Ia supernovae are expected to produce greater quantities of ^{56}Ni and to become transparent to gamma rays earlier than their gravitational counterparts of high masses (SNII, SNIb, SNIc), as was discussed by S. Kumagai. Thus they are good targets of opportunity for gamma ray line astronomy. Indeed, an unusually bright SNIa (SN1991T), located at 13 - 17 Mpc, was already observed by COMPTEL at the edge of the Virgo cluster, close to the detection limit. The amount of ejected ^{56}Ni derived from the observation (higher than 1 M$_\odot$) appears quite unusual, as does SN1991T itself. SN 1998bu, at a distance of 8.2 Mpc, presents certain similarities to SN 1991T. However, contrary to SN 1991T, the mean 847 keV line flux observed by COMPTEL from 5 to 131 days after the appearance of the supernova is somewhat low compared to the predictions of the models.

Concerning the observability of type Ia supernovae by INTEGRAL, S. Kumagai gave a mildly optimistic view. However, recent work (8), that took into account the width of the 847 keV ^{56}Co decay line, tempers this enthusiasm somewhat. If, by chance, an SNIa is captured by INTEGRAL in good conditions, the observation would help to calibrate the explosion models, the ejecta structure and the ^{56}Ni distribution.

C. Novae

The best prospects are the lines resulting from ^{7}Be and ^{22}Na decays. None of these have been observed up to now. Only upper limits exist (2×10^{-8} M$_\odot$) on the ejected ^{22}Na from neon rich novae. The GRO sky survey at 1.275 MeV gives only a marginal excess from South Aquila. M. Hernanz presented detailed nucleosynthesis calculations in nova explosions, employing hydrodynamical models for a variety of CO and ONe white dwarf masses. The low ejected mass of ^{22}Na obtained in the ONe model is consistent with the observational upper limit. Only nearby novae should be captured by INTEGRAL, through the radioactive decay of ^{7}Be (CO novae: 500 pc) and ^{22}Ne (1.5 kpc, (9)). In all models a strong continuum dominates the gamma ray spectrum during the early period of expansion. This short and intense emission could be detected at least up to 3 kpc, during a few hours (10).

3. Non thermal gamma ray lines and associated nucleosynthesis

A broad overview of all aspects of gamma ray lines induced by non thermal particles in various astrophysical sites, including solar flares, was presented by R. Ramaty. The ^{12}C and ^{16}O lines, at 4.438 and 6.129 MeV, can only be produced by non thermal particle interactions, a fact that can be used to distinguish a nonthermal from a nucleosynthetic origin of an observed gamma ray line spectrum. The ^{12}C and ^{16}O lines, as well as many others have been observed from solar flares. The most prominent ones are at 2.223 MeV following neutron capture on H, at 0.511 MeV from positron annihilation (both the neutrons and positrons results from nonthermal particle interactions), at 1.634 MeV from ^{20}Ne and at 0.429 and 0.478 MeV from ^{7}Be and ^{7}Li produced in interactions of fast α particles with He. These lines have provided much new information on particle acceleration as well as on the properties of the solar atmosphere (11, 12). With the withdrawal of the COMPTEL observations of the ^{12}C and ^{16}O lines from Orion, there remains no convincing evidence for such lines from non-solar sites. But the fast particles which produce the lines could have an important role in the origin of some of the light elements, in particular Be (13). This was the subject of a recent conference, the proceedings of which should appear shortly (see 14).

4. Conclusion

The COMPTON GRO mission has provided a wealth of data which has given a strong impetus to nuclear astrophysics. Now a new episode is opening up with INTEGRAL. In this context, F. Lebrun has shown the potential of the INTEGRAL satellite. The high quality spectroscopy of the SPI instrument, between 2 keV and 1 MeV, will shed light on fundamental questions of nucleosyntheis. Lines from ^{44}Ti decay will be observed with both the SPI spectrometer and the IBIS imager. The proceedings of the invited talks and posters are available in the CDROM of the Texas Symposium.

REFERENCES

1. Iyudin, A.F. et al. 1998, Nature, 396, 142.
2. Aschenbach, B. 1998, Nature, 396, 141.
3. Leising, M.D. 1998 Third INTEGRAL Symposium, "The extreme Universe", Taormina, to be published.
4. Knodlseder, J. 1997, Thesis, Toulouse University.
5. van der Meulen, R.P. et al. 1998, Third INTEGRAL Symposium, "The Extreme Universe", Taormina, to be published.
6. Woosley, S.E. and Weaver, T.A. 1995, ApJS, 101, 181.
7. Thielemann, F.K. et al. 1996, ApJ, 460, 108.
8. Isern, J., 1998, Third INTEGRAL Symposium, "The Extreme Universe", Taormina, to be published.
9. José, J. and Hernanz, M. 1998, ApJ, 494, 680.
10. Gomez-Gomar J. et al. 1998, MNRAS, 296, 913.
11. Ramaty, R. et al. 1995, ApJ, 455, L193.
12. Mandzhavidze, N. et al. 1997, ApJ, 489, L99.
13. Vangioni-Flam et al. 1998, AA, 337, 714
14. Ramaty, R. et al. 1999, Publ. Astron. Soc. Pacific, in press).

ELSEVIER

Nuclear Physics B (Proc. Suppl.) 80 (2000) 219–221

NUCLEAR PHYSICS B
**PROCEEDINGS
SUPPLEMENTS**

www.elsevier.nl/locate/npe

Pulsars and Neutron Stars

M. Ali Alpar[a] and Alex Wolszczan[b]

[a]Physics Department, Middle East Technical University, Ankara 06531, Turkey

[b]Department of Astronomy & Astrophysics, The Pennsylvania State University,
525 Davey Laboratory, University Park, PA 16802, USA

Since the last Texas conference, exciting developments with the X-ray astronomy satellite Rossi XTE (RXTE) have brought us into direct contact with dynamics very close to the surface of neutron stars, possible evidence of the last stable orbits of general relativity and thereby strong constraints on neutron star structure. We therefore decided to focus the mini-symposium on "Pulsars and Neutron Stars" on these discoveries. Our invited speakers, Michiel van der Klis and Chris Pethick, reviewed the observational developments and theoretical work on neutron star structure, respectively.

Van der Klis noted that kilohertz QPOs, first discovered shortly after the launch of RXTE in 1996, had been observed in 19 sources (van der Klis et al 1996; for references on the kilohertz QPOS and "burst oscillations" see van der Klis 1999). In 17 of these sources, high frequency QPOs are observed in two different frequency bands simultaneously, as twin peaks in the power spectra. The higher peak frequency has values between 500 and 1200 Hz in the different sources. Both peaks move to higher frequencies as the mass accretion rate $\dot{M}$ increases. The difference between the two peak frequencies remains constant to within 5 or 10 Hz in many sources while both peak frequencies vary over several hundred Hz. There are some sources, notably Sco X-1, for which the peak separation varies by as much as 70 Hz. The range of peak separations observed in the 17 sources is 250-360 Hz. A common interpretation of the two "kilohertz" peaks has been that the higher frequency band reflects Kepler frequencies at the inner edge of the ac-

cretion disk while the lower frequency band is the band of beat frequencies between the Kepler frequency and the rotation frequency of the neutron star. The difference between the two QPO peak frequencies would then be the rotation frequency of the neutron star. This view gains significant support from yet another class of high frequency QPOs. First discovered shortly after the kiloherz QPOs, these "burst oscillations" have been observed from six sources during type 1 X-ray bursts. The frequencies range between 330 and 590 Hz among the different sources. In each source, the frequency is almost constant, drifting by a few Hz while the source luminosity varies. 4U 1636-53 shows a 290 Hz frequency and its 581 Hz harmonic. Four out of these six burst oscillation sources also exhibit twin QPO peaks. The burst oscillation frequency approximately coincides with the peak separation in two of these, with twice the peak separation in one, while in 4U 1636-53, the fundamental 290 Hz burst oscillation coincides with the peak separation to within 15%. But if the peak separation really reflected the rigid rotation rate of the neutron star itself, it would have to be strictly constant for each source, and the burst oscillation frequency itself would not drift even by a few Hz. Models attempt to address these issues by invoking matter at the neutron star surface, differentially rotating with respect to the star, and shifing in frequency as clumps of matter spiral from the disk down to the neutron star surface. More definitive and crucial support for the beat and Kepler frequency assignment to the lower and higher "kilohertz" QPOs could come from observations

of such QPOs and/or the burst oscillations from
the unique millisecond X-ray pulsar SAXJ 1808.4-
3658 (Wijnands & van der Klis 1998).

The requirement that the Keplerian orbital
radius implied by the higher kilohertz frequencies lies outside the neutron star radius already
places tight constraints on the mass and radius
of the neutron star, through the relation $R <
(GM/\Omega_K{}^2)^{1/3}$. It has been proposed that the im-
plied orbit is actually the last stable orbit of general relativity (Kaaret, Ford & Chen 1997, Zhang,
Strohmayer & Swank 1997; Kluzniak 1998). In
addition to being a very exciting prospect per se,
this possibility makes the constraints very tight:
Now the upper kilohertz frequency gives a measurement of the mass, about $2M_\odot$, and the mass
radius relation must give a radius that is less than
the corresponding radius for the last stable orbit,
$R < 6GM/c^2$. The observation by Zhang et al
(1998, see however Mendez et al 1998) that the
upper kilohertz frequency saturates at a constant
frequency at high luminosities supports the last
stable orbit idea. Testing further for this saturation in frequency as the luminosity increases remains a very exciting prospect for the near future.

Van der Klis also pointed out that plotting
the frequencies of various QPOs and features in
the power spectra against other QPO frequencies of the same source in a plot encompassing
many QPO sources yields well defined relation-
ships. The sources sharing these empirical relations include not only low mass X-ray binaries
of both the Z and the atoll types, and the millisecond X-ray pulsar SAXJ 1080.4-3658 but also
black hole candidates. This may be an important
clue to the universality of the dynamics yielding
the QPOs from compact objects, both neutron
stars and black holes.

Chris Pethick reviewed the recent developments on the equation of state. The nuclear
physics in the core regions of neutron stars, at
densities above nuclear saturation density, determines the structure of the neutron star (Pethick
et al 1999). The questions are (i) what are the
constituents of high density neutron star matter, (i.e. the presence of pion, kaon, and quark
phases), (ii) the nuclear interactions, (iii) energy
calculations, and (iv) determination of the equa-

tion of state and the stellar structure. The final answer includes the determination of possible constituents and thus feeds back to (i) for
a consistent stellar model. The development of
an equation of state from the microscopic model
entails many-body theory with well developed
techniques but troubles in the treatment of relativistic corrections, or mean field models which
in turn make questionable assumptions. Pethick
discussed a recent many-body calculation (Akmal
et al 1998) that yields neutron star models consistent with the kilohertz QPO observations, and
may thereby identify the effects that need to be
taken into account in a realistic calculation of the
equation of state. Starting from a new nucleon-
nucleon interaction model, the Argonne v_{18} (A18)
potential, which provides a good fit to nucleon-
nucleon scattering data, Akmal et al incorporate
relativistic corrections and three-nucleon interactions. These effects increase the maximum mass
of the neutron star from $1.67M_\odot$ to $1.8M_\odot$ and
$2.2M_\odot$ respectively. They find that there is a
strong increase in correlations brought about by
the tensor forces, which results in a pion con-
densed phase, not favoured in earlier calculations
because of the repulsive non-tensor correlations.
At the core of the neutron star, a mixed quark-
nucleon phase is expected. The presence of these
phases causes rapid cooling of neutron stars. The
new phases bring in a softening of the equation of
state, with a maximum mass of $2.02M_\odot$. Akmal
et al estimate an upper limit of $2.5M_\odot$ to the maximum neutron star mass by taking the causally
limited equation of state $c_s = c$ at high densities.
The lower limit they estimate for the maximum
mass is $1.91M_\odot$, which is obtained with a bag
model constant of 122 Mev fm^{-3}, and $1.74M_\odot$ if
together with this choice of the bag constant, a
reduced three-nucleon repulsive interaction is assumed. As far as the overall neutron star properties like maximum mass are concerned, their results with the A18 potential plus relativistic corrections and the Urbana UIX model of the three
nucleon interaction are similar to the earlier results of Wiringa, Fiks and Fabrocini using the
A14 + UVII (three-nucleon corrections), and the
U14 + UVII interactions, as far as the overall
neutron star properties like maximum mass are

concerned. The differences are that the A18 provides a superior fit to the nucleon-nucleon scattering data, and with U14 no pion condensation is found. Taking the upper kHz QPOs as Kepler frequencies from the last stable orbit, Kluzniak (1998) and Kaaret, Ford and Chen (1997) found that the neutron star masses were greater than $1.91 M_\odot$, and that the A14+UVII models of Wiringa, Fiks and Fabrocini were practically the only viable neutron star models. As the current updated version of the A14 + UVII, we have at hand the A18 + UIX as a good candidate for a realistic model of the nuclear interactions. Prospects of further observations of strong field general relativity from neutron stars will open up if the attractive possibility of the marginally stable orbit is further supported by astrophysical observations.

Among the contributed talks of this minisymposium, we note here those submitted for publication in the CD Rom proceedings of the conference. Haensel, Lasota and Zdunik report an approximate relation between the maximum surface redshift of a static neutron star and the minimum rotation period, based on a causal EOS. For the maximum measured neutron star mass of 1.442 solar masses they get a minimum rotation period of 0.29 ms. Drago explores the structure of a hybrid neutron star with a quark matter core, reporting a maximum mass of 1.6 solar masses for the nonrotating star. Perez and Leinson report that in the presence of extremely strong (10^{17} G) magnetic fields the direct URCA process can proceed for arbitrary proton concentration, so that such extreme magnetic field neutron stars should cool rapidly. Ramachandran and Portegies-Zwart discuss the detectability condition for neutron stars accelerated in binaries. Several papers (Lattimer; Zavlin; Shearer) discussed implications of soft X-ray and optical observations of likely thermal emission from neutron stars. Shearer et al report an upper limit of 9.5 Km for the radius of Geminga.

REFERENCES

1. Akmal,A., Pandharipande,V.J. & Ravenhall,D.G., 1998, Phys.Rev.C 58, 1804

2. Pethick,C.J., Akmal,A., Pandharipande,V.J., & Ravenhall,D.G., 1999, CD-Rom Proceedings of this meeting

3. Kaaret,P., Ford,E. & Chen, K., 1997, ApJ 480, L27

4. Mendez, M., van der Klis,M., Ford,E.C., Wijnands,R. & van Paradijs,J., 1999, ApJ 511, in press

5. Kluzniak,W., 1998, ApJ 509, L37

6. Van der Klis,M., Swank,J.H., Zhang,W., Jahoda, K., Morgan,E.H., Lewin,W.H.G., Vaughan,B., van Paradijs, J., 1996, ApJ 469, L1

7. Van der Klis,M., 1999, Proc. The Third William Fairbank Meeting "The Lense-Thirring Effect", Rome June 29 - July 4 1998, in Advances in Astrophysics and Cosmology, World Scientific (publ.) Fang Li-Zhi and Remo Ruffini (eds.)

8. Wijnands,R. & van der Klis,M. 1998, Nature 394, 344

9. Zhang,W., Strohmayer,T.E. & Swank,J.H., 1997, ApJ 482, L167

10. Zhang,W., Smale,A.P., Strohmayer, T.E., Swank,J.H., 1998, ApJ 500, L171

ELSEVIER

Nuclear Physics B (Proc. Suppl.) 80 (2000) 223–226

NUCLEAR PHYSICS B
**PROCEEDINGS
SUPPLEMENTS**

www.elsevier.nl/locate/npe

Cosmological parameters

G. Theureau[a] and G.A. Tammann[b]

[a]Osservatorio Astronomico di Capodimonte, I-80131 Naples, Italy
ARPEGES, Observatoire de Paris, F-92195 Meudon Cedex, France

[b]Astronomisches Institut der Universität Basel, Binningen, Switzerland

We report here on the mini-symposium *"Recent results on H_0"*. The session was organized in ten invited reviews and ten poster papers covering a wide range of observational methods: from studies in terms of the classical extragalactic distance ladder, to the S-Z effect in clusters, gravitationally lensed Quasars, and CMB analysis. Two aspects have been stressed: the measurements of H_0 and the estimation of the cosmological constant Λ.

1. Recent results on H_0

1.1. Primary calibration

RR-Lyrae, Cepheids, and supernovae of type Ia are the corner stone of the extragalactic distance scale. By providing accurate distances to nearby galaxies, they enable the determination of zero-points for various secondary distance indicators, some of which can in turn be used to measure the far-field Hubble flow. The Hubble Space Telescope (HST) has increased the range of Cepheid studies by at least a factor five in distance (up to ∼25 Mpc, after its refurbishing), and, even if the result is still provisional, the Hipparcos astrometric satellite has provided for the first time the missing link between extragalactic distances and true geometrical distances.

The state of the Cepheid and RR-Lyrae absolute calibration after Hipparcos is reviewed by **R. Gratton** [1]. A central point is the statistical treatment of few data in the context of large relative errors on parallaxes. The Hipparcos calibration of the Cepheid P-L relation leads to a variety of results for the LMC distance modulus, ranging from 18.4 to 18.7, with quite large uncertainties of 0.1-0.2 mag. RR-Lyrae in LMC clusters and other methods like Baade-Wesselink techniques, main-sequence fitting in clusters, and proper motions lead to the same kind of results. It is shown that part of the discrepancy is also due to the way the metallicity dependence of both P-L Cepheid and RR-Lyrae relations is taken into account. The task appears indeed difficult since color terms, metallicity, and reddening are generally observationally undissociable.

It is important to point out that empirical and theoretical studies do not agree well on the metallicity dependence of the Cepheid P-L relation. In most cases the metallicity effect to be detected is already compensated by the reddening correction method. As a result, some empirical determinations give a positive metallicity term ($\sim +0.4$), while theoretical calculations favor a negative one varying with the period. Progress in the modeling of the evolution of Cepheids along the different crossings of the instability strip in function of metallicity is reviewed by **A. Gautschy** [2] who shows that upper limits on the reaction of the P-L relation on assumptions in evolution computations, pulsation stability analyses, and mappings onto photometric passbands can be estimated now quite reliably.

An empirical treatment is given by **A. Mazumdar** and **D. Narasimha** [3] concerning Galactic Cepheids and HST Cepheids in the Virgo spiral M100. The authors map the instability strip in the surface gravity vs effective temperature plane using multi-color light curves and derive relations between period, color and amplitude. In this way, when flux-limited incompleteness and extinction are properly corrected, they reanalyze the M100 distance and estimate it to be 19.6 ±1.7 Mpc.

Some $\sim$ 30 Cepheid distances of galaxies are now available, among them 8 hosting a type Ia supernova. **A. Saha** [4] reviews and updates the calibration of the peak luminosity of blue SNIa on the basis of these new HST Cepheid measurements. It is examined wether the scatter in the SNIa Hubble diagram can be reduced using additional parameters. Early attempts have shown that residuals around light curve maximum are correlated with decay rate or light curve width (Δm_{15}). In addition, residuals show dependence on observed color, an effect which is not related to extinction. Consequently, the corrected absolute magnitude has the following expression: $M_{corr} = M_{obs} + f(\Delta m_{15}) + \alpha(B_0 - V_0)$. An updated calibration of that maximum is: $\langle M_B^{corr} \rangle=$ -19.45 $\pm$ 0.06 and $\langle M_V^{corr} \rangle=$ -19.44 $\pm$ 0.08. This calibration applied to $\sim$ 50 blue SNIa with $cz \leq$ 30000 km s^{-1} leads to a value of H_0=60 $\pm$ 2 km s^{-1}Mpc^{-1}.

1.2. Secondary distance indicators

The class of secondary indicators comprises a variety of distance criteria generally related to global properties of galaxies. With an accessible range of distances reaching 200 Mpc (h=0.5) and the availability of large and statistically robust samples, both in the field and in galaxy clusters, they allow an estimation of the Hubble constant free from the influence of local peculiar motions. The two main ways are the Tully-Fisher (spirals) and Fundamental Plane (FP) or $D_n - \sigma$ (ellipticals and lenticulars) scaling laws, to which one may add the use of the Globular Cluster Luminosity Function (GCLF), the Tip of the Red Giant Branch (TRGB), and according to some authors the Planetary Nebulae Luminosity Function (PNLF), and Surface Brightness Fluctuations of ellipticals (SBF). All these distance indicators are based, directly or indirectly, on the primary calibration provided by the Cepheid P-L relation.

The Virgo Cluster is one of the very few clusters whose member galaxies have been identified to very faint limits and thus it plays a dominant role for the extragalactic distance scale. Cluster ellipticals and spirals constitute a complete, volume limited sample, which is immune against selection (Malmquist-like) bias. The cluster is in addi-

tion tightly tied to the large-scale expansion field through relative distances to more distant clusters. **G.A. Tammann** [5] reviews the different techniques for determining the Virgo cluster distance. The mean value of the Virgo distance obtained from Cepheids, novae, SNIa, Tully-Fisher, GCLF, and $D_n - \sigma$, is 21.1 $\pm$ 0.9 Mpc ($\langle \mu \rangle$= 31.62 $\pm$ 0.09), leading to a large-scale Hubble constant H_0= 56 $\pm$ 4 (random error) $\pm$ 6 (systematics) km s^{-1}Mpc^{-1}.

The general use of Tully-Fisher distances of field galaxies is reviewed by **G. Theureau** [6]. It is shown that careful completeness analysis of the available samples, computation of reliable Malmquist or selection bias corrections, and extraction of unbiased subsamples are unavoidable steps toward a safe determination of H_0. The results obtained with different forms of the TF relation, and applied to a variety of samples, all converge towards the unbiased value of H_0=55 $\pm$ 5 (random) $\pm$ 5 (external) $^{+0}_{-4}$ (systematics) km s^{-1}Mpc^{-1}. The major source of uncertainty of H_0 comes here from the primary calibration, i.e. from the absolute zero-point calibration of the Cepheid P-L relation.

M. Liu et al. [7] present some new results from K-band surface brightness fluctuations of Fornax and Coma cluster ellipticals. It is shown that K-band SBF are dominated by late-type stars and that IR photometry can potentially measure distances 2-3 times as far as the optical SBF. With a sample depth of $\sim$ 120 Mpc (h=0.5), K-band SBF could yield for individual galaxies values of H_0 which are affected by $\leq$ 10% due to peculiar velocities.

1.3. Gravitational lenses, SZ-effect, and CMB

Thanks to the development of new observational techniques such as VLBI, large field CCD photometry, and space telescopes, we have seen in the last few years the birth of new and completely independent methods allowing an estimation of H_0: 1) distances of gravitationally lensed double quasars can be determined with increasing precision through measurements of time delays and models of the mass distribution of deflector lenses; 2) the mapping of the Sunyaev-

Zeldovich (SZ) effect in galaxy clusters (inverse-Compton scattering of CMB photons by intra-cluster ionised gas) confronted with X-ray contours provides redshift-independent distances up to $z \sim 0.5$; and 3) the Cosmic Microwave Background provides one of the most promising routes for the determination of cosmological parameters, or inversely the best test for the underlying theory.

Gravitational lenses provide a direct way to measure distances in the universe, circumventing the use of the distance ladder, by measuring the time delays between variations of different images of the same source. **R. Blanford** [8] reviews the results obtained from four sources Q0957+561, PG1115+080, B0218+357 and CLASS1608+656, as well as the future prospects with the discovery of e.g. 17 and 30 radio lenses from CLASS and JVAS, respectively. In principle, to measure the Hubble length in a clean way, an adequate model of the lens and an assumed geometry of the universe are required. In particular, dark matter in clusters and groups has to be allowed for (a uniform sheet leads for instance to smaller H_0), while the cosmography and large scale structures introduce an uncertainty of 10-20% and 5-10%, respectively. Using standard models, the present determinations are consistent with a value H_0=60-80 km s^{-1} Mpc^{-1}.

The double quasar Q0957+561A,B is reanalyzed by **A. Oscoz et al.** [9] on the basis of CCD observations in several bands during a period of observation of almost 3 years. A new refined self-consistent test suggests a time delay of 424 days instead of 417, and the measured velocity dispersion of G1 has been found to be σ_l=310 ±20 km s^{-1} instead of the previous value of ~280. The authors analize the present models of Q0957+561 and their consequences on H_0.

The recent progress in the measurement of H_0 from observations of the Sunyaev-Zeldovich (S-Z) effect is reviewed by **Y. Rephaeli** [10]. It is shown that improvements in the implementation of this method to determine H_0 include the realistic description of the temperature distribution of intracluster gas, more exact relativistic calculations of the effect, and a better assessment of some of the observational uncertainties. The

present mean weighted value of H_0 determined from S-Z and X-ray measurements of eight clusters is *roughly* in the range $40 - 70$ km s^{-1} Mpc^{-1}, with most of the uncertainty due to the unknown gas temperature distribution, to the morphology, and to the degree of small-scale clumping. With upcoming detailed interferometric S-Z observations and cluster X-ray images, and observations in the near future using bolometric arrays and the X-ray satellites XMM and Chandra, it is realistic to expect the determination of H_0 by this method to become significantly more accurate.

Some new constraints on H_0 are given by the study of Cosmic Microwave background data in conjunction with information from large scale clustering, high-z SNIa and galaxy clusters. This field is reviewed by **A. Lasenby** [11], together with the substantial improvements expected from the next generation of CMB data from ground, balloon and satellite-based experiments. The author shows that, due to the complementary and orthogonal nature of the combined data sets with respect to the degeneracy of the cosmological parameters, the error bars on H_0, Ω_m and Ω_Λ are dramatically reduced. Assuming an inflationary Cold Dark Matter model with a cosmological constant, in which the initial density perturbations in the universe are adiabatic, it is found that H_0=59 ± 8 km s^{-1} Mpc^{-1}.

2. Cosmological constant and Ω

There is growing evidence that the total energy density of matter in the universe is significantly less than unity and that one should consider the existence of a positive cosmological constant. The most recent argument in this sense comes from the progress of observation of SNIa at high redshifts provided by the two large teams of the "Supernova Cosmology Project" (Berkeley) and the "High-z SN Search" (Harvard).

S. Perlmutter [12] ("Supernova Cosmology Project") presents the analysis of 42 high-redshift SNIa covering the range from z=0.18 to 0.83. The data are shown to be strongly inconsistent with a Λ=0 flat cosmology, the simplest inflationary universe model. The size of the sample allows a variety of statistical tests to check for possible systematic errors and biases. Particular attention has been payed to the control of K-corrections, extinction, Malmquist-like biases, influences of gravitational lenses (see also [15]), and evolution and environment effects on the light curve width-luminosity correction. Their best fit for a flat universe is $\Omega_M = 0.28 \pm 0.1$ leading to an accelerating universe with $q_0 = \Omega_M/2 - \Omega_\Lambda \sim -0.58$.

B. Schmidt [13] presents for the "High-z SN Search" the observations of more than 20 supernovae with reshift $z \geq 0.3$, which have been observed in the restframes B and V. Up to now, 90 SN have been discovered, of which 40 spectroscopically confirmed type Ia. As above, when compared to nearby samples, derived SN distances indicate an accelerating expansion rate with a positive cosmological constant.

These studies stress the importance of SN search programs at low and intermediate redshifts. The EROS collaboration has devoted 10% of its observing time to such a program. **D. Hardin** [14] summarizes the EROS results on the basis of a sample of 30 newly discovered SN. **D. Holz** [15] explores the systematic error due to lensing standard candles. Re-analyzing recent results from high-z supernovae, when most of the matter in the universe is either in the form of compact objects or is continuously distributed in galaxies, it is found that the best-fit model remains unchanged. The confidence contours change size and shape, becoming larger and moving towards higher values of matter density, Ω_M.

Independent constraints on cosmological parameters have been obtained from JVAS and CLASS Gravitational Lens Surveys. On the basis of these data, **H. Helbig** [16] shows that flat-spectrum radio sources are an ideal parent population for unbiased samples of gravitational lenses. New results, particularly in terms of Λ are presented.

L. Gurvits et al. [17] discuss the "angular size-redshift" relation for a sample of compact radio sources distributed over a wide range of redshifts. In the framework of the Friedmann-Robertson-Walker model, when no evolutionary effect is taken into account, their best fit gives $q_0 = 0.21 \pm 0.3$.

A. Starobinsky [18] discusses the reliability of ΛCDM models with an approximately flat ($n \sim 1$) initial spectrum of the adiabatic perturbations. This category of models is in good agreement with all existing observational data for some region of parameter space (H_0, Ω_M). **D. Khokhlov** [19] presents a homogeneous and isotropic model in which background space is a coordinate system of reference. It is shown that the observational constraints (H_0, age, effective shape parameter, CMB and high-z SN data) favor this model. **A. Beesham** [20] re-examines models with variable G and Λ and causal bulk viscosity. The models, when based on the first-order thermodynamic theory of Eckart, are not satisfactory since they permit the propagation of superluminal signals and their equilibrium states are unstable. A possibility to overcome these problems is to use the second-order thermodynamic theory of Israel-Stewart.

REFERENCES

1. R. Gratton, this conference
2. A. Gautschy, this conference
3. A. Mazumdar, D. Narasimha, this conference
4. A. Saha, this conference
5. G.A. Tammann, this conference
6. G. Theureau, this conference
7. M. Liu et al., this conference
8. R. Blandford, this conference
9. A. Oscoz et al., this conference
10. Y. Rephaeli, this conference
11. A. Lasenby, this conference
12. S. Perlmutter, this conference
13. B. Schmidt, this conference
14. D. Hardin, this conference
15. D. Holz, this conference
16. P. Helbig et al., this conference
17. L. Gurvits et al., this conference
18. A. Starobinsky, this conference
19. D. Khokhlov, this conference
20. A. Beesham, this conference

ELSEVIER

Nuclear Physics B (Proc. Suppl.) 80 (2000) 227–228

NUCLEAR PHYSICS B
**PROCEEDINGS
SUPPLEMENTS**

www.elsevier.nl/locate/npe

X-ray Binaries

P.A. Charles[a] and J.-M. Hameury[b]

[a]Astrophysics, University of Oxford, Nuclear and Astrophysics Laboratory
Keble Road, Oxford OX1 3RH, UK

[b]Observatoire Astronomique, 11 rue de l'Université, F-67000 Strasbourg, France

Since its launch at the end of 1995, the Rossi X-ray Timing Explorer (RXTE) has undertaken extensive ground-breaking studies of the full range of X-ray binaries. Many of these results were reviewed and summarized by Jean Swank (GSFC). RXTE has observed 17 black-hole X-ray binaries (BHXRBs) and the All-Sky Monitor (ASM) has produced definitive and well-sampled X-ray light curves of these transient systems. The PCA has given excellent wide-range X-ray spectroscopy of the various transient states, and the timing information led to the discovery of QPOs in BHXRBs, which were compared with those observed in the neutron star systems.

Whilst the twin-peak kHz QPOs in the LMXB X-ray bursters have been interpreted using the "beat-frequency model", there have been difficulties with this model in observations of three systems (Sco X-1, X1735-444, X1608-52) where the difference frequency (which is supposed to represent the actual neutron star spin period) changes smoothly. This led to an alternative proposal in which the second peak arises in the disc due to periastron precession and this was described by Luigi Stella (Rome). If confirmed these observations could lead to important tests of General Relativity as this is a strong field effect that occurs if there is a slight eccentricity in the inner disc.

Observations of two more transients, XTE J2123-058 and XTE J1808-369, were then presented by John Tomsick (Columbia) and Michiel van der Klis (Amsterdam) respectively. The former proved to be an interestingly short period neutron star system at high inclination (further details of which were contained in several poster papers), whilst the latter has been revealed as the long-sought millisecond pulsar in an LMXB. If kHz QPOs could also be detected then this source could confirm or reject current models of both burst and kHz QPOs, and so the next outburst is eagerly awaited.

The oldest BHXRB known is Cyg X-1 and Monika Bałucińska-Church (Birmingham) presented the first comprehensive survey of the X-ray dips in Cyg X-1, observations of which go back to the 1970s. Using archive data covering almost 24 years and a new ephemeris for Cyg X-1 that is valid for this entire period, Church showed that the dips are predominantly close to phase 0, but with a second group at phase 0.6 seen for the first time which is interpreted as evidence for an accretion stream from the OBI companion star.

Neutron star transient systems in quiescence have been detected as weak X-ray sources. Whereas this emission is usually attributed to accretion at a very low rate, Ed Brown (Berkeley) suggested that this could simply be the cooling of the neutron star whose core has been heated by nuclear reactions occurring deep in the crust during the accretion outburst phase. Estimates of the quiescent X-ray luminosity are in good agreement with observations; other predictions of the model, such as the possible absence of metals in the spectra and a large difference between the effective and color temperature will be detectable by future X-ray missions such as AXAF and XMM. A test of this hypothesis is strongly needed, as the determination of the mass accretion rate onto the neutron star is of crucial importance for models of the transient behaviour.

The outbursts in these transient systems are believed to result from a thermal and viscous instability occurring in the accretion disc, in much the same way as in dwarf novae. Guillaume Dubus (Meudon and Amsterdam) presented the first attempt to include self-consistently irradiation of the disc by the central X-ray source, which is observed to play a major role in both the stability of these systems and in the optical emission. He showed that the X-ray source must be extended, or the disc warped, or both, in order to avoid self-screening effects in the disc that would totally suppress irradiation effects, contrary to observations.

The comparison of observations with the predictions of evolutionary models of X-ray binaries can be used to constrain physical processes occurring during the formation of the system. Vicky Kalogera (CfA) showed that the mass of the black hole progenitor must be at least twice as large as the actual BH mass, that the parameter α_{CE} measuring the efficiency of mass ejection during the common envelope phase must be larger than unity, implying energy sources other than the orbit. Andrew King (Leicester) analyzed in detail the case of Cyg X-2, which is a bright persistent X-ray source containing a neutron star. He concluded that this system must have followed early massive case B evolution, during which the neu-

tron star ejected 2 $M_\odot$ that overflowed from a massive star that expanded in a thermal time while crossing the Hertzsprung gap, thus validating the hypothesis of efficient mass ejection in super-Eddington accretion. He also showed that systems like Cyg X-2 must be quite common, since their lifetime are short, and that they could be an important channel for forming millisecond pulsars.

A review of the X-ray sources within globular clusters as observed by Beppo-SAX was given by Matteo Guainazzi (ESTEC). SAX has provided the first simultaneous observations in the range $0.1 - 200$ keV. At low luminosities, hard tails with an exponential cutoff are found, presumably due to Comptonization, together with a softer, thermal component. The origin of both components is still unclear.

A clue to the understanding of accretion flows around black holes comes from a omparison of the properties of galactic and extragalactic sources. Andrzej Zdziarski (Warsaw) found a strong correlation between the intrinsic spectral slope in X-rays and the amount of Compton reflection from a cold medium in Seyfert AGNs and in the hard state of X-ray binaries. Such a correlation requires a coupling between the cold and hot medium, and indicates that the soft photons that are comptonized in the primary X-ray source are emitted by the reflecting medium.

SCIENTIFIC PROGRAM

Venue: Hotel Sofitel Paris-Forum Rive Gauche, Paris, France

Monday, December 14, 1998

Morning Program

OPENING CEREMONY

INVITED TALK

Peter Mészáros (Penn State University): Gamma-ray Bursts and Bursters

HIGHLIGHT TALKS

Dieter Hartmann (Clemson University): Hypernovae
Chryssa Kouveliotou (Marshall Space Flight Center): Magnetars

Afternoon Program

MINI-SYMPOSIA

Gamma-ray Bursts
Quantum Gravity
Recent Results on H_0

Tuesday, December 15, 1998

Morning Program

INVITED TALKS

Gabriele Veneziano (CERN): Probing the pre-Big Bang Universe
Keith Olive (University of Minnesota): Primordial Nucleosynthesis
Alfred Vidal-Madjar (Institut d'Astrophysique de Paris): Cosmic Deuterium

HIGHLIGHT TALKS

Pilar Ruíz-Lapuente (Universidad de Barcelona & Max-Planck Institut für Astrophysik):
  Type Ia Supernovae at High Z
David Smith (Centre d'Etudes Nucléaires de Bordeaux-Gradignan):
  First Detections of 50 GeV Gamma-rays

Afternoon Program

MINI-SYMPOSIA

Early Universe
Nucleosynthesis and Gamma-ray Line Astronomy
Pulsars and Neutron Stars

Wednesday, December 16, 1998

Morning Program

INVITED TALKS

Ramesh Narayan (Center for Astrophysics, Cambridge, USA): Astrophysical Black Holes
Michel Spiro (Service de Physique des Particules, Saclay): Dark Matter
Peter Schneider (Max-Planck Institut für Astrophysik): Gravitational Lenses

HIGHLIGHT TALKS

Lev Titarchuk (Goddard Space Flight Center & George Mason University):
    Signature of Accreting Black Holes
I. Félix Mirabel (Service d'Astrophysique, Saclay): Microquasars in the Galaxy

Afternoon Program

MINI-SYMPOSIA

Dark Matter
Mapping the Universe from Weak Lensing Analysis
X-ray Binaries

Thursday, December 17, 1998

Morning Program

INVITED TALKS

Thibaut Damour (Institut des Hautes Études Scientifiques, Bures-sur-Yvette):
    Experimental Tests of Relativistic Gravity
Jean Heyvaerts (Observatoire de Strasbourg): Astrophysical Jets
Venyamin Berezinsky (Laboratori Nazionali del Gran Sasso): Cosmic Neutrino Oscillations

HIGHLIGHT TALKS

François Vanucci (Laboratoire de Physique Nucléaire et Hautes Energies, Paris): Neutrinos after
    Takayama
Sylvaine Turck-Chièze (Service d'Astrophysique, Saclay): The Solar Neutrino Puzzle

Afternoon Program

MINI-SYMPOSIA

AGN, QSO, and Jets

Gravitational Waves and Numerical Relativity

Large-Scale Structure and Galaxy Clusters

Friday, December 18, 1998

Morning Program

INVITED TALKS

George Efstathiou (Institute of Astronomy, Cambridge, UK):
    Anisotropies in the Cosmic Microwave Background Radiation

Richard Ellis (Institute of Astronomy, Cambridge, UK): Galaxy Formation

James Cronin (Enrico Fermi Institute, Chicago): Ultra-High Energy Cosmic Rays

HIGHLIGHT TALKS

Jean-Loup Puget (Institut d'Astrophysique Spatiale, Orsay): Cosmology in the Infrared

Massimo Tarenghi (European Southern Observatory, Garching): Early Data from the VLT

Afternoon Program

MINI-SYMPOSIA

Cosmic Microwave Background

Galaxy Formation and Evolution

High-Energy Cosmic Rays

CD-ROM CONTENTS

01. AGN, QSO AND JETS

02. COSMIC MICROWAVE BACKGROUND

07. GRAVITATIONAL WAVES AND NUMERICAL RELATIVITY

08. HIGH ENERGY COSMIC RAYS

10. NUCLEOSYNTHESIS AND GAMMA-RAY LINE ASTRONOMY

11. PULSARS AND NEUTRON STARS

CD-ROM contents / Nuclear Physics B (Proc. Suppl.) 80 (2000) 233–246

243

12. QUANTUM GRAVITY AND GENERAL RELATIVITY

13. RECENT RESULTS ON H_0

AUTHOR INDEX

GENERAL INFORMATION

Associate Editorial Board

Nuclear Physics B - Proceedings Supplements (PS) is the premier publication outlet for the proceedings of key conferences on high-energy physics and related areas. The series covers both large international conferences and more specialized topical meetings. Under the guidance of the Editorial Board, the newest discoveries and the latest developments, reported at carefully selected meetings, are published covering experimental as well as theoretical particle physics, hadronic physics, cosmology, astrophysics and gravitation, field theory and statistical systems.

Note to Conference Organizers

Organizers of upcoming meetings who are interested in exploring the possibilities of *Nuclear Physics B - Proceedings Supplements* as the publication outlet of their proceedings are invited to send full details of the planned conference to:

Drs. Carl Schwarz
Publishing Editor
Elsevier Science B.V., P.O. Box 103, 1000 AC Amsterdam, The Netherlands
Telephone: +31 20 485 2355; Telefax: +31 20 485 2580; Telex: 10704.
E-mail: C.SCHWARZ@ELSEVIER.NL

Proceedings will be produced from camera-ready copy and will be published within three months of the Publisher's receipt of the final and complete typescript. Attractive bulk-order arrangements will be offered to conference organizers for participants'copies.

Subscription information

In 2000 eleven volumes of *Proceedings Supplements* will be published.

Subscription prices are available upon request from the Publisher. A combined subscription to Nuclear Physics A volumes 662-678, Nuclear Physics B volumes 564-591, and Nuclear Physics B - Proceedings Supplements volumes 80-90 is available at a reduced rate. Single copies of proceedings published in the *Proceedings Supplements* can be ordered directly from the Publisher.

Subscriptions are accepted on a prepaid basis only and are entered on a calendar year basis. Issues are sent by surface mail except to the following countries where air delivery via SAL is ensured: Argentina, Australia, Brazil, Canada, PR China, Hong Kong, India, Israel, Japan, Malaysia, Mexico, New Zealand, Pakistan, Singapore, South Africa, South Korea, Taiwan, Thailand, USA. For all other countries airmail rates are available upon request.

Advertising Information

Europe and ROW: Rachel Leveson-Gower, The Advertising Department, Elsevier Science Ltd., The Boulevard, Langford Lane, Kidlington, Oxford OX5 1GB, UK; phone: (+44) (1865) 843565; fax: (+44) (1865) 843976;
e-mail: r.leveson-gower@elsevier.co.uk

USA, Canada and South America: Mr Tino de Carlo, The Advertising Department, Elsevier Science Inc., 655 Avenue of the Americas, New York, NY 10010-5107, USA; phone: (+1) (212) 633 3815; fax: (+1) (212) 633 3820; e-mail: t.decarlo@elsevier.com

Japan: The Advertising Department, Elsevier Science K.K., 9-15 Higashi-Azabu, 1-chome, Minato-ku, Tokyo 106-0044, Japan, phone (+81) (3) 5561 5033; fax (+81) (3) 5561 5047

Orders, claims, and product enquiries: please contact the Customer Support Department at the Regional Sales Office nearest you:

New York: Elsevier Science, P.O. Box 945, New York, NY 10159-0945, USA;
phone: (+1) (212) 633 3730 [toll free number for North American customers: 1-888-4ES-INFO (437-4636)]; fax: (+1) (212) 633 3680; e-mail:usinfo-f@elsevier.com

Amsterdam: Elsevier Science, P.O. Box 211, 1000 AE Amsterdam, The Netherlands;
phone: (+31) 20 485 3757; fax: (+31) 20 485 3432; e-mail: nlinfo-f@elsevier.nl

Tokyo: Elsevier Science, 9-15, Higashi-Azabu 1-chome, Minato-ku, Tokyo 106-0044, Japan;
phone: (+81) (3) 5561 5033;
fax: (+81) (3) 5561 5047; e-mail: info@elsevier.co.jp

Singapore: Elsevier Science, No. 1 Temasek Avenue, #17-01 Millenia Tower, Singapore 039192;
phone: (+65) 434 3727; fax: (+65) 337 2230; e-mail: asiainfo@elsevier.com.sg

Rio de Janeiro: Elsevier Science, Rua Sete de Setembro 111/16 Andar, 20050-002 Centro, Rio de Janeiro - RJ, Brazil; phone: (+55) (21) 509 5340; fax (+55) (21) 507 1991;
e-mail: elsevier@campus.com.br
[Note (Latin America): for orders, claims and help desk information, please contact the Regional Sales Office in New York as listed above]